LANDSCAPES OF DEFENCE

LANDSCAPES OF DEFENCE

Edited by
JOHN R. GOLD and GEORGE REVILL

An imprint of **Pearson Education**

Harlow, England · London · New York · Reading, Massachusetts · San Francisco · Toronto · Don Mills, Ontario · Sydney
Tokyo · Singapore · Hong Kong · Seoul · Taipei · Cape Town · Madrid · Mexico City · Amsterdam · Munich · Paris · Milan

Pearson Education Limited
Edinburgh Gate
Harlow
Essex CM20 2JE
England

and Associated Companies throughout the world

Visit us on the World Wide Web at:
http://www.pearsoneduc.com

First published 2000

© Pearson Education Limited 2000

ISBN 0 582 38234 3

British Library Cataloguing-in-Publication Data
A catalogue record for this book is available from the British Library

10 9 8 7 6 5 4 3 2 1
04 03 02 01 00

Typeset by 35 in 10/12pt Times

Transferred to digital print on demand, 2008
Printed and bound in Great Britain by CPI Antony Rowe, Eastbourne

For Maggie and Eleanor

CONTENTS

FIGURES

TABLES

PREFACE

Any book involving more than twenty authors is likely to have some history, and this one is no exception. The idea for this collection came originally from work carried out as part of a three-year funded project on 'Risk, Commercial Insurance and the Urban Environment', which was directed by the editors of this book between 1995 and 1998. We were interested in the role being played by the large insurance firms in shaping the contemporary urban environment, particularly in light of the insurers' response to risk linked to terrorist campaigns. Our involvement in that project made us aware of the large number of other researchers who were working on related topics, especially those whose work touched on investment in high-technology security measures and fortification of buildings, both in Britain and overseas.

As often happens, however, those who work in cognate areas, especially where these involve different academic disciplines, have little contact with one another. To try to learn from each other's experiences, therefore, we held various seminars and symposia on subjects of shared conceptual and empirical concern. Throughout, we used the notion of 'landscape' as an integrating concept: partly because it had an intuitive appeal to us as geographers, but also because it is a term that truly crosses the disciplinary boundaries of academic inquiry. Lawyers, sociologists, criminologists and environmental psychologists as well as geographers all seemed to find it a flexible and worthwhile framework within which to situate discussion of their research – often to their surprise. These contacts led in turn to a one-day symposium at Oxford Brookes University in May 1998 on the subject of what we termed 'landscapes of defence'. The agenda for the meeting was left deliberately open, apart from identifying four initial themes that the symposium might address: risk theory and its application; fear, surveillance and social control; the security of residential areas or central business districts; and financial and institutional aspects of landscapes of defence. However, no restrictions were placed on the empirical contexts that might be explored.

The resulting symposium pleasantly exceeded our expectations, bringing together more than fifty participants from the arts, humanities, built environment professions and security industry. The fascinating range of papers and the lively ensuing discussions convinced us that the event might usefully serve as the basis for a collection of essays. Having said that, it was immediately recognised that such a text must be more than the happenstance expression of papers given at the symposium. Essays for this volume have come, therefore, from a variety of sources. Some are revised and extended versions of spoken precursors. Others were contributed under invitation by an international group of researchers whose work allows us to address themes not covered at the Oxford meeting. The result is a book that has developed and matured in the process of preparation and one which, we believe, goes beyond the limits of case studies to address key themes in the study of contemporary political, environmental, economic and social conflict.

Inevitably, a variety of debts have been incurred in the process of preparation. The Schools of Social Sciences and Law, Planning and Humanities of Oxford Brookes University have all provided finance and other assistance to facilitate our work. Our colleagues in the Geography Department gave us their untiring support. Brian Rivers and Linda Brown cheerfully handled all the financial side of our funded project and the symposia. Tim Blackman and Frank Webster supplied helpful comment and challenging ideas. Gerry Black, David Elsmore and the staff of the university's Computer Centre helped us to iron out the technical problems encountered in the extensive use of e-mail by which we have gathered material and corresponded with contributors. The global village has yet to arrive here. We would particularly like to thank the Landscape Research Group for its support of the Oxford symposium. The breadth of its vision of the scope and importance of landscape has done much to revive this concept as a focus for academic study during the last twenty years. We would like to record our special thanks to Jon Coaffee and Paul Lehane, literally as well as figuratively our fellow travellers between 1994 and 1998. Jon in particular served as the researcher on the urban terrorism project and helped us to develop many of the ideas explored here. Finally, we would like to thank our families for their continuing support and tolerance.

West Ealing and Newthorpe

ACKNOWLEDGEMENTS

Whilst every effort has been made to trace the owners of the copyright material, in a few cases this has proved impossible and we take this opportunity to offer our apologies to any copyright holders whose rights we may have unwittingly infringed.

NOTES ON CONTRIBUTORS

Stuart Aitken is professor of geography at San Diego State University. His research interests focus on the gendered and critical geographies of children, families and communities. His books include *Putting Children in Their Place* (1994) and *Family Fantasies and Community Space* (1998).

Harri Andersson is professor of urban geography at the University of Turku and is currently Programme Director of the Research Programme for Urban Studies at the Academy of Finland. He has research interests in urban regeneration, power structures and changing urban landscapes.

Maoz Azaryahu is a senior lecturer in geography at the University of Haifa in Israel.

John Baxter is chair of the advanced diploma in law course and senior lecturer in law at Oxford Brookes University. His research is mainly on policing and civil liberties. He co-edited *Police: The Constitution and the Community* (1985) and is author of *State Security, Privacy and Information* (1990).

Andrew Blowers is professor of social sciences (planning) at the Open University. His research and publications cover environmental policy, politics and planning, and he specialises in the politics of radioactive wastes. Among his books are *Something in the Air* (1984), *The International Politics of Radioactive Waste* (1991) and *Planning for a Sustainable Environment* (1999). He is a member of the government's Radioactive Waste Management Advisory Committee and has served as an elected councillor on Bedfordshire County Council since 1973.

Stanley D. Brunn is professor of geography, University of Kentucky at Lexington. He is interested in electronic human geographies, scholarly networks and communities, humane geographies, and geographical futures. His many publications include books on social, political and information/communications geography.

Paul Catley is principal lecturer in law at Oxford Brookes University. His teaching interests centre on human rights and the criminal justice system, and issues in policing. His research interests include policing, human rights and sexuality and the law.

Jon Coaffee is a research assistant in the Centre for Urban Technology at the University of Newcastle-upon-Tyne. Having previously undertaken doctoral studies on the institutional response to terrorism in the City of London, his research and teaching interests centre on contemporary developments in the urban environment.

Carl T. Dahlman is a research student at the University of Kentucky at Lexington.

John R. Gold is professor of urban geography in the School of Social Sciences and Law at Oxford Brookes University. His major research interests are in the impact of architectural modernism on urban reconstruction, the production and consumption of urban spectacle, and place promotion. His most recent books are *Place Promotion* (1994), *Imagining Scotland* (1995), *The Experience of Modernism* (1997) and *Cities of Culture* (2001).

Mark Goodwin is professor of human geography in the Institute of Geography and Earth Sciences, University of Wales at Aberystwyth. He has a particular interest in local governance and economic development. These have informed research that he has recently directed for the ESRC, the Welsh Office and the Joseph Rowntree Foundation and a research project on CCTV, which he co-directed.

Phil Hubbard is a lecturer in human geography at Loughborough University. He has published widely on the changing nature of Western cities, with interests that include the geography of heterosexuality, the marketing of place and the social spaces of modernism and postmodernism. His latest books are *Sex and the City: The Geography of Prostitution in the Urban West* (1999) and the co-edited *People and Place: Exploring the Extraordinary Geographies of Everyday Life* (2000).

Craig Johnstone is a lecturer in geography at the University of Wales, Swansea. Previously he worked as a research assistant at the University of Wales at Aberystwyth on the introduction of public-space CCTV to rural towns. His current research is concerned with the governance of crime, focusing specifically on CCTV as a catalyst for local partnership formation.

Christina B. Kennedy is an associate professor of geography and public planning at Northern Arizona University. She teaches cultural geography and courses dealing with the US landscape. Her research foci and publications range from studies on aesthetics in geography and the use of landscape in film to resource management and Native American issues.

Alan A. Lew is a professor of geography and public planning at Northern Arizona University, where he teaches urban planning, the geography of the USA and courses in tourism that contribute to a www-based certificate in international tourism management. He is the editor of the journal *Tourism Geographies* and has edited books on tourism in China, on Native American lands and on the geography of sustainable tourism.

Simon Marshall completed his doctorate as a member of the Urban Morphology Research Group at the University of Birmingham in 1997. He currently works for Crime Concern and is a research associate of the International Centre for Prison Studies, King's College, London. He has previously worked for HM Prison Service and is a regular contributor to the *Prisons Handbook*.

Taner Oc is Director of the Institute of Urban Planning at the University of Nottingham. He has degrees in architecture, planning and social sciences from Middle East Technical University, Ankara, the University of Chicago and the University of Pennsylvania. His recent publications include *Urban Design: Ornament and Decoration* (1995), *Revitalising Historic Urban Quarters* (1996) and *Safer City Centres: Reviving the Public Realm* (1997).

Martin Phillips is a lecturer in human geography at the University of Leicester. He has research interests in rural social and cultural geographies. He has written numerous articles and book chapters and is also a co-author of *Writing the Rural* (1994) and *Society and Exploitation through Nature* (2000).

George Revill is a senior lecturer in geography in the School of Social Sciences and Law at Oxford Brookes University. His research interests are concerned with landscape and with the cultural geography of the UK. His most recent books are *The Place of Music* (1998) and *Pathologies of Travel* (2000).

Peter Shirlow is a senior lecturer in geography in the School of Environmental Studies at the University of Ulster in Coleraine, Northern Ireland. His most recent books include *Development Ireland: Contemporary Issues* (1995) and *Who Are 'The People'?: Unionism, Protestantism and Loyalism in Northern Ireland* (1997).

Steven Tiesdell is a senior lecturer in the Department of Land Economy at the University of Aberdeen. He is a chartered architect and town planner whose current research interests are in urban design, urban revitalisation and planning for safer city centres. He is co-author of three books: *Urban Design: Ornament and Decoration* (1995), *Revitalising Historic Urban Quarters* (1996) and *Safer City Centres: Reviving the Public Realm* (1997).

Kate Williams is a lecturer in law at the University of Wales, Aberystwyth. Taking a socio-legal approach she researches into criminal justice, technology and human rights. These issues informed the research project on CCTV, which she co-directed. She is currently co-directing a research project on 'Devolution and Human Rights'.

CHAPTER 1

LANDSCAPE, DEFENCE AND THE STUDY OF CONFLICT

John R. Gold and George Revill

Some time ago, one of the editors of this book visited the newly opened 'Broadway Centre' of an outer suburban borough in West London. The centre itself was an interesting mixed development, finished externally in the neo-vernacular style of postmodern architecture that flourished in the early 1980s. Inside it contained the typical retail chain outlets and restaurants found in any British shopping mall but, somewhat unusually, it also incorporated a new public library, an outdoor meeting area with seating called the Town Square, fountains, an interior space that included a bandstand, and generous statuary. It seemed the right place to illustrate a planned lecture that could show that such places might play a role in future urban life if only it was possible to find the right blend of public and private, civic and commercial. With this, ironically sympathetic thought in mind, he took out his camera.

It was not the wisest move. Before he had done more than squint through the camera's viewfinder, two security guards appeared on the scene, seemingly from nowhere. They proceeded to demand not just that he cease taking photographs – which he had not yet done – but also that he open the camera and hand over the film. A heated discussion ensued of the sort that probably convinced passers-by that a dangerous felon had been detained. When asked why photography was banned, the security guards stated that the architects had given 'strict instructions' that no one was to do so – presumably to ensure that only favourable images appeared in the local or professional press. 'What you have to remember,' said one, 'is that this is private property.' When challenged that this could scarcely be so when the complex now contained the borough's main public library, he replied: 'that is only the inside of the building. They can let you take photographs if they want, but the outside *belongs* to us.'

In the event, all ended reasonably amiably in the centre manager's office. The photographer was allowed to keep his film, albeit at the expense of receiving a short lecture about 'correct behaviour' and the need to go through 'the proper channels' if he wanted to take pictures in future. There was, however, no change in the underlying attitude. Although it was admitted that appearances might be deceptive, it was emphasised that this was private space owned by a large financial corporation. The management reserved the right to control access. It was free to exclude any person or usage deemed unsuitable without the need for further justification. Taking photographs apparently placed the offender in that category.

1

All then was not as it seemed. A private corporation had provided space for a town square and a public library but had blurred the lines between private and public space. The new facilities gave every impression of being meeting places of the type that are standard parts of the civic life of most British towns and cities, but the owners retained the right to regard them as space within their control, policed by their own security staff. This control even extended to making attempts to restrict the way in which the external appearance of the buildings was represented, despite the fact that this had few implications for the security of tenants, proprietors or users.

Revisiting the Broadway Centre in 1999, we found that, while it retains its character as defended space, it is no longer an isolated island of security. Steel pillars, sometimes cast in the shape of upturned cannon but sometimes just functional square-sectioned posts, block the approaches to stores in the adjacent High Street at risk from ram-raiders. The gates of a local Church of England school carry notices to warn potential intruders that the metal has been treated with special non-drying paint that will adhere to anyone attempting to climb over. Ungainly iron frames found in surrounding streets await the return of the rubbish containers that were removed during the early 1990s to reduce the threat of parcel bombs left by terrorists. More significantly, the centre itself now lies in the heart of a large area of shops and offices over which networks of closed-circuit television cameras keep watch. These networks include both security video cameras belonging to private firms and a highly visible public network. The latter employs cameras placed at elevated points on buildings or mounted at the top of 10-metre posts, the massive reinforced bases of which look strong enough to guard against accidental or deliberate damage by a passing articulated lorry. Very occasionally the cameras silently swivel to change their angle of regard – the only tangible signs that they are in use. It is unclear though whether they simply feed images into video recorders for potential use in case of need or whether observers actively monitor them.

This all seems to convey an atmosphere of danger, but local people report that they do not feel any more or less secure than they did prior to the introduction of security equipment in the central area. Normal life proceeds much as before. Nor for that matter are there significant signs that surrounding areas, unprotected by the paraphernalia of high-technology security equipment, are suffering grave insecurity. The tree-lined residential streets have front gardens bounded by traditional low fences or neatly clipped hedges. Relatively few have burglar alarms or security lights, and there are no grilles on windows. There are few signs of graffiti or of wilful damage to road signs, bus shelters or other street furniture. Indeed, most people scarcely seem to have noticed the increase in surveillance and control that has been implemented in the adjoining town centre. The security equipment found there is regarded as unexceptional: commonplace features in the landscape that are taken for granted by those who use the town centre. They have simply become part of the reliable background for everyday life.

These small examples hint at wider themes. Whatever one's assessment of the long-term prospects for global security after the end of the Cold War, other indications suggest that little has occurred to reduce the role of security considerations at other levels. The new, often fiercely contested, boundaries established after the internal partitions of Yugoslavia, Czechoslovakia and the former Soviet Union have offset the erosion of national boundaries within the European Union. The military continues to hold huge tracts of land in Europe and North America for training and weapons testing. Indeed, it may be argued that concerns about security, control and reduction of risk have become *increasingly* important influences in the conduct of everyday life. Close observers of cities, suburbia and rural areas throughout the Western world would argue that the private domain is steadily expanding through the agency of defensive strategies that seek to create surveillance over, exert control over and partition the public domain. In the process, we find the growth of 'landscapes of defence', defined here as landscapes shaped or otherwise materially affected by formal or informal strategies designed to reduce the risk of crime, or deter intrusion, or cope with actual or perceived threats to the security of the area's occupants.

Our aim in this book is to bring together a set of essays that offer critical analyses, explanations and some understanding of landscapes of defence in their wide diversity. Through a series of case studies focusing substantially on Anglo-American experience, our authors examine issues of ethnicity, nationalism and landscape symbolism, land rights, resource use, globalisation and environmental risk, privatisation and public space, surveillance, policing, and citizenship. They find defensive landscapes in suburbia and the inner city; in the financial heart of the City of London and the remotest rural regions; at international airports and in the design of prisons. They find expression of landscapes of defence in the fears of parents for their children and in the schemes of planners and architects; in the cultural politics of community activists and in the rhetoric of ethnic separatists. They show that even in an age apparently characterised by globalisation and transnational developments, defensive landscapes continue to proliferate, addressing new sources of conflict and uncertainty as well as reflecting longstanding sources of social, cultural and political conflict.

However, with such a multiplicity of phenomena potentially offering themselves for inclusion within the scope of this text, it is important to explore carefully the rich meanings of our central term 'landscapes of defence'. Certainly, when used independently, the terms 'defence' and 'landscape' both have complex patterns of usage which have an important bearing on any term that expresses their conjunction. The task in the remainder of this chapter, therefore, is to explore the nature of landscapes of defence, sample something of their diversity and begin to ask what it is that brings these different phenomena together.

DEFINING TERMS

Defence, as the extensive entries under this heading in the *Oxford English Dictionary* make clear, is a word with rich connotations. It can apply equally to the prevention of physical attack, a response to symbolic threat and the presentation of a response to legal prosecution. It is conceptually related to security and control, to minimisation of risk and maintenance of privacy, to preservation of status and projection of power. It can apply equally to the actions of an individual in warding off direct aggression or the operation of culturally derived rules that regulate grounds for dispute so thoroughly that conflict is scarcely noticed in practice. 'Defence' can imply creating divisions between insider and outsider, civilisation and wilderness, order and chaos, self and other: but the meaning of each of these dichotomies can and does change over time. Depending on the prevailing circumstances, the hard-fought battlefield of one era can easily become a forgotten name on a changing political map, an ecological treasure, a mythologised site for rallying nationalism or a heritage centre for a booming tourist industry (Gold and Gold 2000). Certainly, analysis of the different ways in which the term 'defence' is used to give legitimacy to physical and social interventions can supply a potent starting point from which to scrutinise a wide range of social processes and practices.

Strategies to achieve defence against other people or against nature have long figured prominently in shaping the human environment. Although it is always dangerous to infer too much from the archaeological record, the evidence clearly shows that defence in its broadest sense was intrinsic to the ordering of early civilisations. Dykes and other works designed to defend communities from flood-ing were implemented as much as 6000 years ago to allow settlements to flourish in the otherwise hazardous bottom lands of the Tigris, Euphrates, Nile, Indus and Yangtze valleys. Field evidence also provides ample evidence of military defence. The survival of relic walls, ditches, ramparts, castles and garrisons, sometimes in a well-preserved state of repair, reflects the effort once expended upon their con-struction. Some were undoubtedly designed to protect communities from their natural or human enemies, yet concentrating the populations in fortified centres also served the political goals of ruling elites. Throughout Europe, for instance, the feudal regimes of the medieval period created a network of fortified towns situated in planned and regulated countryside (Baker and Harley 1973; Ashworth 1987; Dodgshon 1987; Roberts 1987; Morris 1994). These, as European history clearly shows, were necessary to meet defensive needs, but defence also justified the concentration of population into a space where *they* could be controlled. Later, during the period of colonial expansion, many defensive constructions served the function of protecting the *interests* of an imperial power, symbolically serving to establish a presence and create an image of power that might impress indigenous populations or rival colonists. For example, the planned settlements of the British Empire – ranging in architectural style through Italian Renaissance and mock Tudor to pastiche Mogul – both translated European culture into alien environments and represented an imaginative reconciliation with an exotic colonised 'other' (King 1976). Over time, 'defence' encompassed an ever wider range of features. Access to scarce resources, ethnic separatism and the security needs of private financial corporations, among others, presented new demands for defensive measures giving rise, as we shall see later in this chapter, to a new repertoire of secure structures and forms of control.

Landscape, our second key term, also provides a challenge to the lexicographer. Originating in the late sixteenth century as a technical term used by painters to describe the depiction of natural inland scenery, it quickly took on wider meanings and connotations. *Inter alia*, landscape became associated with styles of gardening, artistic representation, topographic interpretation and photographic convention. From the standpoint of the present inquiry, however, it is important to recognise that there are firm grounds for associating 'landscape' with notions concerning 'defence'. It may be convincingly argued that the very term 'landscape' is itself intimately bound up with activities associated with defence, security and protection. Warnke (1994), for example, began his study of landscape painting by considering the representation of frontiers in both language and painting. He argued that it is only relatively recently that the borders between European nations have been reduced to mere lines. Hence the border and all that this represented in terms of cultural, political and economic difference was conveyed in a vocabulary of landscape. Moreover, he noted (*ibid.*: 11) that the common German word for 'frontier' until well into the sixteenth century was *Landmarke*, the second element of which is cognate with the English *marche* (as in the 'Welsh Marches'):

> Borders were thus constituted by marches – woods, mountain crests, wildernesses, steppes, swamps, moors, lakes or rivers . . . To describe such boundaries as 'natural frontiers' tends to obscure the fact that nature was left artificially intact so that it could function as a boundary and so serve a cultural purpose.

In the same vein, Olwig (1996) showed how the medieval German term *Landschaft* referred to a restricted or administratively bounded piece of ground (see also Muir 1999: 3). Other authors claim that the development of landscape as a set of aesthetic practices and as a way of seeing are closely tied into landscape's defensive functions. In its most general formulation, W.J.T. Mitchell (1994: 9) related the aesthetic practices of landscape to militaristic cultures widely ranged across time and space, arguing that:

> Two facts about Chinese landscape bear special emphasis: one is that it flourished most notably at the height of Chinese imperial power and began to decline in the eighteenth century as China became itself the object of English fascination and appropriation at the moment when England was beginning to experience itself as an imperial power. Is it possible that landscape, understood as the historical 'invention' of a new visual/pictorial medium, is integrally connected with imperialism? Certainly the roll call of major 'originating' movements in landscape painting – China, Japan, Rome, seventeenth-century Holland and France, eighteenth- and nineteenth-century Britain – makes the question hard to avoid.

The various historical aspects of this assertion have been examined in a range of contexts. Typically, these connect landscape aesthetics to the strategic practices of surveying, cartography and civil engineering. Schama (1995: 10) suggested that landscape as a pictorial aesthetic developed from the culture of defensive landscapes in the seventeenth-century Netherlands:

> *Landshap*, like its German root, *Landschaft*, signified a unit of human occupation, indeed a jurisdiction, as much as anything that might be a pleasing object of depiction. So it was surely not accidental that in the Netherlandish flood-fields, itself the site of formidable human engineering, a community developed the idea of a *landschap*, which in the colloquial English of the time became a 'landskip'.

Cosgrove (1984; 1985) has shown how the origins of pictorial perspective derive from the rediscovery and reworking of geometric theory in Renaissance Italy. The development of visual perspective is therefore linked to the practices of map making, land surveying, cartography and navigation. These technologies were fundamental to the development of the Venetian state as a maritime power and trading nation, to the making of agricultural land and the consolidation of wealth in the Venetian hinterland. Sullivan (1998) argued that issues concerning defence, security and national integration in early modern England were discursively bound into depictions of land and landscape in the theatre, themselves echoing the discourses of politics, estate management, surveying and cartography. In this regard, the work of Shakespeare and his contemporaries amplifies the arguments of the time concerning moral economy of national defence and security in the transformation from feudal relations of authority and security to capitalist relations of absolute property (*ibid.*; see also Helgerson 1986).

It is also true that in Britain during the eighteenth and nineteenth centuries there was an important relationship between topographical drawing and military surveying. This was the case overseas, where sketching and watercolour painting supported topographical and geological surveys for imperial purposes (Livingstone 1992; Godlewska and Smith 1994; Gregory 1994). However, it was also important within Britain itself, where the aesthetics of landscape painting were undergoing radical transformations – as exemplified by the rise to prominence of John Constable and J.M.W. Turner. In this context, for example, Daniels (1993) has argued that the mapping of Britain was part of a broader revision of the country by travel writers, antiquarians, landscape gardeners and landscape painters, linked to the threat of invasion during the Napoleonic Wars. Perhaps the most notable artist in this regard was Paul Sandby, who began his career as a draughtsman employed on the military survey of Scotland. His art continued to be highly influenced by military drawing and map making, and later he became a renowned topographical painter, credited with popularising the sites and scenes of the Highlands of Scotland (*ibid.*: 61). Indeed, Sandby's work, with its underpinnings in military need and topographic representation, was typical of much that helped to shape imaginative geographies of Scotland during the late eighteenth century (Gold and Gold 1995: 39–41).

INSTINCT AND AFFECT

Ideas that are deeply rooted in human history often invite analyses that give prominence to instinct theory. Given the deeply engrained relationship in Western culture between landscape and defence, it is not surprising that a variety of theoretical perspectives have sought to essentialise this relationship as one derived from instinctive behaviour. To Jay Appleton (1990; 1996; see also Orians 1986; Bourassa 1991; Kaplan 1992; Muir 1999: 255–8), for instance, our cultural preferences for designed parkland landscapes and for landscape paintings based on theatrical rules of perspective and visual closure are derived from our ancient ancestry as savannah-dwelling hunter-gatherers. In his view, the conventions of both the designed parkscape and landscape painting are merely expressions of an innate desire to locate ourselves at a hidden, defended and protected vantage point. Appleton's 'habitat theory' and ideas of 'prospect' and 'refuge' assert that the landscapes we find most attractive are those 'that once would have afforded us, as individuals involved in the struggle for survival, the opportunity to see without being seen, to eat without being eaten, to produce offspring that survive' (Greenbie 1992: 65; Muir 1999: 249).

Although such biologically founded theories have had little impact on mainstream landscape theory (see criticisms by Cosgrove 1978; Walmsley and Lewis 1993), ideas of instinctive behaviour have also entered thinking on landscape through another route, namely theories of human territoriality. Notions of territoriality developed from ethological research (e.g. Moffat 1903; Howard 1920; Mayr 1935). The existence of similar patterns in human and animal behaviours has given rise to a body of theory that maintains that human beings, like other species, are biologically predisposed towards spatially defensive behaviour (e.g. Ardrey 1966; Malmberg 1980).

The behaviourist legacy of territoriality is perhaps most pervasive in studies of 'defensible space'. Although Oscar Newman (1972; 1996), the prime mover in such research, is by no means convinced of the homologous nature of human and animal behaviour (e.g. Newman and Franck 1982), the application of territorial theory by others has associated the failings of public-sector housing estates and issues of crime with control over space. For Newman, the abnormally high crime rates associated with high-rise public housing estates could be traced to factors of layout and design. Such buildings denied residents the opportunity to exercise territorial control over the area around their dwellings. In particular, attention was drawn to the fact that the design of high-rise estates gave rise to 'secondary territories', such as stairways and lobbies, that were open to all but under the control of no specific resident. Newman argued that layout and design should be changed to give residents opportunities to have 'defensible space', using real and symbolic boundaries to create perceived territorial zones and increased surveillance opportunities over the semi-public interior spaces. These measures would bring about a residential environment in which the latent territoriality and sense of community of the inhabitants could be harnessed to bring about a safe and well-maintained living space. In one formulation or another this theory has remained attractive to planners, architects and policy makers (see, for example, Perkins and Taylor 1996; Perkins *et al.* 1996; Donnelly and Kimble 1997; Tijerino 1998; Ham-Rowbottom *et al.* 1999; Saleh 1999).

Theories founded on the premises of human ethology or sociobiology have proved neither historically nor socially robust when used as conceptual schemata. Appleton has to recognise that the conventions of landscape are not necessarily shared cross-culturally, least of all in their supposed savannah source regions. Newman and his fellow workers need to account for the fact that the domestic landscape of housing estates is shaped by many social factors; indeed, that high-rise public housing might actually suit some socio-economic and cultural groups rather than others. Aggression and competition are not necessary components of human

behaviour (Callan 1970; Eisenberg 1972), and there is little to support the contention that contemporary human society simply perpetuates the behaviour patterns of
atavistic groups. Human territoriality is complicated by the existence of complex
cultural constructions such as property laws. Animals may defend territory, but
there is no reason to suggest that they view it as their own property, certainly not in
the sense that it can be passed on to others of their species (Gold 1982; Gold and
Revill 1999).

These reservations point to the cultural particularity of human territorial behaviour and the historical specificity of defensive landscapes rather than their genesis
in transhistorical behavioural laws. The concept of landscape, however, does not
require these universalising tendencies to offer critical insight into issues involving
defence, security and conflict. Indeed, it has a number of dimensions that, for
primarily cultural reasons, place it centrally within many conflicts and make it
fundamental to the expression of defensive strategies in a wide range of empirical
contexts.

One dimension, as has already been suggested, is that landscape expresses conflicts over the value of land. Conflicts can arise over the material value of the land,
or its emotional and symbolic value, or sometimes the conjunction of both. In
material terms, struggle for the control of land and the natural resources it contains
has been and continues to be a major source of strife throughout the world (e.g. see
Homer-Dixon 1999). Nevertheless, the role of environmental scarcity in precipitat-
ing violence is complex and indirect. Settlers in the Middle East, for example, have
laid claim to land in the Golan Heights and Palestine occupied since the 1967 war.
These were lands that they regarded as originally unproductive or neglected and
which they have transformed by dint of their labour and application of scientific
management, especially with regard to water resources (Cosgrove and Petts 1990;
Kliot 1994). In the process, the transformed landscape has become endowed with
new material value as productive farmland. Nevertheless, the settlers' demands
transcend ones of compensation for material loss; they also reflect their emotional
investment in the landscape. To the settlers, the transformation in the landscape was
part of their own biographies and identity, and some would also invoke biblical
justification for their actions and claims. Understandably, such arguments carry
little weight with those whose families left the area through conflict or whose lands
were appropriated by military or civil authorities. The population that owned the
land prior to 1967 sees its restitution as equally intrinsic to their identity. They are
predictably unmoved by claims based on readings of theology or by the material
and emotional investment in the land by those who were not its rightful owners.
They may also not see the change in the agricultural status of the land as a positive
step, since it may well disrupt other, culturally valued, patterns of rural life. Small
wonder, then, that the conflict aroused by settlers' claims has proved so enduring,
as the protracted peace negotiations on this point have shown.

Conflict due to the emotional value of landscape is a theme that must be emphasised. In both Western and non-Western societies, for example, landscapes often
have worth as sacred space fundamental to the legitimation of social hierarchies
and cultural organisation (Tuan 1974). In the wake of nineteenth-century romanticism,

the landscape has taken on a secular, and partly mystical, role as a source of psychic health and physical renewal. Places that afford spectacular scenery and the opportunity for solitude are viewed as places of retreat from the ills and threats of modern life, where emotional and bodily health are restored. The national parks movement, campaigners for recreational open space policy and the activities of the Sierra Club in the USA and the Council for the Preservation of Rural England have helped to enshrine and codify this romantic conception of landscape as both a defence and a necessity for active participation in the modern world (Nash 1967; Shoard 1987). In this way, cultural constructions of landscape become inexorably fused with the imagined vitality of ways of life (Wright 1985). Any threat to a valued landscape becomes reified as a threat both to the way of life it symbolises and to the very idea of landscape – itself now only conceivable within the terms of romanticism (Wilson 1992; Mitchell 1994: 20).

Landscapes at all scales from the continental to the domestic can symbolise the social ideals of defence and security at the same time that their material presence demonstrates physical effort and tangible worth. In this context, the private garden, denoted by culturally accepted rather than physically impregnable boundary markers, has long constituted an archetypal example of a landscape of defence. It is the object of active labour, in terms of horticulture and husbandry, and provides a measurable source of physical sustenance, but it also occupies a culturally defined position of psychic security between country and city, wilderness and civilisation. The garden at once represents ideas of home and domesticity, privacy, intimacy, nurturing, private wealth and public respectability (Williams 1973; Marx 1967; Pugh 1988; 1990).

At a very different scale, landscape is also fundamental to the nation-state as a functional unit. As A.D. Smith has shown, territory is not only vital as living space and a source of economic wealth but it is also fundamental to the imagining of the nation for the purpose of forging a national identity. For Smith, specific landscapes, or features in the landscape, are evoked to constitute the historic home of the people. These may include natural features such as lakes and mountains, idealised landscapes, edenic gardens or sacred cities, which often refer back to an imagined national 'golden age'. These landscapes can 'all be turned into symbols of popular virtues and "authentic" national experience' (Smith 1991; 1997). In the twentieth century, when mechanised warfare has both been directed against and involved entire populations, authors have examined the symbolic role of landscapes of national remembrance and personal loss (Heffernan 1995; Gold and Gold 2000), the use of landscape symbolism in civilian propaganda (Harries 1983; Foot 1990; McCormick and Hamilton 1991) and the aesthetics of totalitarian regimes (Golomstock 1990; al Khalil 1991; Affron and Antiff 1997).

Perhaps the broadest studies of the affective and imaginative aspects of landscape are to be found in the pioneering work of Yi-Fu Tuan, particularly his study *Landscapes of Fear* (1979). In this book, Tuan explored the values and attitudes bound up with specific landscapes and conceptions of landscape. He then used them to investigate human response to a wide range of threats and uncertainties. Like our notion of landscapes of defence, Tuan's 'landscape of fear' was an enabling metaphor, facilitating study of imaginative landscapes from children's fairy tales to the perception of natural hazards. He focused on landscape because it drew together human attitudes, values and our physical responses and interventions in the world. He remarked: '"Landscape", as the term has been used since the seventeenth century, is a construct of the mind as well as a physical and measurable entity. "Landscapes of fear" refers both to psychological states and to tangible environments' (*ibid.*: 6).

As a site in which material practices and human values are brought together, the concept of landscape fits easily with more recent versions of territoriality (see above). Robert Sack (1986), a critic of the behaviourist legacy bound up with the concept of territoriality, argued that the spaces associated with this concept are intimately related to how people use the land, how they organise themselves in space and how they give meaning to place. These are functional qualities that could equally well be applied to the concept of landscape. Although Sack (*ibid.*: 2) stated the need for a historically and geographically sensitive approach, his own abstract functionalist 'ten tendencies of territoriality' fell foul of this call. The register of defining attributes becomes merely a set of empty and ahistorical categories when divorced from specific historical settings (*ibid.*: 28–51).

By comparison, landscape is a much more powerful vehicle for accessing the processes that produce conflict. Social scientists might infer that a particular group of people are behaving territorially even though they may well not think of or articulate their own behaviour in this way. However, people in a wide variety of historical and geographical locations do actively create landscapes, even though this certainly means different things in different times and places. Moreover, they recognise themselves as doing so, in the process producing a range of evidence from the economic to the aesthetic. Together, the discourses and practices of landscape bring together spheres of existence otherwise held apart (Daniels 1987; Matless 1998). In this sense, Matless used the language of 'actor-network theory' to call landscape an 'immutable mobile', an entity whose simultaneously imaginal and material constitution is held in uneasy and contingent unity by its usage in society. As Barbara Bender (1998: 5) commented in her study of Stonehenge as a landscape of conflict:

> Landscape – people's engagement with the material world – not only works between fields of knowledge, but also incorporates everyday life and contemporary politics, and highlights institutional constraints. It offers a holism: parts of life that are often compartmentalised spill over in a satisfactorily untidy way. And once you start to think your way into landscape you discover that it permeates everything. Our language is saturated with landscape metaphors. Look at this last paragraph: 'sign-posts', 'boundaries', 'divides', 'wideopen' [*sic*], 'exploration' and more ... We talk and walk landscape every moment of our lives.

Unfortunately Bender's enthusiasm for the landscape may have had the inadvertent consequence of rendering the concept less rather than more meaningful through ubiquity. However, like Tuan and others, there is no doubting the importance of her view that landscape is a powerful means of investigating interaction between a range of social dynamics and processes because of the way it blends imaginal and material qualities. Most importantly, as a repository for material and emotional value, landscape forms an important arena in which conflict between sets of values and investments takes place. Significant here, for our purposes, is the assertion that landscape and landscaping are an expression of power, of material and symbolic capital. As Zukin (1991: 16) remarked in her analysis of the impact of capitalist relations on the urban landscapes of North America: 'In a narrow sense, *landscape* represents the architecture of social class, gender, and race relations imposed by powerful institutions.'

Nevertheless, as she also noted (*ibid.*), this is only one dimension to landscape in a broader sense:

> it connotes the entire panorama that we see: both the landscape of the powerful – cathedrals, factories, and skyscrapers – and the subordinate, resistant, or expressive vernacular of the powerless – village chapels, shantytowns, and tenements . . . As the opposition within vernacular implies, powerful institutions have a preeminent capacity to impose their view on the landscape – weakening, reshaping, and displacing the view from the vernacular.

ARENAS OF CONFLICT

Landscape therefore denotes a contentious, compromised product of society, a feature highlighted by the prominence recently given to the role of landscape as a site of contestation. As Mitchell (1993: 11) remarked, we can examine the production of landscape in the way that social groups with contrasting access to power, varying financial and social resources, and differing ideological legitimacy struggle over issues of production and reproduction in place. The form of the landscape emerges from these contests (Anderson 1987; Mitchell 1993: 11).

For some authors (Rees and Borzello 1986; Rose 1992; 1993), the history of landscape is almost synonymous with processes by which the wishes of the powerful come to be imposed on the powerless. This perspective, for example, has informed studies of elite landscaping practices in eighteenth-century England (Pugh 1988; 1990; Williamson 1998), urban landscapes of consumer and financial capital (Domosh 1992; 1996; McDowell 1997) and the activities of European colonisation (Smith 1984; Hughes 1997; W.J.T. Mitchell 1994). A particular focus of interest lies in the relations between colonists and indigenous peoples during the period of European colonisation. Encounters between colonisers and indigenous peoples repeatedly threw up different interpretations of the relationships between people and land. Colonists often regarded land primarily as a tradeable commodity with value primarily to the extent that it was a source of expropriable material wealth. By contrast, the indigenous landscape ethic might well have given priority to spiritual values, stewardship, communal use and inviolable rights to dwell (Pawson 1992; Young 1992; Hirsch and O'Hanlon 1995; Ellen and Fukui 1996; Strang 1997). Such cultural differences over landscape became institutionalised in treaties and agreements, particularly those establishing the enforced concentration of colonised peoples into homelands and 'reservations'. From those beginnings, the contested nature of landscape led to continuing conflict. As the settlement history of North America, South Africa, New Zealand and Australia shows, culturally specific conceptions of landscape and the social and material inequalities that these imply are subject to defence and challenge.

A related point can be made by returning to the exchange between Barbara Bender and Sharon Zukin. In contrasting high culture with the vernacular, Zukin actually had much in common with Bender and others who interpret landscape anthropologically. In this sense landscape is the outcome of material practices, the product of human activity traced across and marked out in and on the land (Ingold 1996). Like Zukin, Bender contrasted the formal, reflexively articulated landscapes of authority with the landscapes of everyday practice. These may be formed from the slow accretion of dull routine or the fleeting marks of deliberately counter-hegemonic activity. This perspective offers a way of examining landscapes of defence as a means of resistance to formal authority. For instance, Bender (1998: 3–4) quoted Pred (1990), who, when trying to assess opposition to the increasingly controlled landscape of the later Neolithic, had written: 'in trudging a well trodden path, in creating short cuts and detours to interrupt the humdrum and the accustomed, in self-imposing obstacles and intentionally avoiding accessible pathways, in making . . . [a person makes] a statement.'

This formulation echoes Michel de Certeau's (1984) account of walking in the city as a counter-cultural activity of making landscape. For de Certeau, all human activity is inscriptive. Human actions become stories, consciously and unconsciously written on the world and its inhabitants. Each narrative maps out a landscape. The trajectory becomes a reflexive justification of activity, which creates the 'landscapes' in which we live. To de Certeau (*ibid.*: 123): 'everyday description is more than a fixation, [it is] a culturally creative act.' He described two forms of spatiality, which he termed 'strategies' and 'tactics', a dichotomy that recognises the differing power and knowledge capacities of different forms of discourse. This in itself reflects the way that landscape brings together that which is normally held apart.

To elaborate, 'strategies' are spatialised practices, formalised in writing, that give strategic activity a technocratic purchase on the world. In this context, de Certeau's use of the word 'landscape' is not uninformed, since it relates to writing that is scientific, logical, progressive and taxonomic with implications of linear perspective. Conversely, 'tactics' are opportunistic actions that weave in and out, temporarily colonising the space opened up by the sciences of formally organised writing. By referring to 'tactics' as opportunistic, de Certeau associates them with risk taking, struggle and achievement against the odds. He is concerned with the processes situating the individual in society and characterises this relationship as one between 'the sciences of space' and the 'arts of living'. Although romantic, this does set out a plan for exploring the relationships between dominant ideology and counterculture that stresses the ways in which they are mutually interdependent in the production of landscape. Rather than one simply feeding off the other, all inscriptive activity not only remakes and reproduces the material of which it is constituted, it also opens the space necessary for further writing – at once, both emancipatory and imprisoning.

In this sense, de Certeau's formulation is reflected in the criticism that Mitchell made of Zukin's work. Mitchell argued that Zukin was right to recognise both dominant and subaltern landscapes. Nevertheless, her argument suggested the two categories, 'high culture' and 'vernacular', as being separate and complete in themselves. Mitchell (1993: 10) suggested that Zukin and others had forgotten that landscape is both process and product, a means of representing and something represented:

> Landscape is best understood . . . as a certain kind of produced, lived, and represented space constructed out of the struggles, compromises, and temporally settled relations of competing and cooperating social actors: it is both a thing and a social 'process', at once solidly material and ever-changing. Like a commodity, however, the evident (that is temporarily stabilised) form of landscape often masks the facts of production.

Landscape is then an aesthetic object, the product of representation, unified and stabilised. It disguises and smoothes over the clashes of values, aspirations and ideals that characterise landscape as a site of contestation.

This view was set out most usefully by Daniels (1987) in an essay on the 'duplicity of landscape'. For Daniels, landscape was an ambiguous synthesis that could neither be completely reified as an authentic object in the world nor thoroughly discounted as an ideological mirage. Obfuscating the divide between culture and nature, and the power relationships suggested by this division, landscape as pictorial perspective hid the very processes and practices that create it. Its aesthetic and political triumph was the process of naturalisation, an ability to separate the sign from the practices that produce signification. This is an aspect of landscape central to our understanding of its role as a defensive strategy.

Hence, in light of this discussion, we might now be able to understand Appleton's 'predatory' view of defensive landscapes as one that is complicit in landscape's ability to conceal its own artifice. Assuming without question the apparent perceptual naturalism of landscape, Appleton consequently attributed this to the effects of instinct on perception (Mitchell 1993: 16–17). To take another example, Allen (1999) shows how landscape's ideological role in the naturalisation of social and productive relations was important for the representation of contemporary ethnic conflicts in such areas as Liberia, Sudan and the former Yugoslavia. He argued that the rise of new nationalisms, and the intra-state conflict that this brings, have placed current conflicts beyond the realm of the conventional and familiar means for representing wars between states (*ibid.*: 12). The frequent result is that Western media and academics resort to metaphors that suggest regression from civilised order into an atavistic primitivism. While the origins of wars such as the First World War or the American Civil War are normally studied in terms of contemporary economic and political processes, the origins of war itself are frequently explored with reference to archetypal models derived from literature on groups like the Yanomami of Brazil. The purpose is to suggest models of primordial behaviour and to explore the possibility that human beings need to wage war, either due to an innate propensity or because of the manner in which they cohere and survive as social groups (*ibid.*: 15). In this formulation, aggression is linked causally to people who live 'close to the land'. Most frequently nationalist groups, their controlling elites and propaganda readily supply the Western media with a discourse of blood, soil and ancient history, supported by images of burning farms, ransacked villages and fetishised emblems of ethnic homeland. Together these establish the idea that particular landscapes are natural sources of conflict, which in turn is legitimate as an expression of basic instincts or ethnic birthright.

Despite the different context, there are important parallels with Neil Smith's (1996) arguments about gentrification in the cities of North America. Smith described the way that discourses of urban renewal from the mid-1960s came to represent urban redevelopment in language derived from the American landscape undergoing European colonisation in the eighteenth and nineteenth centuries. The city was represented as a wilderness and its zones of reconstruction and rehabilitation as the urban frontier. In the language of gentrification, the appeal to frontier imagery became exact: 'urban pioneers, urban homesteaders and urban cowboys became the new folk heroes of the urban frontier.' During the 1980s (*ibid.*: xiv), 'the real estate magazines even talked about "urban scouts" whose job it was to scout out the flanks of gentrifying neighbourhoods, check the landscape for profitable reinvestment, and, at the same time, to report home about how friendly the natives were.' Smith concluded: 'In the end and this is the important conclusion, the frontier serves to rationalise and legitimate a process of conquest, whether in the eighteenth and nineteenth century West, or in the late-twentieth-century inner city' (*ibid.*: xv).

As an arena of conflict, and in particular one that contrasts elite pictorial order with vernacular networks of social routine, we may begin to understand the role of landscape in counter-hegemonic practices – the practices that endeavour to break the ideological stranglehold of those defensive landscapes by which dominant groups hold subordinate groups in place. Often this is couched in a language of mobility and stability that recognises the role of dominant landscapes in the establishment of social and visual order. In this context, philosophers, and social and political theorists argue that 'nomadic', mobile or anti-territorial behaviour can be politically revolutionary, undermining the taken-for-granted basis of society (see also Virilio 1986; Deleuze and Guattari 1987). Applying these ideas, for example, researchers have examined the destabilising effect on conventional British notions of rural landscape of new age travellers, and anti-nuclear and anti-road protesters (Cresswell 1996; Hetherington 1997). Here, too, we might begin to understand the landscape interventions of environmental protest groups such as 'Reclaim the Streets'. Their encouragement of active reappropriation of road space, legally defined as thoroughfare, for enjoyment, socialisation and children's play, consciously work in opposition to the current conventions of urban life. Their actions deliberately hark back to visions of past life in the city in order to demonstrate the extent to which government policy and contemporary capitalism have impoverished the lives of city dwellers. The conflicts between activists, residents and the authorities are couched in terms of defence: as freedom of movement and the commercial life of the nation versus protection of children's lives and the viability and vitality of community. At the same time, each enshrines differing conceptions of landscape: on the one hand as classified, compartmentalised and an economic utility; on the other hand as sociable, communal and multi-use.

This brings home a central problem. We might think of individual 'landscapes' as being compromised, partial, contested and only provisionally stable as modes of ordering the world and our engagement with it. If so, this suggests that we should not think of individual landscapes as discrete pieces of territory because they are supported by, and help to sustain, the interests of mere sections of any given society. Alternatively, we might think of landscapes as being formed only in relation to other landscapes and conceptions of landscape. In that case, perhaps also we should base our analysis in terms of the interconnectedness of landscape, its links with other landscapes, other geographies. For the study of landscapes of defence this suggests that we examine how its discourses and practices create landscapes that ensure security for some while legitimating insecurity for others. The cultural historian John Barrell recognised this while studying depictions of the English rural poor during the worst excesses of parliamentary enclosure. Barrell referred to the 'dark side to the landscape'. This dark side is something that is not merely mythic, not simply a feature of the regressive instinctual drives associated with non-human 'nature'. Rather, it is a moral, ideological and political darkness that shrouds landscape as a producer of inequality. Don Mitchell (1994: 11; see also Harvey 1985), in part of his commentary on Zukin, remarked that:

> If, as Zukin [1991: 14] ... claims, a 'landscape mediates, both symbolically and materially, between the socio-spatial differentiation of capital implied by *market* and the socio-spatial homogeneity of labour suggested by *place*', then we must begin to see shantytowns, homeless shelters, housing projects, tract home developments, and for that matter, nature or the countryside, as the *necessary* obverse of the 'urbanisation of capital'.

In that case, for example, we may better understand the justification of urban security measures that displace people from 'trouble spots' to less sensitive locations without regard for their inhabitants and users. It may help us to understand the activities of national agencies that wish to export their toxic waste to countries that have 'a pollution deficit'. Landscapes that express power and privilege are always the flip side of landscapes of exploitation and disadvantage. To understand landscape as a defensive strategy is to recognise these processes at work.

ORGANISATION AND STRUCTURE

The chapters in this volume have been grouped together in three thematic categories, which, while far from mutually exclusive, draw out issues central to defence and security in the contemporary landscape. The first concerns *ontological security and the distribution of risk*, in other words, issues involving individuals' basic sense of self, their knowledge of the world and their place in society. In the opening chapter in this section, Andrew Blowers examines landscapes of nuclear risk. The threats posed by nuclear power constitute a defining case for the type of socially produced risk analysed by Ulrich Beck, his followers and associates. As a product of human science and technology, such 'mega-risks' lie at the heart of contemporary mistrust of systems of expert knowledge, which contribute to the formation of ontological insecurity. At the same time, the case of the nuclear oases brings into focus a contemporary politics of inequality based on the distribution of risks rather than resources. In this context, Blowers examines the future for the nuclear industry at four sites: Hanford in the USA, Cap de la Hague (France), Sellafield (United Kingdom) and Gorleben (Germany). He considers popular acceptance of high-level risk within these localities in terms of a variety of geographical factors, their remoteness, economic marginality, the social and political powerlessness of local communities, environmental degradation, and local cultural acceptance of the nuclear industry. Blowers argues that these communities are not so marginal or defensive as might first appear. Although they may be dependent on the nuclear industry, the industry is also dependent on them because of the formidable obstacles it faces in the search for suitable sites.

Stuart Aitken's chapter moves the study of risk to a neighbourhood and domestic scale. He examines the fears and anxieties of young parents in San Diego, southern California. Using questionnaires and a series of in-depth interviews, he examines how the experience of having children changes people's perceptions of urban risk and security. He argues that the birth of a child promotes certain attitudes and values, which translate into strategic choices. These, in turn, bear heavily on the construction of landscapes of defence. He then shows how these fears are both consolidated and ameliorated by a parent's reactions to the urban landscape. He highlights parents' and children's perceptions of the scale of the local community, their active relocation to a 'safer' area and involvement in community activism as key factors. However, these are in themselves exclusionary gambits by which individualised and privatised family units seek to protect themselves against an unknown, demonised 'other'. In the process, they help to perpetuate precisely the form of ontological insecurity that lies at the core of such exclusionary activity.

The subsequent chapter by Stanley Brunn and his associates focuses on the creation of landscapes of defence as an expression of social power. They produce a wide-ranging survey of the physical forms that landscapes of defence may take. They argue that the social practices producing individual, residential, corporate, military and political spaces are identified and marked by subtle landscape elements and features that separate, integrate and unify individuals, groups and communities. These landscape features are in essence formed from subtle and overt symbols of power, which, they suggest, are basically protective and defensive in nature. The authors develop a typology of defensive landscapes that connects the built form of defensive landscapes, along with their organisational and technological features, to the sources of insecurity that typically produce or represent such landscapes. In doing so, they usefully summarise many connections between the physical form of defensive landscapes and their symbolic meanings.

The second of our organising themes is that of *boundaries, identity and symbolic landscapes*. This theme, which is addressed in the next four chapters, sees authors adopting approaches that link both to notions of landscape as a unified and ordered pictorial image and to ideas of landscape in an anthropological sense as a set of meaningful lived routines or practices. These chapters indicate the importance of landscapes of defence as simultaneously material and imaginal entities.

Peter Shirlow's paper on 'fundamentalist loyalism' is based around a case study of the Loyalist Volunteer Force in Northern Ireland. It examines the ways in which the ideological formations of fundamentalist loyalism in Northern Ireland are expressed, reinforced and amplified in an ethno-sectarian sense of landscape. Shirlow shows how particular historical and mythical resources are marshalled and translated into an effective focus for community action. Here, landscape symbolism produces a sense of homogeneity and community against which to contrast an objectified 'other' as enemy. He concludes that religion, as an instrument in the creation of landscapes and the performative enactment of violence, is crucial in terms of understanding the rise of militia groups. There are clear parallels to be draw to ethnic conflict and the growth of violent religious movements worldwide.

Maoz Azaryahu's chapter on Israeli 'securityscapes' examines the pervasiveness of 'security' in the cultural milieu of daily life for Israel's Jewish population. His discussion lends weight to Shirlow's assertion of the widespread importance of ideologically anchored social spatialisations in the maintenance of ethnic violence. Azaryahu believes that concern about, and quest for, security in Israel resonates with an ingrained ideology of national survival. Landscapes of defence, therefore, are symbolically central to Israeli identity and therefore transcend many political divisions within Jewish-Israeli society. Azaryahu finds the symbolic reminders of the Israeli condition at a variety of spatial scales: a planned fence between the West Bank and Israel; the towers and stockades of Jewish *kibbutzim* planted on occupied territory; routine patrols by the military and notices on buses warning against abandoned parcels.

His emphasis on a morphological approach is also followed by Jon Coaffee in his chapter on the impact of terrorism on the landscape of the City of London in the 1990s. He traces the strategies used by the security agencies, the local authority, the business community, insurers and government in response to this threat to a high-profile financial and banking centre. Coaffee highlights three urban security strategies that emerge from this analysis. These are management, where a series of spatial and temporal regulations or laws are enforced; fortification, with the introduction of physical defensive features such as walls, barriers and gates; and surveillance, through increased policing and electronic means, particularly closed-circuit television. He shows how a complex series of security measures was instigated that, he argues, had both a practical and a highly symbolic role in maintaining confidence in the City of London as a world financial centre.

Martin Phillips's study of exurban private communities in North Carolina shares Stuart Aitken's concern for constructions of risk and security at the domestic level. Phillips examines ideas concerning the 'militarisation' of residential space in the USA through a case study of 'gated' or 'walled' communities. He examines the range of physical defensive measures used to fortify such residential developments and the discourses of defence and exclusivity used to market them. He argues that features of the defensive landscape such as gatehouses may be more important for the symbolic construction of exclusivity than a feature of security. Conversely, the marketing of leisure pursuits in such residential developments, ostensibly a mark of status, may also serve a defensive function simply because they also produce socio-economic exclusivity.

It is clear that processes of inclusion and exclusion are fundamental both to the creation of landscapes of defence and to the identity of those who are excluded or included. The operation of these processes highlights the need for explicit discussion of issues of *citizenship and governance* – our third major theme and one that forms a central thread in the remaining chapters. Each of these chapters asks, in its own way, who is responsible for these landscapes of defence, who benefits, who loses and who makes the decisions.

Christina Kennedy and Alan Lew examine issues of authority, control and self-determination for Native American peoples in their homelands. They show how the defensive role of the homelands has been transformed since their foundation in nineteenth century. Originally established to protect non-Indians from the military threat of 'captive nations', they are now cultural reference points fundamental to the defence of Native American ways of life. Kennedy and Lew show how many Native American peoples are taking active steps to counteract attacks on their homelands and culture through strategies related to citizenship and governance. Many homelands have their own laws and police forces. Increasingly, national groups actively work to gain greater input into the management and protection of places that are sacred to them. The resurgence in cultural pride and increasingly politicised senses of national identity are reflected in an increased emphasis on sovereignty and a determination to gain control of resources as well as policies to limit outside access to homelands, and improve education and the dissemination of traditional knowledge.

In the ensuing chapter, Kate Williams and her colleagues consider current policing and crime prevention strategies in the United Kingdom, focusing particularly on CCTV. By contrast with Kennedy and Lew, their exploration of new modes of partnership in governance between people, civil institutions and government bodies indicates the negative implications and the asymmetries of power disguised in a rhetoric of empowerment. Their paper opens by considering CCTV's role in the strategies of crime prevention and policing, locating the development of CCTV within a broader social context, which includes an increasingly authoritarian approach to tackling crime in the United Kingdom, and an increasing emphasis on collaboration between a range of public and private institutions in local government. They conclude that the cameras often do not live up to the claims made for their use by government, such as deterrence, detection and reduction of fear. Instead, they argue that it is the police themselves who have most to gain from the installation of CCTV systems, not least because funding comes from local government rather than from the police budget. Williams and her colleagues conclude that other agencies are largely excluded from the formation of CCTV policy in terms of the fundamentally important issues of location and operation. Although both of the principal partners are fulfilling the formal requirements placed on them, neither is really working in a true partnership. They also point to the symbolic role played by

CCTV in the development of local entepreneurialism. In this context, the fact that CCTV does not deliver crime reduction or ease the fear of crime does not necessarily matter, since it is being used more as a symbolic tool to stimulate economic development.

Taner Oc and Steven Tiesdell focus on related matters. Their concern is with reducing the fear of crime in town centres as part of a strategy of urban regeneration. Like Williams and her associates, they recognise that the defensive landscapes so created are as much a matter of urban marketing as they are policies for crime prevention. They outline four design strategies that might be adopted to create safer urban spaces. These range from the target hardening measures of the 'fortress' approach to the 'animated' approach, which holds that safety comes from creating an urban environment that is lively and welcoming to people. Strategies for the city centre as a focus for a leisure and entertainment economy can help to provide a form of 'natural' surveillance well into the night. Oc and Tiesdell argue that the seductive, albeit short-term, appeal of fortification is that something is being seen to be done, but while the city centre may become safer for some, it may become increasingly unsafe for others. Alternative ways of making the landscape of the city centre feel safer offer more expansive and positive notions of public space and, indeed, city centres generally.

The remaining three papers explicitly address issues of policing. Like the previous three chapters, issues of citizenship and governance are again in the foreground. John Baxter and Paul Catley examine the role of the British 'bobby' policing the street. They stress the importance of the constable's visible presence to the traditional effectiveness of British policing and examine the history of consensual policing as part of a mutually agreed matrix of authority and control. Like Williams *et al.*, they trace the change to a more authoritarian form of policing back to the political climate of the early 1980s. Baxter and Catley examine the model of 'zero tolerance' policing as it was translated from New York to Cleveland in Britain. They examine the range of problems that the end of discretionary policing implies, including concern for civil liberties, the exacerbation of racial tensions and an increased confusion of the distinction between civil and criminal law. In the worst-case scenarios, they argue this may place the police in day-to-day supervision of even the most routine social interactions, ensuring good citizenship and negotiating urban governance rather more centrally than either they or anyone else would wish.

By contrast, Simon Marshall's chapter examines the design issues of policing and control within a closed and clearly defined environment. Impinging on debates concerning the role of prison as variously a punishment, an act of social containment and a place of rehabilitation, prisons as much as schools or streets are sites for the definition of citizenship. For example, Marshall is keen to assert that prisons are not landscapes of power but landscapes of *authority*. He argues that strong physical security will not ensure control; indeed, evidence suggests that it may in fact incite disorder. Cooperation, a set of shared values and an agreed code of behaviour are clearly fundamental to the routine functioning of such institutions. Like Brunn *et al.*, Marshall is concerned with the analysis of built form, like Blowers he is concerned with an assessment of risk, yet like Oc and Tiesdell he is ultimately sceptical of the power of built form alone to solve the problems of these defensive landscapes. Marshall uses a number of perspectives on urban form to analyse the built form of UK category C prisons. He concludes that the importance of the built environment of prisons should be neither underestimated nor overstressed. All too often since the days of Jeremy Bentham, penal research has fallen back upon environmental determinism when confronted with unexplainable differences that arguably require a social solution.

The final chapter deals with the citizenship and governance theme yet also returns us to the issues of risk and ontological security raised in the first section of the book. In his analysis of the policing of street prostitution, Phil Hubbard describes a landscape of contestation, exclusion and resistance that echoes the discussion in Shirlow's contribution. Focusing on defensive landscapes as a form of social exclusion, he argues that attempts to purify residential spaces by removing prostitution represent part of a more general process in which threatening 'others' have their access to public space challenged. As Hubbard, Shirlow and others in this volume show, it is a process that may force exclusions based on gender, race or class. Street prostitutes are scapegoats, identified as a polluting influence detrimental to local amenity and quality of life. He concludes that it is perhaps not surprising that there have been a number of high-profile protests against prostitutes in many British red-light districts and elsewhere. People's desire for 'ontological security', for a sense of well-being and identity, is increasingly manifest in defensive protests designed to maintain social and spatial ordering.

CHAPTER 2

LANDSCAPES OF RISK: CONFLICT AND CHANGE IN NUCLEAR OASES

Andrew Blowers

Risk from radioactivity is present at all stages in the nuclear cycle, from uranium mining through enrichment and fuel fabrication, energy production and reprocessing to the management of radioactive wastes. Nuclear energy is produced in over thirty nations, the biggest concentration being in Europe (including the former Soviet Union) with over 200 reactors, North America with 130 and Japan with fifty-one (Table 2.1). Reprocessing spent fuel to recover plutonium was originally developed for military purposes. There are five so-called nuclear weapons states – USA, Russia, China, France and the United Kingdom – together with India, Pakistan and Israel, which are known to possess nuclear weapons. Others have achieved the technical basis for making them. With the ending of the Cold War, military reprocessing has largely ceased, although the capability remains, and decommissioning of weapons has added to the burden of nuclear wastes. Reprocessing to produce materials for nuclear fuel for the civil nuclear programme is carried out in France and the United Kingdom, with Russia continuing to reprocess fuel from reactors in Eastern Europe and the former Soviet Union.

All these activities produce nuclear waste in the form of spent fuel, high-level waste (HLW) from reprocessing, intermediate-level wastes (ILW) from reactor operations and voluminous low-level wastes (LLW) from a multiplicity of operations. In addition, decommissioned plants leave a legacy of clean-up, restoration and waste management that will take decades to complete. With the construction of nuclear power plants effectively over in most Western countries, the key sites for conflict over the nuclear industry have shifted to the centres of reprocessing and waste storage. As Zonabend (1993: 128) observed: 'The popular opposition that was once in evidence against the building of nuclear power stations is now directed against the construction of these storage sites.'

The setting of these sites is distinctive. They are in remote locations, peripheral to the main centres of population and economic activity, and consequently have been termed 'nuclear oases' (Blowers *et al.* 1991). In both social and environmental terms they are 'landscapes of risk'. The basic argument of this chapter is that these areas provide the key to the future of the nuclear industry, notably the problem of managing the burden of nuclear wastes into the far future. In particular, they reveal the characteristics of peripheral communities where social changes define and

Table 2.1   Nuclear power reactors and generating capacity by country

	In operation		Under construction
	Number of units	Total net MWe	Number of units
Argentina	2	935	1
Armenia	1	376	
Belgium	7	5,527	
Brazil	1	626	1
Bulgaria	6	3,538	
Canada	21	14,907	
China	3	2,167	
Czech Republic	4	1,648	2
Finland	4	2,310	
France	56	58,493	4
Germany	20	22,017	
Hungary	4	1,729	
India	10	1,695	4
Iran	–	–	2
Japan	51	39,917	3
Kazakhstan	1	70	
Korea, Rep. of	11	9,120	5
Lithuania	2	2,370	
Mexico	2	1,308	
Netherlands	2	504	
Pakistan	1	125	1
Romania	–	–	2
Russian Federation	29	19,843	4
South Africa	2	1,842	
Slovak Republic	4	1,632	4
Slovenia	1	632	
Spain	9	7,124	
Sweden	12	10,002	
Switzerland	5	3,050	
United Kingdom	35	12,908	
Ukraine	16	13,629	5
United States	109	98,784	1
World total *	437	343,712	39

* The total includes Taiwan, where six reactors totalling 4884 MWe are in operation. *Notes to table*: data are subject to revision. During 1995 two reactors were shut down (including Bruce-2 in Canada, which could restart in the future); seven reactors were connected to the grid, and the construction of three reactors was temporarily suspended (in Romania).
Source: *IAEA Bulletin*, 1/1996, p. 53.

constrain the power of the nuclear industry. These changes need to be understood if acceptable and lasting solutions to the problem of nuclear waste are to be found.

LANDSCAPES OF RISK

Nuclear oases are in areas that experience actual or potential radioactive contamination and whose inhabitants are exposed to the possibility of impacts from radioactivity. They are the defining case of landscapes of risk. The impacts arising from the deliberate, accidental or routine releases of radioactivity generate greater anxiety than perhaps any other socially induced risk in the modern world. These impacts include the chronic effects of exposure resulting in cancers, reproductive failure, birth or genetic defects, and the wholesale death and destruction that can result from a major nuclear accident or a thermonuclear conflict. Radioactivity presents a low-probability/high-consequence risk that, once in the accessible environment, spreads via pathways of water and atmosphere covering wide areas and crossing national frontiers.

In common with certain other 'mega-risks', such as petrochemicals and genetic engineering, nuclear risk is a product of science and high technology where knowledge of causes, impacts and solutions is in the hands of experts (Boehmer-Christiansen and Skea 1991; Boehmer-Christiansen 1994). Experts exert enormous power over decision makers despite the problems of empirical uncertainty and theoretical controversy on which their pronouncements and proposals may be based (Buttel *et al.* 1990).

In certain respects, nuclear risk is a distinctive, perhaps unique, case. In the first place it is a 'dread risk' (Slovic *et al.* 1980), invisible and engendering greater and more widespread fear than any other form of mega-risk. Second, it is feared whatever the source and whatever the level of radioactivity (Bertell 1985). Third, the nuclear industry is identified with its military origins, and the civil and military components of the industry are related to a greater or lesser degree. Fourth, the scale of impact may vary. Unlike global warming, radioactive emissions do not inevitably pose a global threat; they may be limited in scale covering an area within the vicinity of a release. Yet once a major release occurs, its widespread impact is unavoidable and irreversible. Radioactivity presents a risk of regional and potentially global reach ultimately posing a threat to the environment and to people for which there is no effective defence. In the ultimate case, a reactor meltdown or a nuclear war, the impact will almost certainly be catastrophic. The impact is sudden, total and instantaneous. As Beck (1992: 38) put it so cogently: 'The effect only exists when it occurs, and when it occurs, it no longer exists, because nothing exists any more.' Fifth, nuclear risk is intergenerational. Its impacts can be gradual and long-lasting, perhaps not manifested for many generations. This is the hazard posed by nuclear waste, which contains some radionuclides, which will remain highly dangerous with half-lives extending to hundreds of thousands or even millions of years. For example, plutonium, one of the most dangerous substances known, has a half-life of 24,000 years and will remain dangerous for an unimaginable time span.

THE GEOGRAPHY OF LANDSCAPES OF NUCLEAR RISK

Landscapes of nuclear risk occur anywhere where there are nuclear facilities or on the routes over which nuclear materials are transported. In some cases, major accidents have contaminated large areas. The most extensive examples are the area around Kyshtym near Chelyabinsk in the Urals, where a probable explosion at a reprocessing plant left a substantial area of devastation and contamination (Medvedev 1979), and the reactor meltdown at Chernobyl in the Ukraine in 1986, which affected a substantial surrounding area (Medvedev 1990). Other accidents at Chalk River, Canada (1952), Windscale (now Sellafield) in the United Kingdom in 1957 and Three Mile Island in the USA in 1979 also contributed to a growing perception of the dangers associated with nuclear facilities. Around many nuclear sites the contamination from routine releases or from poorly managed waste facilities has left a legacy of risk and a problem of cleaning up, especially where the rear-end operations of the nuclear industry are located (Figure 2.1). In addition, there are those areas of potential risk where the nuclear industry is seeking to develop facilities in greenfield locations.

In the USA, landscapes of risk occur in the various military plants scattered across the country. Of these, the largest is the Hanford nuclear reservation in Washington state, where operations began in 1943 to produce the plutonium used in the first atomic bombs and where a clean-up is now a priority to prevent leakage of high-level wastes into the Columbia River (Gerber 1992). The Idaho National Engineering Laboratory (INEL) and the Savannah River Plant in South Carolina, both developed in the 1950s, are also engaged in clean-up activities. Problems of contamination have also occurred at the Rocky Flats bomb-trigger plant near Denver; at the Oak Ridge, Tennessee, research centre; at the bomb assembly Pantex plant at Amarillo, Texas, and at a number of other sites. Decommissioning is taking place at a former civil reprocessing works at West Valley in New York state and at several closed nuclear waste facilities (Sheffield, Illinois; Maxey Flats, Kentucky; Beatty, Nevada). There are also two greenfield locations. One is at Carlsbad, New Mexico, a deep repository in salt opened in 1999 for the disposal of military wastes, the other at Yucca Mountain, Nevada, the site of the controversial deep-disposal repository for the country's spent fuel from the civil nuclear programme.

In Western Europe, the significant landscapes of risk are the reprocessing centres in the United Kingdom at Sellafield (begun 1947) and Dounreay (early 1950s, now largely devoted to decommissioning activities) and the French plants at Marcoule (1958) and Cap de la Hague (1966). In Germany, Gorleben in Lower Saxony is the proposed greenfield location for a nuclear waste repository and the site of a storage facility and associated processing plant. In Japan, Roakkashomura is the proposed location for both a reprocessing plant and a storage site for high-level wastes.

Russia has a number of seriously contaminated landscapes of risk arising from both military and civil nuclear operations. On the coast of the Kola Peninsula in the far north there are decaying reactors from nuclear submarines. Across the country there are twenty-nine power reactors, including one fast breeder, located at nine sites, mainly in European Russia; these under-maintained and aging plants constitute a significant risk. Reprocessing and a mixed oxide (MOX) fuel-fabrication plant are

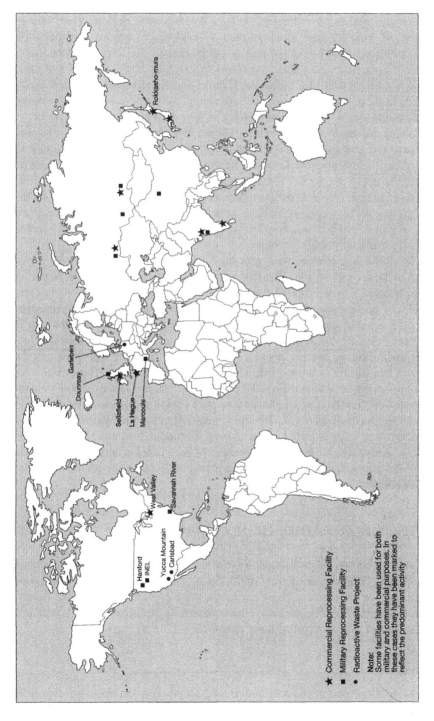

Figure 2.1   Location of reprocessing plants and nuclear waste facilities

concentrated at Chelyabinsk (Mayak). Tomsk and Krasnoyarsk are Siberian centres of plutonium production and reprocessing with associated nuclear waste storage facilities. At each of these locations there is evidence of serious contamination (WISE 1999). In Eastern Europe and in other parts of the former Soviet Union there is a legacy of aging plant, problems of clean-up and a lack of sufficient finance, notably in Bulgaria, Lithuania, Ukraine and Kazakhstan. Elsewhere in the world there is an established nuclear industry in India and a rapidly developing nuclear investment in the Far East, notably in China, South Korea and Taiwan.

Although the landscapes of nuclear risk are spread across countries in several continents, this chapter will concentrate on examples from the Western world. In these countries it is possible to observe the social changes that are beginning to shape the future of the nuclear industry and, in particular, the prospects for nuclear waste management. While experience here may not be translated easily into different social and political contexts in other parts of the world, it does provide evidence of the sources of conflict and the possible means of resolution of what has become a difficult, if not intractable, problem.

In the following section, four sites have been selected to explore the social context of landscapes of risk: Hanford in the United States, Cap de la Hague in France, Sellafield in the United Kingdom and Gorleben in Germany. They are examples of nuclear oases at different points on a continuum moving from established to greenfield sites. Moving along the continuum from Hanford, through la Hague and Sellafield to Gorleben there is increasing conflict over the nuclear industry. In the following section, the social changes in these communities will be examined empirically in terms of their peripheral characteristics. The existence of integrating processes within them will be revealed. This leads on to a more theoretical discussion that challenges certain assumptions about the nature of modern society and suggests that traditional, and more modern, forms of community integration are a significant component of social change. In the final section, I shall suggest how an understanding of the social context of nuclear oases is necessary in order to identify the principles upon which solutions may be found to the problem of managing radioactive wastes.

THE PERIPHERAL NATURE OF NUCLEAR OASES

The four sites chosen are each examples of what have been described as 'peripheral communities' (Blowers and Leroy 1994). Such communities can be characterised in terms of five dimensions. These are remoteness, economic marginality, powerlessness, a culture of acceptance and environmental degradation. Each site will now be examined in turn in terms of these dimensions.

Hanford

Hanford is at one end of the continuum, long established and relatively free from internal conflict over the nuclear industry. The pre-existing agricultural community was displaced by an adventitious population working in the various plants primarily dedicated to the production of plutonium for nuclear weapons. The reasons for its location are fairly obvious. The Columbia River could provide the vast quantities of water needed to cool the reactors, and there was a plentiful supply of hydroelectricity from the dams along its length. In addition: 'There were few people to relocate, few eyes, ears, and lips to compromise the high degree of secrecy that needed to be maintained, and few potential casualties in the event of an accident' (Dunlap *et al.* 1993: 137).

Remoteness Thus, Hanford was certainly remote, and it also occupied a very large site (640 square miles). It was around 150 miles from Spokane, the nearest large city, and 250 miles from Seattle and Portland, the major cities across the Cascade Mountains to the west. Surrounded by river and marshes and away from major roads and airports, its isolation was complete. It has remained so, although the cities of Richland, Kennewick and Pasco now have a combined population of around 100,000, and they are heavily dependent on the nuclear reservation, thus constituting the archetype nuclear oasis (Figures 2.2 and 2.3).

Economic marginality This dependence renders Hanford economically marginal. The nuclear reservation was selected in 1942. Within a decade, 'the Columbia Basin, chosen as the site of the Hanford project partly because it was sparsely inhabited, had been transformed by that very endeavour into the third most populous region in Washington State' (Gerber 1992: 56). The sheer scale of the Hanford operations created a monocultural community displacing the earlier population and dominating both the landscape and the life of the surrounding area.

Powerlessness The power of the nuclear industry in Hanford resulted in the powerlessness of the community. As part of the US military complex, Hanford is subject to the impacts of decisions taken elsewhere. During and after the Second World War it experienced vigorous growth, but in recent times, with the ending of the Cold War and the mounting problems of cleaning up the site, the role of the complex has changed profoundly. Moreover, Hanford and its local region are effectively a huge company town in a remote part of a state that is generally hostile to the nuclear industry. Opposition has been fanned both by a general and growing anti-nuclear sentiment and by a number of incidents at the Hanford site. A state referendum taken in 1986, at a time when Hanford was a possible candidate for a high-level waste repository, indicated that over 80 percent of the state's population was opposed to the project. This and other findings indicate that, from being a dynamic and confident focal point of high technology, Hanford gradually became a beleaguered bastion increasingly vulnerable to external pressures.

Cultural acceptance This powerlessness and economic dependence fostered a cultural fatalism, defensiveness and acquiescence. Hanford expressed what Loeb (1986), in a fascinating portrait of the community, has called a 'nuclear culture'. It was 'a demonic fountainhead, a symbol of a broader war culture characterised as a

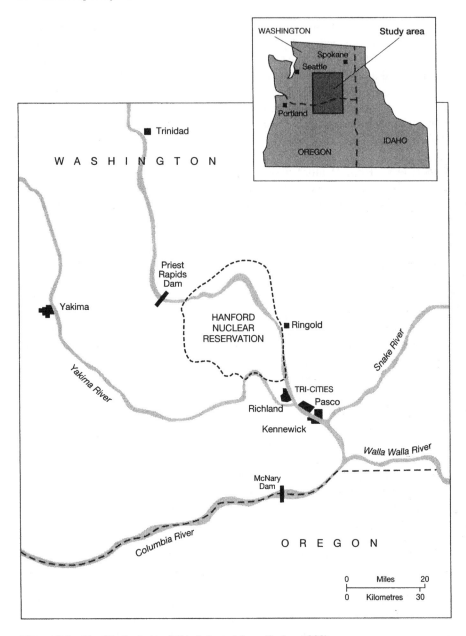

Figure 2.2   The Hanford site, USA (adapted from Gerber 1992)

frontier mentality, aggressively defensive against a hostile world.' The optimism engendered in the early years later shifted to a reluctant recognition of a diminished role. The community has had to adjust to the knowledge that the nuclear reservation is no longer engaged in vital military activities but is a centre for a long-term clean-up programme.

Figure 2.3   Hanford

Environmental degradation   The Hanford site contains areas that are heavily contaminated. This is 'the unfortunate legacy of the rapid and intensive production rate' (*ibid.*: 45) during the war and subsequent years, when speed and short cuts were of the essence. Airborne releases covered large areas downwind of the plant and leakage from tanks entered the groundwater of the Columbia basin. The release of information was parsimonious, but banning of hunting, limitations on the use of pastureland and control of milk indicated concerns about the possible, but little understood, effects of radioactivity on livestock and crops. A number of incidents contributed to growing concerns about contamination. The most notorious was the unannounced and uncontrolled Green Run in 1949 – apparently an experiment to test monitoring methodology. This resulted in the release of 5500 curies of iodine-131 and an inventory of other fission products downwind from Hanford, giving levels of radioactivity far exceeding accepted exposure standards (Gerber 1992; Blowers *et al.* 1991). Other incidents, such as leakage from a waste storage tank in 1973, contributed to the growing opposition elsewhere in the state. The threat to aquifers poses a major clean-up problem, which has now become the major activity at the site.

Hanford, therefore, expresses the changing rhythms of the quintessential nuclear community. Its early history of rapid, almost cavalier expansion, the very symbol of modern high technology, has been eclipsed by its contemporary image as the icon of a declining industry with its problems of environmental degradation and its search for a new social identity. It is a community created by the exigencies of war but later marooned as geopolitical forces changed to remove its *raison d'être*. The environmental legacy that remains ensures that Hanford's continuing role is one of cleaning up the landscape of risk that is the consequence of its former glory.

Cap de la Hague

Cap de la Hague in Normandy, France, is the largest civil nuclear reprocessing complex in the Western world. The industry has not entirely supplanted the pre-existing farming and fishing community but has become the dominant element of a nuclear triangle that also includes the Arsenal at Cherbourg (the production base for nuclear submarines) and the power plant down the coast at Flammanville. This juxtaposition of the modern and traditional indicates that la Hague lies further along the continuum than Hanford. Although the nuclear industry is well established it has not displaced the traditional community and, while it is free of local conflict, it has been the focus of wider concerns about the nuclear industry.

Remoteness   'A peninsula on the end of a peninsula, la Hague at the western extremity of the Cotentin is one of those in-between places. It neither belongs completely to the sea, nor is it wholly attached to the land' (Zonabend 1993: 13). Like Hanford, it was selected as an area sufficiently remote for the secretive nuclear industry to develop reprocessing plants with sufficient capacity to handle spent fuel from the French nuclear industry (second only in size to the United States) and from foreign customers, notably Germany and Japan, which had opted for a closed nuclear cycle but had not yet developed their own reprocessing facilities. La Hague was also selected for environmental reasons in that strong winds and powerful offshore currents would provide dispersion and dilution of gases and liquids from the plant (Figures 2.4 and 2.5).

Economic marginality   La Hague has become the dominant employer in a relat-ively backward area. The nuclear industry directly employs about 10,000 people here, a fifth of the area's total, and they tend to earn more than those in traditional occupations. The industry supports an infrastructure – a four-lane highway and an electrified rail route to Paris serve the peninsula. It builds up the local tax base and provides a range of social facilities. The reprocessing company, COGEMA (Compagnie Générale des Matières Nucléaires), has a profound economic impact in an area that attracts few other activities.

Powerlessness   In this economic context it has proved difficult to exert 'a countervailing power to that automatically and somewhat mysteriously exerted by COGEMA purely by virtue of its dominance over the regional economy' (*ibid.*: 60). For the present, with France remaining strongly committed to the nuclear industry and to reprocessing, support for the industry remains strong among its workers. Local opposition to the industry, more vigorous in the past when the industry was developing, is today muted. The opposition that is manifested is

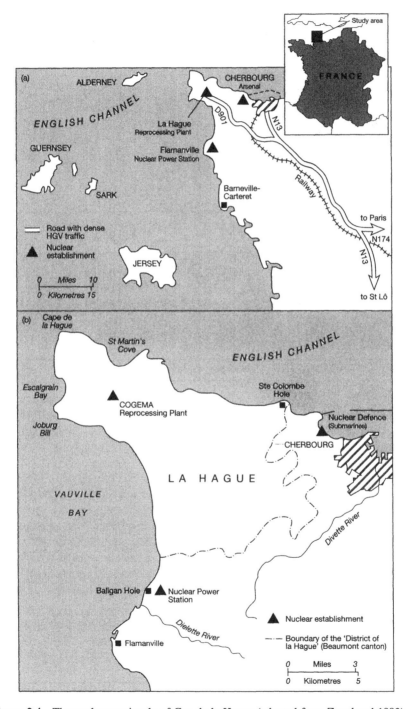

Figure 2.4 The nuclear peninsula of Cap de la Hague (adapted from Zonabend 1993)

Figure 2.5   La Hague (courtesy COGEMA, photographer Gerard Hallary)

directed at the trade in nuclear materials and is largely fomented by environmental groups acting at a national or even international level. Though powerlessness undoubtedly exists it appears to be an acceptable condition for the prosperity the plant brings to the area.

Cultural acceptance   The cultural adaptation to the impact of the plant at Cap de la Hague is complex. A modern industry has been accommodated, if not wholly accepted, into this traditional community. Its economic benefits are welcomed, but the attendant risks (economic and environmental) must also be borne. Hence there is ambivalence in attitudes, expressed to me by one trade unionist as follows: 'The industry is not necessarily popular, but it is necessary.' Zonabend (1993: 120) described the feeling as one of 'repressed anxiety', a 'selective blindness' that 'helps the nuclear worker of today to resign himself to a fate that has always existed, a destiny that none can escape.' The industry and its workers are mutually dependent. There is widespread, if passive and sometimes ambiguous, support in the community at large. As a local deputy put it: 'I don't say it's good – but it is a fact.'

Environmental degradation   This acquiescence is also reflected in the relatively low level of local concern about the environmental implications of the plant. This is partly a result of COGEMA's efforts to provide information, create greater openness and engage in dialogue with the local community. La Hague has also been free of any major incidents that might arouse local concerns. Most of the environmental concern has come from outside, from the nearby Channel Islands and from environmental groups based elsewhere in Normandy. However, la Hague is the focus of international protests led by Greenpeace against the trade in nuclear materials with Japan and Germany. A shipment of plutonium in 1992–93 and of high-level wastes two years later were dogged by international protests along the route from Cherbourg to Japan. Similarly, transport of nuclear materials to the Gorleben store in Germany beginning in 1998 has provoked opposition in both countries. This trade 'until the present day has attracted little attention in France: the nation is *exporting* more and more nuclear services but is *importing* inherent risks of many types and in particular the problem of waste management' (WISE 1994: 23).

Cap de la Hague provides an example of a nuclear oasis within a more traditional community. It remains a dynamic production centre reprocessing fuels for the French and foreign industries. Conflicts in its earlier years have subsided and, although la Hague has captured some international attention, there is evidence of a stable, integrated community in which traditional and modern coexist. Although tensions and anxieties remain they are generally submerged in a common acceptance of economic necessity.

Sellafield

With Sellafield (formerly known as Windscale) we move further along the continuum, encountering much greater local conflict than in Hanford or Cap de la Hague. Although the plant has been established since 1947, it has continued to arouse opposition both within the local community and at a broader national and international level. It is a complex site. Originally constructed for the production of plutonium for military purposes at the Calder Hall and nearby Chapel Cross reactors, it subsequently developed reprocessing facilities for the United Kingdom's original Magnox reactors. More recently, with the opening of the thermal oxide reprocessing plant (THORP), it has engaged in reprocessing foreign fuels. In addition, it manages the bulk of the United Kingdom's radioactive wastes, disposing of low-level wastes at the Drigg repository and storing intermediate- and high-level wastes on site. Sellafield is the heart of the United Kingdom's nuclear fuel cycle and the national focus of contemporary conflict over the industry.

Remoteness   Sellafield is in 'a remote and relatively undeveloped corner of the north of England' (Wynne *et al.* 1993: 7). It is on the west coast of Cumbria, around 330 miles from London and connected to the main routes by inadequate roads and railways (Figures 2.6 and 2.7). As with Hanford and Cap de la Hague this remoteness was seen as an asset. It was in an area of economic decline where 'Nuclear power was seen by many as part of the golden post-war era. These people were not to be deterred by the unknown risks of the industry nor by any rumours of the harm radiation could do' (McSorley 1990: 25–6).

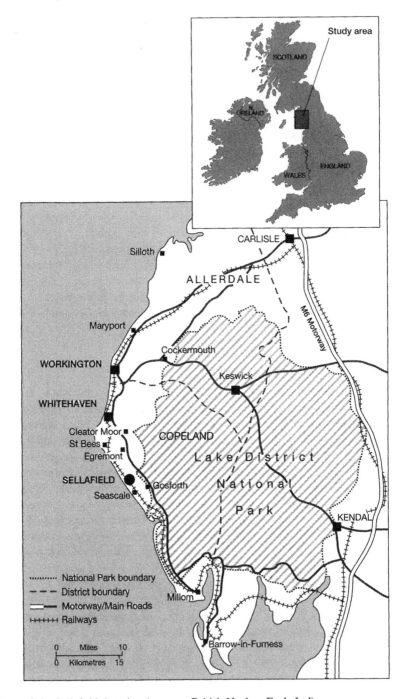

Figure 2.6 Sellafield, location (courtesy British Nuclear Fuels Ltd)

Figure 2.7 Sellafield

Economic marginality The nuclear industry in Sellafield was introduced into an area of traditional farming and fishing and, just to the north, a declining coalfield and iron and steel industry. It was an area of economic risk experiencing relatively high unemployment and identified as an area for special assistance under various regional development initiatives. The nuclear industry at Sellafield, now largely controlled by British Nuclear Fuels (BNFL), has become the dominant employer, accounting directly for about 7000 jobs or a fifth of the total in west Cumbria. Indirectly, it provides income and capital investment for the whole sub-region. Although there have been efforts at diversification, 'this may be interpreted as a recognition of the vulnerabilities of the local economy to nuclear domination' (Wynne *et al.* 1993: 13).

Powerlessness As with other nuclear oases, the fate of Sellafield seems largely to be controlled by external forces. The very power of the nuclear industry, with its high wages and investment, and the stigma that attaches to its activities may have acted as a disincentive to inward investment by other firms. However, dependence on the industry has proved a powerful reason for investment in its continuing development. Despite the opposition to reprocessing on both economic grounds (e.g. Sadnicki *et al.* 1999) and environmental considerations (e.g. CORE n.d.), the need to sustain the west Cumbrian economy was the prime reason for the commissioning of THORP, with its guarantee of continuing employment. Similarly, local employment is a major consideration in the development of the MOX plant. Thus, the ability to defend the interests of Sellafield suggests a form of power that derives from the powerlessness of dependence.

Cultural acceptance   While Sellafield has proved remarkably resilient in securing its position, among its workers there still remains a sense of resignation at being tied irrevocably to a stigmatised industry. As Wynne *et al.* (1993: 37) noted: 'People combined a sense of fatalism about the situation with indications of guilt and even shame at being a "community" which allowed itself to be dictated to, not just by the nuclear industry but by outside forces as well.' Sellafield lacks the cultural homogeneity of Hanford, or even of Cap de la Hague. The surrounding area 'is one of the few locations of relative solidarity and distinctive traditional cultural identity left in industrial Britain' (Wynne 1992: 21). In this context, the nuclear industry has proved divisive, supported by its dependents but resented by other parts of the community. For example, Wynne's study of the hill sheep farmers revealed deeply held animosity towards an industry that could contaminate their environment and affect their economy (*ibid.*). Similarly, the emissions into the Irish Sea have provoked fears in the fishing industry. In addition, the area, which is on the edge of the Lake District, one of England's national parks, has attracted in-migrants who resent the blight and risk associated with the nuclear industry.

The tensions within and around Sellafield have become palpable from time to time, the most notable recent example being the conflict over the proposed Nirex underground laboratory. Although this project was unconnected with the reprocessing industry, the choice of Sellafield had depended on local support for the nuclear industry (Blowers *et al.* 1991). Although the local district council was. generally supportive, Cumbria County Council opposed the proposal, and local and national environmental groups campaigned against it. Ultimately, the proposal was rejected by the Secretary of State on a variety of grounds, including the failure 'to select the site in an objective and methodical manner' (decision letter, 17 March 1987, para. 6xxiii). The Nirex laboratory project had revealed the fractures inherent in a community that has accommodated but not assimilated the nuclear industry.

Environmental degradation   Over the years there have been a series of incidents, most notably the Windscale fire of 1957 and the release of radioactive 'crud' into the Irish Sea in 1983. Routine releases remain a cause of concern, and the stockpile of historic wastes and the decommissioning of plant represent a major future clean-up problem. Some of this work is now in progress. Sellafield, like la Hague, has attracted international attention and demands for a reduction in its emissions to the sea. These reached a peak in the 1970s and have since been reduced with the UK agreeing at the 1998 Ospar Convention (Convention on the Protection of the Environment of the North-East Atlantic) to reduce them to close to zero by 2020.

Sellafield, then, presents a somewhat different case to Hanford or Cap de la Hague. Although it has been developed vigorously in recent years, it remains a source of conflict, which occludes its long-term future. The community is divided, and even support for the nuclear industry, which seems to be strong among its workers, may be 'less solid than is usually assumed, and based more upon fatalistic resignation than positive active espousal' (Wynne *et al.* 1993: 54). The industry continues to be supported because it is firmly established. Where the industry is seeking to establish itself, it is likely to be resisted, as the fourth example shows.

Gorleben

Gorleben in Lower Saxony, Germany, lies at the opposite end of the continuum to Hanford. It is a greenfield location where the industry is not yet fully established and has been the centre of vigorous and sustained conflict over the future of the nuclear industry both locally and in Germany itself (Blowers and Lowry 1996; 1997). Since the 1970s, Gorleben has been identified as the potential centre for the management and disposal of radioactive wastes from the German nuclear programme. Indeed, in the early years it was selected as the location for the 'Integrierte Entsorgungskonzept' (Integrated Waste Management Concept) a combined reprocessing, fuel-fabrication, fast reactor, waste-conditioning, packaging and disposal complex. The reprocessing plans were abandoned at the end of the 1970s. Since then, waste-conditioning and storage facilities have been developed on the site, and a nearby salt dome has been the focus of investigation for a possible deep-disposal facility.

Remoteness   Gorleben is located in the centre of Germany on the River Elbe about 80 miles southeast of Hamburg. It is in the district of Wendland, which, when the site was selected, was a border area surrounded on three sides by East Germany. It was an area of high unemployment and sparse population, on the edge of Western Europe relatively remote from major centres of population (Figure 2.8).

Economic marginality   Wendland is an area with a traditional and conservative rural culture distinguished by its *Rundlings-Dorfer* (round villages). Its economy is predominantly agricultural, with a modest tourist industry. It was an area where it was felt the economic investment represented by the nuclear industry would be welcomed and where resistance would be relatively weak. However, the proposals for an integrated nuclear complex were strongly resisted by local groups, supported by groups from outside the area. Mass demonstrations and the use of tractors by the local farmers to block roads became the typical form of resistance, which has been repeated from time to time. After an international review the reprocessing plans were dropped in 1979 and attention turned to the exploratory drilling work at the

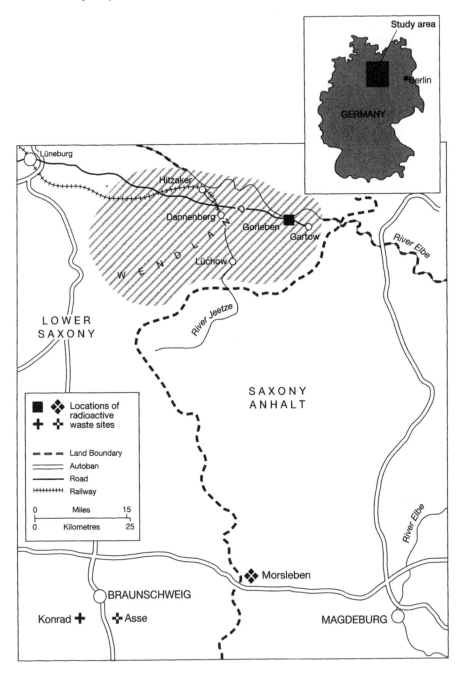

Figure 2.8   Location of Gorleben

Gorleben salt dome, where further protests hampered operations in subsequent years. Meanwhile, the high-level waste store and conditioning plant were constructed to receive spent fuel from German power stations and vitrified wastes returned from overseas (France and the United Kingdom). Opposition to the transfer of wastes resulted in a series of protests during 1995–96, culminating in 1997 in a confrontation with an estimated 30,000 police and border troops in one of the country's largest internal security operations in peacetime.

The level of opposition indicates the difficulties in establishing a greenfield site, even in an economically marginal area, where the risks and blight are perceived to far outweigh the benefits of investment and employment. In any case, without reprocessing the waste management functions of Gorleben will produce only a modest number of employees, and thus the nuclear industry is unlikely to achieve the economic dominance that has occurred in Hanford, Cap de la Hague and Sellafield.

Powerlessness The local community has been unable wholly to resist the presence of the nuclear industry. However, the industry itself has been besieged and nuclear workers, often recruited from outside the area, have expressed fears of violence to themselves and property by the protesters. The local community has been mobilised by the *Burgerinitiativ* and constitutes a 'a broad church that breaks the barriers of age, class, wealth and political affiliation' (*The Guardian*, 8 May 1996). It is supported by environmental groups based outside the area, whose objective is often the destruction of the nuclear industry itself. Gorleben has become a key issue in the contest over the future of the German nuclear industry. Decisions taken, and enforced, at federal level have secured the industry's foothold in the area. At the same time, Gorleben has also become embroiled in a conflict between the federal government and the government of Lower Saxony, which has become the location of the major radioactive waste projects at Gorleben and at Asse and Konrad near Braunschweig (see Figure 2.8). These various dimensions of conflict indicate that the local community can draw on resources of power from outside in order to defend its territory.

Cultural acceptance Gorleben is a community where the nuclear industry and its opponents are defending different sets of values. The vigour of the protests indicates a surviving sense of community able to draw on its rural values to defend a way of life that has not yet been overwhelmed by the intruding nuclear industry. Resistance rather than acceptance is the prevailing condition. Cultural defensiveness here is not fatalistic; rather, it adopts an aggressive determination to sustain traditional values against the social trends of modern society (Blowers and Leroy 1994).

Environmental degradation Apart from the Gorleben salt dome (where headworks are ringed around by razor wire, water cannon and watchtowers) and the pilot conditioning plant and waste store, the nuclear industry has had little visible impact on the area. The railway depot at Dannenberg and the 12-mile road route to the store have been the focus of attention during demonstrations. It is the blight caused by the nuclear industry and the possible impact on health and the hazards to future generations that are the cause for concern.

Gorleben is a nascent nuclear oasis. In terms of remoteness and economic marginality it is on the periphery, but it has not acquired the other social, political and environmental characteristics of peripheral nuclear communities. It is a place where traditional values have been activated during a time of conflict. The outcomes of that conflict will have a much wider impact influencing the course of nuclear politics in Germany.

Nuclear oases: similarities and contrasts

In terms of peripheral characteristics the four communities show both similarities and contrasts. They are all remote, although the geographical isolation of Hanford is pronounced while Cap de la Hague and Sellafield are on the geographical margins of their countries. This was also true of Gorleben, although reunification has placed it in a more central location. All four communities are economically marginal. Hanford's *raison d'être* is the nuclear industry, while in Cap de la Hague and Sellafield the industry is the dominant employer. In Gorleben, the industry has secured its base but has not supplanted the existing economy.

The third characteristic, powerlessness, is most pronounced at Hanford, which is largely at the mercy of external forces. Although Cap de la Hague is also dependent on external forces, in this case the French commitment to the nuclear industry provides a sense of economic security. By contrast, Sellafield has long been the focus of conflict over the nuclear industry, but its importance to the local economy has provided powerful leverage in ensuring its continuing development. In Gorleben, local community power has been directed at resisting the industry and this has been a major element in preventing reprocessing, delaying waste transfers and halting progress of the repository project.

Evidence of cultural defensiveness is present in all four locations. At Hanford it is a form of aggressive defence of its nuclear culture, whereas at Cap de la Hague there is greater ambiguity, captured in Zonabend's phrase as a 'selective blindness' on the part of those working in the plant and a passive acceptance of the plant's existence by the surrounding community. At Sellafield there is conflict, on the one hand between nuclear employees defending their livelihood and, on the other hand, the traditional and adventitious communities defending their ways of life. Conflict is most pronounced at Gorleben, where differences of occupation, class and politics have been submerged in a combined resistance to the industry.

In terms of environmental degradation, Hanford exhibits evidence of serious contamination requiring a massive clean-up operation. At Sellafield the historic wastes are being cleaned up, and there have been longstanding international as well as local concerns about contamination of the landscape and the Irish Sea, which have been fuelled by incidents from time to time. Cap de la Hague has provoked rather less concern about its emissions (although local groups and the Channel

Islanders have been alert to potential hazards), but the trade in nuclear materials has given it international prominence. At Gorleben it is the risks to health and future generations and blight from a developing industry rather than the presence of contamination that have mobilised local resistance.

COMMUNITY INTEGRATION: THE TRADITIONAL AND THE MODERN

In each of these communities there is a juxtaposition of the traditional and the modern. To an extent the two have become intertwined as the modern, represented by the nuclear industry, has itself begun to seem outmoded as the dynamics of postmodern society have developed in other directions. As Wynne *et al.* (1993: 52), reflecting on Sellafield, put it:

> Whilst the rest of society has supposedly been learning to swim in the less determin-
> istic and less all-controlling, more exhilarating currents of post-modern diversifica-
> tion, areas like that around Sellafield have remained, monolithically dependent upon
> an unreconstructed icon of modernism, looking out it seems with some sense of regret
> or loss at missing out.

One of the interesting features of the nuclear industry is its ability to inspire those traditional integrating institutions supposedly overwhelmed by modernisation. Thus, resistance to the nuclear industry is motivated by a sense of communal identity. Likewise, it is values of solidarity that provide the defensive support for the industry in the nuclear oases.

Moreover, traditional values are complemented by quite untraditional cross-cutting alliances acting to defend the *status quo*. Capital and labour, middle and working class, radical and conservative, traditional and modern subordinate their perpetual differences to a wider common interest in the survival of their community. Whether, as in Hanford, the community is entrenched in its dependence on the industry or, as in Gorleben, embattled against the intrusion of the industry, there is evidence of integration within the community in a battle against hostile external forces. At the same time they draw on external support. At Sellafield, and to a lesser extent la Hague, both types of integration are present, creating conflict between communities supporting and opposing the nuclear industry. Thus, the relationships both within these communities and between them and the outside world are complex. The dynamics of these relationships shape the course of conflict over the nuclear industry and its outcomes.

This presence of processes that integrate communities in nuclear oases contrasts with the picture of increasing fragmentation in modern society that is present in some theoretical writings on social change in an era of ecological risk. In his earlier work on the idea of the 'risk society', Ulrich Beck (1992; 1995) argued that society is increasingly exposed to risks from technology, including nuclear technology, which imperil our very survival. The complexity of the systems and the uncertainty and disagreement among experts make them technologies that ultimately may be beyond control. A process of 'individualisation' compounds the anxiety engendered by the awesome potential hazards of technology. This is the result of the insecurity felt by individuals in a world of economic uncertainty, withdrawal of the support structures of the welfare state and dislocation in personal lives brought about by the breakdown of integrating institutions. Individualisation results in a fatalistic acceptance of risks for which all are responsible but which individually we are unable to control.

In this unrelentingly bleak portrayal of the malaise of modern society, Beck *et al.* (1994) offered a small ray of hope in the form of reflexivity. This is the condition whereby individuals are consciously aware of the reality of their situation and able to undergo the self-reflection and self-criticism that leads on to self-transformation. As Giddens (1991: 20) put it: 'Modernity's reflexivity refers to the susceptibility of most aspects of social activity, and material relations with nature, to chronic revision in the light of new information or knowledge.'

Individualisation and reflexivity may seem rather abstract notions, not empirically verified. The evidence from the nuclear communities studied here suggests a rather different situation. Individualisation, a consequence of modernisation, is hardly a universal condition since latent integrating feelings of community, which transcend conventional divisions, can be mobilised. In all four cases we have studied, it is the sense of integration through an expression of community that is most evident rather than the fragmentation brought about by individualisation. It is the prospect of economic harm (of job loss in the nuclear oases) or of blight and risk (in those communities resisting the nuclear industry) that is the motivating force.

In this context the notion of reflexivity, if it exists, is not an outcome of greater consciousness arising from individualisation but a communal response to external threat. To put it more simply, a motivating force in nuclear communities comes from a recognition of their situation and a determination to respond to that knowledge. However, this response may take different forms, as our examples have shown. It may take the form of fatalistic acceptance, as seems to be the case at Hanford, where the community is powerless to determine its own future. Alternatively, it may result in the kind of cultural defensiveness that is expressed at la Hague and among the nuclear workers at Sellafield or may provoke a more active and aggressive response, as at Gorleben. In each case there appears to be a recognition within the community of the realities of the situation and hence of the possibilities for dealing with it.

In these nuclear communities the pre-existing forms of traditional integration have been accompanied by more modern forms of integration arising from the development of the nuclear industry, which in a sense have themselves become traditional. Indeed, Beck (1998) suggested that industrial society has created its own traditional forms. The integration of the interests of the industry and its workers or the integration of diverse elements of society combining in opposition are indicative of combinations acting together for a common purpose. This is distinctively different from the social fragmentation that has supposedly become one of the hallmarks of modernism. Moreover, this integration has a local context; it is the sense of local community that binds people together to defend or oppose the nuclear industry.

In a broader spatial context, the survival of nuclear communities suggests a relatively stable pattern as a declining industry is resisted in all but its nuclear oases. However, on examination this is far from the case. The nuclear communities examined are each, though in different ways, vulnerable and, as a result, unstable. This is least obvious in Hanford, where dependence is most evident, although, even here, stability is achieved only through continuing external support. At la Hague, too, it is the commitment by the state to the nuclear industry that has been a necessary condition for the acceptance of the wider local community. At Sellafield, the situation is more evidently unstable, giving rise to conflict and the need for continuing investment to shore up the industry's position. Gorleben presents a case where the industry is introducing an unstable situation into a hitherto quiescent community. These communities, though peripheral, are not so powerless, marginal or defensive as might at first appear. They may be dependent (except for Gorleben) on the nuclear industry, but the industry is itself dependent on them. This provides them with resources of power that enable them to exert leverage on such issues as waste clean-up (Hanford), opposition to a repository (Sellafield) or on the very presence of the industry (Gorleben).

NUCLEAR OASES IN A GLOBAL CONTEXT

The shift in power relationships within the nuclear oases is linked to the much wider conflict over the nuclear industry that is now engaged in most Western countries. On the one hand, the nuclear industry still commands considerable resources of power through its privileged access in its relationship to the state. It is powerful as a result of its mere presence and reputation for power (Crenson 1971; Bachrach and Baratz 1970; Lukes 1974; Blowers 1984), but such power is not merely passive. It can be activated when challenged, as we have seen in the case of the nuclear oases. On the other hand, this power has been increasingly and successfully undermined by the antagonism of environmental movements feeding from and replenishing local opposition, combining traditional and modern values and, in the process, opening up the decision-making process.

This broader context further exposes the increasing weakness of the industry in the four cases examined. While the industry is embedded in Hanford, it faces considerable state-wide opposition in Washington state, notably in the main cities of Seattle and Spokane. La Hague has, in recent years, become a focus for opposition to the international trade in nuclear materials destined for Japan and Germany. At Sellafield, too, the reprocessing industry has attracted international attention, while the Nirex proposals for a potential deep repository met an opposition combining local (county) and national elements. Gorleben has been at the heart of protests against the nuclear industry across Germany.

The intrinsic changes within the nuclear communities suggest a greater (reflexive) awareness of these areas as places that take a disproportionate share of risk for present and future generations. They are similarly perceived by surrounding populations concerned to ensure that the risks remain confined to those areas where they already exist. This combination of internal and external social perception suggests two consequences. The first is that it is likely to prove very difficult to establish nuclear facilities that impose risks into the far future in any but the existing nuclear oases. The second is that, even in the nuclear oases, the introduction of further risk-creating activities may be impossible unless forms of compensation are introduced that are sufficient to overcome resistance. These consequences are having to be faced in dealing with the nuclear industry's most intractable problem – the long-term management of radioactive wastes.

PROSPECTS FOR SOLUTIONS

In the four cases studied radioactive waste is a key issue. During the 1980s, when Hanford was put forward as a possible candidate site for a deep repository (later abandoned on technical grounds), there was only a narrow majority in favour of the project posing the question: 'If a region such as the Tri-Cities that depends so heavily on the nuclear industry . . . is not hospitable . . . where will a suitable host site be found? (Dunlap *et al.* 1993: 167). The controversy surrounding the selected host site at Yucca Mountain in Nevada demonstrates the point. In France, while la Hague is the main storage site for nuclear waste there are no proposals for final disposal there. Attempts in the 1980s to find disposal sites for high-level waste provoked such opposition that a complete re-evaluation of the approach to siting took place. The French have now adopted a more measured, comparative approach based on seeking volunteers and offering generous compensation. In the UK, after attempts to find sites during the 1980s were abandoned, Sellafield was chosen as the preferred site for a repository for intermediate-level wastes. The proposal by Nirex to locate an underground laboratory there was strenuously opposed and ultimately turned down on grounds that it was technically inadequate and 'seriously premature' (Secretary of State's decision letter, 17 March 1997). In Germany, fears about the future safety of a repository in the Gorleben salt dome have been the fount of opposition both locally and in the state of Lower Saxony (Niedersachsen 1993). In all these cases, and elsewhere with the single exception of the Carlsbad repository, proposals for deep disposal of long-lived wastes have yet to get beyond the exploratory stage.

In the search for suitable and publicly acceptable locations for managing nuclear waste the industry faces formidable obstacles. If solutions are to be found they will have to be accommodated to the changing social context that is encountered in the nuclear oases that has been explored in this chapter. In particular, solutions will need to recognise three fundamental requirements: the need for a greater credibility of science; the development of more participative forms of decision making; and a commitment to the principle of equity.

Credibility of science

It may seem axiomatic that any acceptable solution must be scientifically robust and achieve the highest technical standards. However, the nuclear industry has an uneven record in meeting this requirement. The Nirex proposal for Sellafield ultimately failed on scientific grounds. It was described as an 'ill-considered exploratory development within a promising site', and the geology and hydrogeology were 'more complex than would be expected of a choice based principally on scientific and technical grounds' (decision letter paras 6xxiii and 7). The proposed Yucca Mountain site has also been criticised on scientific grounds, especially its location in an area of seismic activity. Similarly, scientific doubts have been raised about Gorleben. The scientific problem has been compounded by a past tendency to make optimistic predictions based on limited evidence and to provide confident assertions about safety that are impossible to verify. The credibility of expertise has been weakened by the failure to confess the limitations of science.

The need for informed scientific debate has been widely recognised. In practical terms this means a rigorous programme of peer review (RWMAC 1997; 1999; Nirex 1997), *in situ* experiment and a measured programme of research that is committed to openness and disseminated through accessible public information. In order to achieve public confidence in policy, a consensus extending beyond the community of experts will be required. In short, a social consensus must be constructed embracing all those groups and communities with a stake in the outcomes. In areas of great uncertainty and long-term risk such as radioactivity, conventional approaches to science are inadequate. As Funtowicz and Ravetz (1990: 20) have argued, there is a need for a 'post-normal science that is neither value-free nor ethically neutral but which takes into account people's beliefs and feelings lest they become totally alienated and mistrustful.' Confidence in science is a precondition for finding acceptable solutions. It is more likely to be achieved if the uncertainties are fully acknowledged.

Participative forms of decision making

This involves openness, early engagement with all those affected and a deliberative, accessible style of decision making. In terms of openness, the ground rules need to be established at the outset, and the process itself needs to be staged so that decision making is measured and progressive. Attempting to find sites for radioactive waste progressively involves eliminating possible candidates according to technical and other (agreed) criteria. Such a measured process has been described in advice to the UK government (for example, RWMAC/ACSNI 1995; RWMAC 1998; 1999; House of Lords 1999). This approach is far different from the secretive identification of sites that are thought least likely to resist, which has resulted in defeat for the Nirex proposals in the UK and the current controversy over Yucca Mountain.

Various methods of participative decision making have been developed to enable wider involvement and to ensure that the process is accessible to a wider public. These include public hearings, community consultation, mediation exercises and consensus conferences (Joss 1998; UKCEED 1999). They provide an opportunity for defining areas of agreement and disagreement that assist the process of consensus building on which an agreement to proceed can be built. Given the scientific uncertainties and the increasing power of nuclear communities, the decision-making process cannot be hurried. If it is to be successful there must be adequate time for resolution of problems.

Equity

The existence of landscapes of nuclear risk represents a particular form of inequality. It suggests that those areas which support the nuclear industry or those that may be identified as future sites should be compensated in some way for the burden they bear on behalf of present and future generations. The reflexive awareness in these nuclear oases indicates that proposed solutions based on principles of equity are not only morally desirable but politically necessary if they are to prove acceptable and effective.

One way of approaching the problem of site selection for nuclear wastes has been the identification of 'volunteer' communities. Various types of compensation have been offered, including the provision of infrastructure and other facilities, investment capital, economic diversification and forms of tax relief. Such measures have sometimes been regarded as bribes, a way of finding willing victims from among the disadvantaged peripheral nuclear communities. Nevertheless, the need to find suitable sites provides potential hosts with considerable bargaining power. For example in Canada, Deep River, the community adjacent to the country's Chalk River nuclear research facilities, volunteered after a referendum to take the country's historic wastes in return for a package of compensatory measures. However, the community also placed a condition that employment levels at the Chalk River plant should be sustained at 1995 levels until 2010, a condition the federal government refused to meet (Siting Task Force 1995a,b).

This example suggests the increasing leverage available to communities in negotiations over the future siting of radioactive waste facilities. The scope of compensation will have to be broadened, not only to take into account economic sustainability but also to embrace a genuine participation, perhaps even a power of veto. This would necessitate the negotiation of agreements over siting, waste volumes, routing of wastes and the timing of the stages of development. Whatever the specific forms of compensation ultimately agreed, the principle of equity implies that the terms are mutually acceptable and should be agreed in principle prior to any process of negotiation over specific sites.

Contemporary approaches to finding sites for managing nuclear waste have, so far, failed to yield acceptable solutions. This is partly a result of the lack of scientific credibility, a failure to engage in participative decision making and an absence of agreed and acceptable compensation programmes. In each of these areas, improvements have been made. However, it is not simply a matter of devising the right decision-making criteria; a solution to the problem of nuclear waste management also depends on an understanding of changing social relationships, particularly in those peripheral communities where the nuclear industry is already established. In these areas, both traditional and modern processes of integration are encountered, which provides both identity and power that enables them increasingly to challenge the nuclear industry. This power is weakest in an isolated monocultural nuclear community such as Hanford. Yet where the nuclear industry cohabits with a traditional community, as in Cap de la Hague, or where the community is itself divided, as is the case at Sellafield, the industry, though powerful, experiences greater constraint. Where the local community is united in opposition, as in Gorleben, the industry's foothold is precarious. In each community, an acceptable solution to the industry's future development will depend on an ability to adapt to local processes of social change.

As we have seen, each of the communities studied is embedded in wider social processes that provide support for both the nuclear industry and for its opponents. In addition to finding locally acceptable solutions, decisions must also reflect a wider social consensus on questions of nuclear power, reprocessing, decommissioning and nuclear waste. The interaction of local and more global social processes will determine the nature, timing and location of these decisions. Since much of the radioactive waste is already located in the nuclear oases, these landscapes of risk will hold the key to the future of the nuclear industry.

CHAPTER 3

FEAR, LOATHING AND SPACE FOR CHILDREN

Stuart C. Aitken

I was raised in a small town. We had a huge yard, not very many people. That would be ideal for Keenu, living in a smaller area with a large yard. I'm not really comfortable with the size. The lots are small here, not a lot of room to play. I don't feel comfortable with cars driving on the street at all. In fact, we're having a community gate. I'm pretty sure it's going in. If that happens. I'll be much, much happier.

Dong-Mei, first interview

I first met Dong-Mei and her husband Peter when they agreed to be part of a longitudinal study that focused on household changes around the birth of a first child and how those changes are contextualised in San Diego communities. Dong-Mei's words are excerpted from our second interview, about six months after the birth of their son, Keenu. The couple moved into a neo-traditional development when they learned that Dong-Mei was pregnant with Keenu, about two months prior to our first interview. The couple's home was about two years old on our first meeting, with two storeys, three bedrooms and a large deck extending the house into a small back yard. Relatively narrow streets, at least by southern California standards, connect the sixty houses that comprise the community. The community's design and layout are fairly typical of the neo-traditional style that is currently in vogue in southern California. Geographers and other academics criticise neo-traditional developments as the apex of commodified values that valorise nostalgia for traditional and mythic community values, the privatisation of local government, the establishment of exclusionary territories and the continued social segmentation of the urban social fabric (Till 1993; McKenzie 1994; McCann 1995). These landscapes are described by some as fortresses, citadels and carceral spaces (Blakely and Snyder 1997; Dear and Flusty 1998). Yet regardless of this critique, neo-traditional communities represented the largest proportion of housing starts in southern California in 1998. It is estimated that gated communities house eight million people in the USA (Blakely and Snyder 1995a).

In this chapter, I want to raise questions that relate to how the fears and anxieties of young parents are contextualised in the built environment. I argue that these southern Californian communities – and other similar kinds of exclusive residential space such as community interest developments (McKenzie 1994) partly arise from a perception that urban America is an unsafe place for children and families. Indeed, Robyn Dowling (1998) argues that these perceptions are part of a larger cultural geography of exclusion that is part of contemporary suburban ethics that uphold neo-traditional values even when those values are not designed into the community. In communities designed specifically with neo-traditionalism in mind, developers such as Andres Duany in Florida and Peter Calthorpe in California promote their developments as a return to old-fashioned family and community values. Peter's and Dong-Mei's feelings about their community are constituted primarily by their perceived needs for security and their fears for Keenu's safety. For example, one of our study's concerns was how often young couples interact with their neighbours and, in particular, how this relates to the raising of children. Peter told us they hardly ever saw their neighbours and would never consider leaving Keenu with them except in a dire emergency. Dong-Mei admits that although the developers suggested their home would be part of a community with traditional family values and this had been attractive to her at the time, her lifestyle after Keenu's birth does not allow the time for community socialising:

> I know that there are some babies around but I never see them. We really don't spend a lot of time mingling with the community. Just to maybe get the mail, and this summer we went to the pool about twice . . . And I'm busy working [so I] don't want to spend a whole lot of time, and certainly [I] don't have time to socialize much with neighbours. (Dong-Mei, second interview)

In this chapter, I draw upon empirical data that were created through questionnaire surveys and in-depth interviews in San Diego between 1991 and 1996 (see Aitken 1998) to present an analysis of how the fears and anxieties of young parents are contextualised by the built environment (Figure 3.1). In it, I use as far as possible the voices of the carers who participated in the study to argue that, despite what the developers of neo-traditional communities offer, young parents often settle for little more than the vestiges of a landscape of defence replete with such things as speed bumps, gates and security personnel. The chapter touches on issues that relate to neo-traditionalism, but it is not meant as a universal criticism of these residential developments. I introduced the chapter with part of Dong-Mei's and Peter's story because it emphasises several recurrent perceptions among the young parents in the study, notably nostalgia for past residential landscapes, concern for the safety of children on public streets and a focus on security. I am not suggesting anything new by noting that fear for children's safety in residential environments is a critical concern for most of the families that we surveyed and interviewed. Rather, with what follows I want to focus on the ways carers' fears are exaggerated by very specific forms of loathing and abjection, and how those relate to issues of difference and the prospect of creating appropriate kinds of space for children.

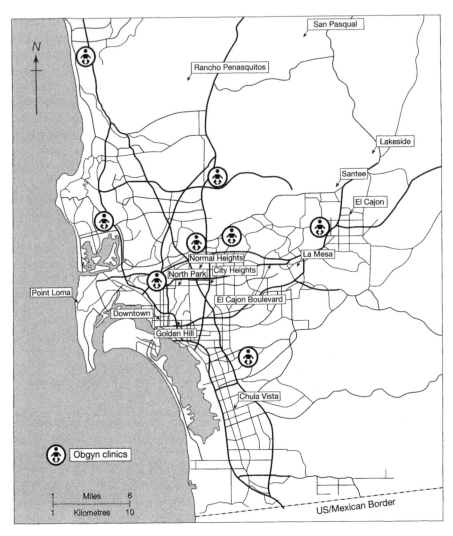

Figure 3.1 The study area, with place names mentioned in the text (from Aitken 1998)

FEAR

Although the majority of those surveyed in the questionnaire felt that their neighbourhood was a good place to raise their new child (Figure 3.2), our interview transcripts are replete with discussions of participants' fears for their children in public spaces:

> Well you definitely get the sense that there are some neighbourhoods that are, that you just feel safer in than others. You know there are some neighbourhoods that just have too many strangers walking around. (young mother, first interview)

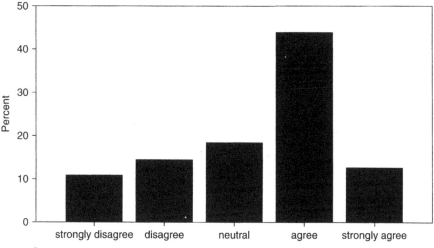

Statement: My neighbourhood is a good place to raise children.

Figure 3.2   A good place to raise children

It's not kind of . . . you know you wouldn't want . . . If you had an older child that was able to go out and ride their bike or something, I wouldn't feel at all safe about a kid going out and wandering around. The biggest fear is somebody is going to come by and grab 'em. (young father, second interview)

In the next section, I will deal specifically with what fear of strangers may mean in terms of racism and the rejection of difference, but it is important to begin by simply outlining the ways that parents articulated these fears. One of the more interesting aspects of the project was monitoring the way in which some parents' views on neighbourhood safety changed after the birth of their child. Like the young father above, some envision how the neighbourhood might accommodate their child when older. Alternatively, several new mothers commented that they felt more vulnerable when out with their infants. 'There are some weirdos around here,' noted one young mother, 'and I never noticed before the baby was born. I notice these things now because before I felt I could take care of myself but now I have another life to protect. I sometimes get nervous when I see a suspicious character walking towards me and will sometime take the buggy and cross the street. Before I would never have dreamed of doing that.' In addition to some parents voicing concerns about strangers in the neighbourhood, an overwhelming majority of respondents to the second questionnaire survey noted some considerable trepidation about leaving their children with neighbours (Figure 3.3).

Fears over strangers, neighbourhood crime and a lack of trust in neighbours stimulate different kinds of response from parents. In what follows I discuss how these fears are mitigated by scale, relocation and community activism. It should be clear that the strategies I choose to discuss are not exhaustive, and the fears upon which they are based relate to a complex set of perceptions that I cover more fully in the next section. Some residents perceive southern Californian cities to be crime-ridden

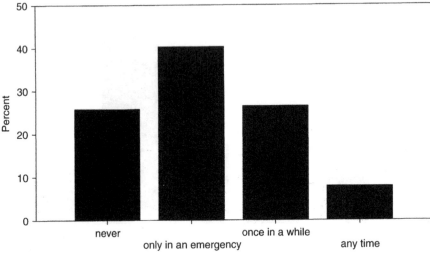

Q: Would you let neighbours look after your child?

Figure 3.3   Neighbours looking after children

and gang-infested. This is a generalisation that covers a substantial geographical area, but it nonetheless transfers general fears of crime on to local streets. Some of this transference is attributable to the influence of past crimes on people's memories (Baker *et al.* 1983; Grogger and Weatherford 1995), but a large part is undoubtedly due to distortion and inflation of particularly violent crimes by the media (Davey 1995). As Justin, whose story will be articulated more fully later, puts it:

> I lived in east San Diego, which is considered dangerous, for a year and never saw anything suspicious. And east San Diego is a much worse neighbourhood compared to here ... You hear about things now and then, but that's the media. There's crime *everywhere*. I think the media sometimes creates more fear ... [We need to be] in touch with more of the reality. (Justin, third interview)

Geographically, media distortions serve to spread perceptions of crime to a neighbourhood when only a street or an apartment is the focus of crime. This is clearly a scale issue.

A second point about scale relates to a segmentation of social space that often results in poor and affluent neighbourhoods abutting each other, raising issues for residents who view crime nearby as seeping into their local areas. These fears help to create geographies of exclusion, which are exacerbated by the hard borders of gated communities and the middle-class exclusivity of neo-traditional communities. I want to articulate some of the parameters of those exclusionary perceptions through the words of Margot, a new mother who uses scale to define what she is comfortable with in her neighbourhood:

> I really like this neighbourhood but the extended neighbourhood I really don't like. The neighbourhood itself where we live is great, the people are wonderful. There are

a lot of people our age, a lot of people with kids. In fact, I think there are going to be five kids born in this neighbourhood this year. It's great. As far as the extended neighbourhood . . . I don't know . . . I wish I had more sense of community.

A year later, Margot still seemed to lack a 'sense of community', but she clearly had a stronger grasp of the scale relations between the one or two blocks that she lived on and the 'extended community':

Um . . . the *little* neighbourhood: the little block or two-block area is nice, because it's, you know, it's a cul-de-sac horseshoe. The traffic's not bad. So those things are good. But if you get a little further out from that . . . Um, the safety factor is a big issue. So the answer is on the micro-level sort of yes. It is a great neighbourhood to raise kids because of the people of the, you know, sort of *closed* little neighbourhood. But on a *bigger* level I would say it's *not* the greatest . . . I mean you're in one of the worst areas in San Diego.

Margot and her husband were planning to relocate to a 'more middle-class' community, but they were waiting until their daughter was ready for school. I will return to Margot's reasons for moving later in the chapter because they have some bearing on how she characterises the people who live in her 'extended neighbourhood'. The point is that people differ considerably in what they perceive to be the geographical extent of their neighbourhood. Some may perceive their neighbourhood to be quite extensive while for others, it is a block or two or simply the house or apartment complex in which they live. In previous work, I showed how those perceptions related to residents' familiarity and experience of local public places (Aitken and Prosser 1990; Aitken 1992). Yet such explanations are partial because they neglect the politics that surround geographies of exclusion and the fact that some residents shrink their worlds away from landscapes that are indefensible.

Jack and his family help to illustrate this latter point. They live in an area that Margot would describe as 'the worst in San Diego'. During our first interview, Jack noted a particularly frightening incident that occurred just after they had arrived home from vacation: 'We got home from Yosemite and somebody drove by the house and shot a gun in the air, right in front here. There [were] shell casings in front of the car.' I probed this last statement a little further. He noted that the perpetrators were 'kids mostly, mostly high school kids'. I then asked if they were kids who lived in the neighbourhood, to which he replied: 'Well it depends on what you mean by the "neighbourhood".' I asked him how he would define the neighbourhood, and he asked: 'Do you mean City Heights? Oh yeah they live in City Heights' (Figure 3.4). This was curious to me because I thought that Jack also lived in City Heights, so I questioned him further as to what he thought of as his neighbourhood, to which he responded: 'I see it as just a couple of blocks. They don't live within a couple of blocks. Some of them come over though.' Like Margot, Jack's strategy for managing a safe environment for himself and his family is to re-scale his neighbourhood until it feels comfortable. Jack's response highlights an intriguing insight into how defendable landscapes contract to exclude hostile elements.

Another strategy for countering fear of crime is involvement in activism and local planning. At the time of his second interview, Jack outlined several goals:

Figure 3.4 City Heights

I know something I'd like to see. We're working towards a neighbourhood patrol. We're starting one up actually ... [it is] a good deterrent to crime because people are out there with their eyes ... There are several new laws been passed ... there are loitering laws in places where there are common drug deals. They can arrest you and take you downtown. [These laws] are going to be enforced real soon ... I hate to say this, it may sound crazy, but people have too many rights. It's like in areas like this, drastic problems require drastic measures. I wouldn't mind house to house searches for weapons; illegal automatic unregistered weapons, I don't have a problem with that ... We got a place over here shut down. A guy was selling motorbikes out of his front yard. Just had it totally filled up. Then I noticed hookers going in and out, looked like they were buying their crystal. Then I saw the cops there two days in a row going through everything. And now they're not there. There are more liquor stores combined in our little community [than anywhere else] which has the highest crime. I tell you, it is crazy but we're breaking it up. We're not allowing it. We've had a couple of liquor licenses pulled because they sold to anybody. You can't get a liquor license in this area, we won't give them to anybody. We don't want any more liquor stores here. They go hand in hand with crime, drugs and prostitution ... We're doing everything we can to cut down on the crime problem. It finally became drastic and that's what it took. People being shot and stuff like that.

The 'we' to whom Jack refers to is the community-planning group of which he is an active participant. Jack's fourth strategy, at least in the long term, is to leave the area: 'We don't plan on staying here for very long, four–five years is kind of our

Figure 3.5 City of El Cajon

starting point. Maybe buy another house around here and then eventually move up to a better area.'

Eddie and Christy live in El Cajon, a city located about 10 miles east of downtown San Diego (Figure 3.5). They moved three times within El Cajon during the life of the project but never seemed to improve their residential environment: 'I'm very dissatisfied with this neighbourhood. You got crime, you got drugs, you have people picking through trash cans. Umm . . . it's not a good environment to raise kids.' When asked whether he planned to move again, he answered: 'Sure yes, but every time we move to a new place, it's ok for awhile and then the place gets bad' (Eddie, second interview).

Eddie's and Christy's annual household income is approximately $50,000, and they have no day-care expenses because Christy's mother lives with them and looks after Joe during the day. They were among a number of participants in the study who perceived their residential environment to be not only unsuitable for raising children but also 'indefensible'. Some sociologists argue that places are often labelled high-crime areas because of people's perceptions rather than any meaningful crime statistics (Skogan 1990). Physical signs of 'incivility' such as graffiti, litter, beggars and so forth (LaGrange *et al.* 1992) engender fears in these areas. For Christy and Eddie, crime is not based simply on perception or how they feel about signs of crime in the physical environment. Crime, for them, is tangible:

When our personal space gets broken into. We've been broken into like three times. You know, the child's clothes have been stolen and people are making drug deals in our driveway. We just don't want to be in an environment like that. We don't want to raise a child in that environment. Our neighbours back behind: the police are here every other day in case they beat their child or each other. I mean if we had our choice we wouldn't live in that type of environment.

When asked whether they would you like to change anything in their neighbourhood to make life for them and their family easier, they answered: 'About this particular neighbourhood? I don't think there's one or two things that need to change. I think a general move would make things better for us. I just think there's some areas . . . that should be forgotten about' (Christy, second interview).

When asked if he would ever leave his child with neighbours, Eddie responded: 'Never, ever. I don't trust anyone around here. He stays with the family and nobody else.' The interviewer probed this response a little further by asking Eddie what he would do if an event occurred that required his full attention and that of his wife. He replied: 'Well I wouldn't. I would sacrifice whatever it was first so that the child will never be watched by anybody else but family.' The example suggests that when residential landscapes are perceived as indefensible, families may implode in on themselves.

A large part of the foregoing discussion concerns general fears of crime and some specific responses of parents who feel that they live in neighbourhoods that are prone to crime and violence. There are other parental fears that I chose not to raise, such as concerns over the amount and speed of traffic. Nevertheless, for some parents, even these fears contain some elements of concern over people who are different: 'This is a nice neighbourhood but the younger people who live on the street tend to drive too fast up and down the street.' In what follows, I try to take this premise and explore it further with strategies of defence that relate to concerns over difference. These strategies focus on the creation of exclusionary geographies and attendant feelings of abjection and loathing. The creation of physical landscapes of defence helps to drive desires, usually only realised by those whose lifestyle is middle- or upper-class, for the seemingly safe environments of gated and neo-traditional communities or the security of knowing that you are surrounded by people like yourself in community interest developments (CIDs).

LOATHING

Interviewing in the mid-1990s coincided with the extension of the San Diego trolley network (light rail transit system) from its original configuration between downtown and the Mexican border to areas east and north of downtown San Diego. The network's original configuration was highly successful in terms of box-fare recovery only partially because it enabled tourists to get to the border with considerable ease. By far the greatest benefit of the trolley was to those who lived in working-class communities south of downtown, who regularly used the system as a means of commuting. A sizeable number of our respondents lived in areas that would come under the influence of an extended trolley line, but few of them voiced appreciation of new commuting opportunities:

> The neighbourhood is not as safe as it could be. The trolley brings in people that would not usually be in this neighbourhood. Other than that it seems to be great. (a young father, noted on his survey)

> The only thing I don't want to see is the trolley coming through here. It should be here in another two years. Because I think we are going to get all walks of life through here that we don't get through here now [wife interrupting from the other room: 'A lot of crime.'] Some people are going to figure out that Santee is here that don't even know about it. (young father, first interview)

Tricia and Russell (their story is elaborated in Aitken 1998) moved prior to Savannah's birth partly because a new trolley line was built quite close to their old neighbourhood:

> We needed to buy a house … Santee's getting scary with the trolley coming in. Because I used to work in La Mesa where they just put a trolley station, and the crime rate right there is out of control. I mean stabbings, cars being stolen, *in La Mesa*. It's not a good thing. (Tricia, first interview)

Anxiety over outsiders in the neighbourhood is also a concern for this young mother, who lives in Point Loma, an upper middle-class community near the coast, but her concerns revolve around youths getting bused into what she considers to be one of the best high schools in town:

> I wish they would do something about it … those school boards … this is gonna sound really terrible, but when I'm living in this area, and I'm paying these property taxes, and my husband's really involved in the community … and my child won't be able to go to one of the better schools that is literally around the corner because they're shipping in all these kids from 30 miles away. That's [stifles response] … I don't agree with that. Those kids, there are a lot of negative influences. At lunch time the crime rate goes up over there because they have off-campus lunches. You've seen them? I mean there's riff-raff. These kids don't want a better education. They have no business being over there running around the neighbourhood at lunchtime, you know … It really bugs me.

Issues regarding 'negative influences' were clearly delineated by some of the respondents, suggesting that, for them, family life required a geography of exclusion.

Julia Kristeva's (1982; 1991) notion of abjection has been discussed recently by geographers interested in the ways racial tension is linked to loathing and exclusion (Sibley 1995; Aitken 1999). Kristeva describes as 'abject' that which we attempt to exclude radically but cannot. David Sibley (1995: 8) points out that the urge to make separations and to expel the abject is encouraged in Western cultures, creating feelings of anxiety because such separations can never be achieved. That a trolley system gives local access to 'people that would not usually be in this neighbourhood' is clearly a source of anxiety for some participants in the study. This anxiety translates into loathing people who are different and wanting to exclude them from local access.

A large part of the loathing that parents feel may be attributable not so much to their own experiences but to expectations based on what they perceive. Tina, for example, is particularly concerned about local gang cultures:

> I think there's a lot of gangster kids around here. Whether they are or not they look like it. How can you, I mean your kid comes home and he's got his head shaved and his pants hanging below his butt with his boxers showing [laughs]. The people who live in El Cajon, they're like so different from anybody else. The way they act and things like that. I don't know. La Mesa to me is just really cleaner and just safer. The worst parts of La Mesa are not as bad as the worst parts of El Cajon.

The issue I want to raise in the above quote relates to difference and perceptions of gang activity based solely on appearances. Tina's husband, Frank, talks about how his youth was much safer:

> I grew up in a small, small town. The nearest place was 20 or 30 miles, maybe more. It took an hour to get there. In that town it was just like La Mesa, even more peaceful than that. It was really, really small and you didn't have to worry about crime. It was there, but you never had to worry about the gangs, it wasn't like they were in the neighbourhood and they were like sitting out and just watching. It was just real safe.

When asked if he thought he could find a less threatening neighbourhood to live in, Frank replied: 'A die-hard gated community would be nice. But yesterday we looked at an apartment in a gated community and a guy just walked through the gate. It has to be really gated with a guard to feel good.'

Constance is a single mother who lives with her mother in Chula Vista. They have lived on the same street for seventeen years. Constance's father (who left the family some years ago) is African-American and her mother is Mexican. Constance studies at City College but intends to transfer to a college in Atlanta with her fiancé. Although she grew up on the street in which she currently lives, she does not like what the neighbourhood has become. She notes that although the level of gang activity has decreased in recent years, drug use seems to be about the same. She does not like her apartment complex and, when probed further, suggested that this was because they felt isolated among the Asian families who now live there:

> I won't let Cassandra go out to play here with kids who are rude. We are the only black family left in the complex. When I was a kid it was all black or Mexican, but now it is primarily Asian and there is tension. The last black family moved out when their kid was picked on and called a nigger. The kids in the complex are outside cussin' and unattended at 8am. This is a terrible place to bring up kids. Everybody keeps to themselves. They may talk on the street but there is significant racial tension. The Filipinos are particular bad at . . . [a local grocery store run by Filipinos]. They get particularly abusive. I prefer to go shop at places where I am treated with a little respect like Vons or J.C. Penney's.

Eleanor lives in a new condominium in Rancho Penasquitos, a predominantly white, middle-class suburban neighbourhood located about 20 miles north of downtown San Diego. She is concerned about her child's future in the neighbourhood, which relates, specifically, to anxieties over the changing ethnic structure of the area:

Figure 3.6 El Cajon Boulevard

> We – this will come out wrong – but we live right next door to low-income housing, and some of the kids aren't ... [shrugs and voice trails off]. They probably want to steal and that kind of stuff. For his age it's fine, I just think we'd be concerned overall with crime as everyone would be! English is almost a second language, and he might get lost in the shuffle.

I have discussed the complexities in Eleanor's prejudices elsewhere (Aitken 1998: 115), but here I want to compare her anxieties with those of Margot, whose contrivance of neighbourhood scale has already been discussed. Like Eleanor, Margot is concerned about ethnic diversity. She says that the problem is the 'ethnically diverse south of El Cajon Boulevard element' (El Cajon Boulevard being a street that developed in the 1920s as the main route east out of San Diego but which went into decline after Interstate 8 was built in the 1950s). This 'element', she maintained, has resulted in increased crime in the area (Figure 3.6):

> the *south* of El Cajon [Boulevard] element in general, I wish it would go away ... I mean you're in one of the worst areas in San Diego. And that's within quarter of a mile straight from our home. And so that I hate. I find it very scary. Um, it is *not* a, you know, ethnic thing; it's an economic thing. It's people who are very poor and people who are very desperate and who are very uneducated and they live very, very, very *close*. And I *hate* that.

Importantly for Margot, the 'south of El Cajon Boulevard element' has made the public schools extremely poor. Later in the interview, however, she contradicts her previous admonition that it was purely an 'economic thing': 'I understand that at [the local high school] they teach, um, you know, maybe six or seven different languages, you know. I don't think that is appropriate myself.' Margot is upset about this and grieves the loss of several families from her street, which she attributes to the state of the local schools:

> In our neighbourhood we have seven families that . . . have children that are about the same age, within a year or so. And of them, two have already moved, um, mainly because of the schools in the neighbourhood. And three are definitely planning to.

When asked about what kind of places they were moving to, she replied: 'Um, away from El Cajon Boulevard basically [laughs], away from south of El Cajon Boulevard. Just a more middle-class neighbourhood.'

I selected interview transcriptions in this section to make the point that sometimes parents' anxieties ferment into a loathing of people who are different. Many of the respondents would be upset that I make inferences of this kind from their words. Some are quite sensitive to issues of racism and often qualify what they say with phrases such as 'this will sound terrible, but . . .' The point I want to make is that these prejudices are subtle. They take the form of loathing someone who is different and unfamiliar, and they are based on a socially constructed form of abjection. Moreover, they comprise a basis for the finer sorting of the urban social fabric that ultimately creates landscapes of defence. In the last section of the paper, I consider how strategies of defence play out in the creation of space for children.

SPACE FOR CHILDREN

> Ya, if this neighbourhood had more kids in it, I think that would calm it down [laughs] . . . although it's not like a wild party neighbourhood or anything, it's a pretty nice neighbourhood . . . if, if it just had more families.
>
> young mother, first interview

> We were looking for a nice neighbourhood, we figured wherever you have a nice neighbourhood and nice houses, that you're probably going to have children in the area.
>
> young father, first interview

At one level, parents want a safe and secure landscape in which to raise their children. A corollary to this axiom is that landscapes with evidence of children are often thought to be family-oriented and, as a consequence, safer and more secure. Ironically, a noted concern for American family landscapes is the lack of children in public streets and the seeming implosion of the family in upon itself (Katz 1993; Aitken 1998; 1999; Davis 1998). Cindi Katz (1993) attributes this in part to media 'terror tales' of child molestation and abuse. Parents are understandably upset by gruesome reports of abductions and, from an early age, warn their children about 'stranger danger'. In reality, most of these abductions are by adults who are known to the child, and 95 percent of child abuse occurs in the home. There are obviously many reasons why some American families seem to be imploding in on themselves, and discussions on this topic include the commodification of childhood experiences (Davis 1997; Aitken 1999), the structuring of children's recreational time (Gagen 2000) and the responsibilities felt by parents (Hewlett and West 1998). Clearly, the perceived lack of safety in public spaces constitutes a large part of parents' anxieties. Gill Valentine (1997) has focused on this anxiety as it relates to the context of older children's experiences of space, but there has been little work on how such fears play out among the parents of very young children.

Derek and Shelley live in Normal Heights, a neighbourhood that other respondents in the study thought was fun and child-friendly. Here is what Shelley had to say about their landscape:

> I hate it. There's lots of graffiti and vandalism and noise . . . The area is awful. It's mostly on Friday and Saturday nights people get rowdy. Adams [Avenue] is strange because there are a lot of antique stores and nice places but then, intermixed with it, there's a lot of low cost housing. There's a lot of rowdiness. We get police cars, you know, going up and down the street looking for people . . . The crime, that's really what we object to. Well it's also an ugly neighbourhood, there's obviously no building codes. It's just ugly buildings . . . ugly apartment buildings with cinder blocks. But we can live with it. We like our house. We're very happy with that.

Derek elaborates more fully on how their neighbourhood context affects their family life:

> I mean, I am not unsatisfied with the neighbourhood, but it's not what you'd call safe. I mean it's alright and we're pretty secure here, but we don't spend a lot of time doing stuff outside. Inside the house it's fine, and hanging out in the yard and stuff, but I don't think it's really safe walking around the neighbourhood or even walking to the store. When it gets dark out, I don't think I'd even walk to the store . . . We had some guy held up in the alley, stuff like that, you know. I guess it happens everywhere.

Many couples in the study actively sought out a place that seemed child-friendly, or they created a space within which they felt comfortable raising their children. I conclude this chapter with the stories of two couples who exemplify these two perspectives. Camille and Chuck wanted to move to a different neighbourhood when Camille was pregnant with Robert. The move was precipitated with, to use Camille's words, 'the whole family thing in mind . . . we want to buy into an area with lots of kids and an appropriate structure . . . and tons of kids.' Both Camille and Chuck hold full-time jobs with a combined income of around $85,000, so they were able to move to the relatively affluent community of San Pasqual, about 30 miles north of downtown San Diego (Figure 3.7). Marcia and Justin live in Golden Hill, a downtown neighbourhood with a diverse population of African-Americans, Hispanics and whites (Figure 3.8). It is partially gentrified and contains a wide range of income groups. Neither Marcia nor Justin would call themselves gentrifiers. He works full-time and she part-time, with a combined income of less than $50,000. They are particularly happy with the neighbourhood because they feel it offers them the freedom of what they call a 'real community'. I compare these two

Figure 3.7 San Pasqual

Figure 3.8 Golden Hill

families (they are the same ethnicity) because each describes their neighbourhood as a 'real community', but they reside in startlingly different residential landscapes. Both may be described as landscapes of defence, but the defensive strategies of each family are quite different.

Camille and Chuck

Camille and Chuck moved into San Pasqual just before Robert's birth. Here is how Camille describes the landscape:

> It's pretty private, because unless you are coming in this neighbourhood you don't have a reason to go up the hill 'cause it doesn't – it's not a through street or anything. So we kind of like the fact that there's less traffic coming through here. Um, and we like the size of the neighbourhood; it's about a hundred homes so there's enough to where there *is* a real *community* feeling but not too huge where you really don't know people. The neighbours are wonderful, when he was born a lot of them we had only met, you know, just as acquaintances kind of thing. And a lot of them brought baby gifts and they really welcomed us into the neighbourhood . . . There's a lot of kids in the neighbourhood, [it's] about ten years old so there was the initial buyers that had their babies, you know, ten years ago, and now its kind of the second cycle.

Chuck also felt that they had found the perfect place for young kids:

> The couple across the street have two kids, and the girl comes over sometimes and watches Junior, the boy mows our yard and . . . then there's Peter and Annette next door. As a matter of fact, Annette's daughter baby-sits for us.

A home owners' association (HOA), which regulates and directs many different aspects of the community, controls the area. The following quote from Camille engenders a sense of how the HOA is used as a resource for residents:

> I'm going to our home owners' meeting tomorrow night to propose . . . to do, um, a neighbourhood directory. 'Cause where I grew up we had a directory and in the directory there was a place where kids could list if they did yard-work or pet-sitting or baby-sitting . . . I thought of starting with some of the other moms some sort of system where we had tickets or, you know, points . . . something that you could trade for baby-sitting services.

Cooperative baby-sitting systems were quite common among families who participated in the project. The point I want to make is that in Camille's and Chuck's context the procedure is formalised and contractual, requiring the approval of the HOA.

Although there are 'tons of kids' in Chuck's and Camille's neighbourhood, they were not necessarily evident in the landscape. One interviewer noted in their field-log that 'this is a very quiet neighbourhood, not a single person seen outside.' Towards the end of one interview, Camille's sense of community was probed further, and when asked how she would improve the neighbourhood she noted that:

> We have that real sense of community but, um, it seems that people don't know their neighbours as well anymore because everyone's so busy. That's one reason why I'm going to volunteer to do [the directory] because I thought that might kind of encourage

people to interact a little bit more . . . if they see, you know, their neighbours offering something they wanna take advantage of.

When pressed further about leaving Robert with neighbours, Camille confessed that they have only had a neighbour's teenager look after him twice: 'We've really not left him with people here . . . We'd choose to leave him with family members first. We also have several other friends that don't live in this exact neighbourhood but that have babies that we do things with, and we kind of trade off watching.'

It may be argued that exclusionary tactics coupled with contractual environments for children define the safe, social structure of southern California's newer residential landscapes. Evan McKenzie (1994) calls these landscapes 'privatopias', noting that they are increasingly held under the control of a local form of private governance. In what follows, Camille suggests the kinds of control exerted by the HOA:

We have an association because we have a common area and we have the pool and the recreation centre. So we pay, you know, fifty bucks a month to maintain all the common areas. And its just the, you know, typical California community where we have restrictions like we can't paint our house purple or anything like that. You know, everything has to be *approved* which is basically good for the value of everyone. I mean, you don't have people who have, you know, RVs parked out front or stuff like that, 'cause there's rules.

When asked if that ever affected her, Camille stated:

No . . . I mean for us it is basically *positive*. Um, you know, we did get a reprimand once 'cause our garage door needed painting but, you know, actually our door wasn't even that bad, but they kind of enforced the standards of the neighbourhood.

Camille said that maintenance of these rules was not as strict as it might be in some other communities, but her tone of voice suggested that it was more strict: 'In fact, after we had to paint our garage door, we went around and we said "Hey! You know, theirs is a lot worse than ours," you know. *Sometimes*, there's people who just don't comply, you know.'

When asked what happened to those people, she replied: 'Um, well, they can fine you if you don't, you know, keep your house up to a certain standard, and ultimately they can put a lien on your house. So, they have some power and grounds for enforcing it.' When the researcher commented: 'So, this isn't a gated community but it kind of functions like one,' Camille replied sharply: 'No, it is *not* gated, but you'll find very few communities these days that don't have some form of common area, and therefore you *have* to have an association.'

There are several inferences that I want to draw from Camille's extended quotes. First, she is adamant that hers is not a gated community and snaps back at the interviewer when he suggests that it functions like one. Second, it is important to note that although not gated, this landscape defends a certain kind of lifestyle that effectively bars certain kinds of people from being part of it. Not only do you need sufficient money to buy into the community (rental of property is forbidden by the HOA), but you must also be willing to pay HOA dues and maintain your property to a set of standards. You must also own a car, because the nearest retail store is over a mile away and the HOA has the power to keep mass transit systems out of the area. As Chuck notes, 'We're secluded here from Escondido [where the nearest bus stop is] . . . we're just a little too far to walk, really.'

Marcia and Justin

There are no home owners' associations where Marcia and Justin live but, like Chuck's and Camille's community, there are some facilities that provide common gathering places such as the public park and several local restaurants. Of those we interviewed in inner city locations, many felt that their local park was too danger-ous because of gangs, prostitution or drug dealing, but this is not how Marcia and Justin feel. In the first set of interviews, when Marcia was pregnant, she said that they go to their local park and feel it is safe because the 'whole neighbourhood actively utilises it and is therefore constantly claiming it for themselves.' After Stella was born, Marcia became a little more aware of the crime in the area: 'It's not Rancho Santa Fe [an exclusive upper-income neighbourhood] or something like that. It's kind of violent. It's got a little of that violent feel to it.'

When asked whether she felt unsafe or threatened, she replied that the things that worried her were 'not gangs', but 'Hearing sirens a lot. Helicopters. Not a lot. It's quiet like this 95 percent of the time.' The interviewer then asked: 'What about walking? You mentioned in the last interview you liked to walk.' Marcia replied: 'Yes, I walk to some local restaurants and the park. That's not bad. It is not perfect but I do feel a little nervous at times.'

Marcia went on to qualify this statement by pointing out that she still loves the people in Golden Hills and its diversity. 'My life has changed since having Stella,' she concludes with a laugh, 'I've changed a lot of diapers. People, people are so different. People are so friendly and helpful, and so it's pretty wonderful. It's a whole new experience. Everybody around here is great.' One of the changes with Stella is the couple's need for a larger house, but Marcia is adamant that she would rather put up with a less than ideal house in order to remain in the neighbourhood. Justin works as a machinist in a shop in the neighbourhood. In addition, most of his family lives and works in the area ('my shop is five minutes walk from here'), but nonetheless some fears similar to Marcia's arose during the second interview: 'I'd say I'm very satisfied [with the neighbourhood]. Sometimes I'm slightly less satisfied. Generally it's a real nice neighbourhood. There might be something of a crime problem. I really haven't seen it myself.'

A year later we interviewed Marcia and Justin again and it seemed that their sense of community had grown. Neither brought up issues of violence or crime in the area but, rather, focused on their developing feelings for the area. Justin put it like this:

Well (in the last year), um, a lot of the families in the block have come closer. We do things together, um, and we had a block party last year where we blocked the streets . . . it was really nice. Um, some old friends of mine have moved in right across the street. Um, we pretty much know just about everyone on the block, and there's a lot of different kinds of people on the block. It's a really good feeling, watching out for each other.

When asked about whether he would leave Stella with neighbours Justin replied: 'We have, and we've watched neighbours' kids from time to time. It feels very comfortable. And it's different from other places I've lived. I mean neighbourhoods I've lived in where I haven't really known the people.' When asked specifically if he thought Golden Hill was a safe place to raise a child, he replied 'I think so.' When probed further his answer took a surprising turn, none of which focused on violence in the area, and crime was dismissed with 'We don't have much of that in the area':

> Primarily it's because she gets to mix with people. Um, I think it's a good place to raise a child, but I don't know how to put it into words. Ya, diversity! I'd *much* rather be here than in a community with a lot *less* diversity and a lot more problems *with* diversity. I think people have bought into this idea that kids are supposed to grow up in suburbs. I wouldn't mind being somewhere more in the country. I love, love being in the country but I don't think the suburbs are the country.

It seems appropriate to conclude with a quote from Justin in response to being asked what advice he would give to young parents who are stressed out and live in fear for themselves and their children:

> Taking more part in your community. There's a lot of like-minded people out there that have a lot of knowledge about things that you don't know about. I think it helps a lot [to share your fears and ask for help]. When you're just a little island unto yourself things can get pretty, pretty heavy . . . It's hard to take the first step. So many people are so afraid.

IN DEFENCE OF DIFFERENCE: A CONCLUDING COMMENT

With this chapter, I highlight landscapes of defence and landscapes of siege as sustained and encountered by a variety of families with young children living in different kinds of community in San Diego. For many young families in our study, fear formed around an anxiety over strangers. If, as Kristeva suggests, this is the rejection of difference then it is also about feelings of loathing and abjection because we cannot, ultimately, rid ourselves of difference. Neo-traditional communities, CIDs, HOAs and 'privatopias' comprise yet another series of exclusionary gambits by that part of American society (and it constitutes a large swathe of that society) that wishes to defend itself against the unknown 'other'. With this chapter I have tried to demonstrate that parents' fears and anxieties are as complex as their responses and strategies. That said, the birth of a child crystallises certain attitudes and values, which translate into strategic choices that bear heavily on the construction of landscapes of defence.

A number of examples from the study suggest that young parents believe that only a specific kind of landscape can provide their children with a safe and secure space. These landscapes are carceral and exclusionary, catering to a problematic political identity that focuses on being surrounded by people who look and think the same. Iris Marion Young (1990: 124) argues that rather than seeking a 'wholeness of the self, we who are subjects of this plural and complex society should affirm the otherness in ourselves, acknowledge that as subjects we are heterogeneous and multiple in our affiliations and desires.' She notes that such a perception entails a revolution of subjectivity. A revolution of this kind would require embracing difference as the most appropriate strategy for procuring a 'real community' within which to raise a child.

ACKNOWLEDGEMENTS

Research for this chapter was supported in part by grant SES-9113062 from the National Science Foundation. Special thanks go to all the students who were employed through this grant. In addition, I would like to thank the families in San Diego who filled out our questionnaires and agreed to be interviewed as part of the study that drives large parts of the arguments in this chapter. Opinions, findings and conclusions expressed in this chapter are mine and do not necessarily reflect the views of the National Science Foundation, San Diego State University, or the students and families involved in this project.

CHAPTER 4

LANDSCAPING FOR POWER AND DEFENCE

Stanley D. Brunn, Harri Andersson and Carl T. Dahlman

Social practices that produce individual, residential, corporate, military and political spaces are identified and marked by distinct and subtle landscape elements and features that separate, integrate and unify individuals, groups and communities. Among those 'dividing' features that are familiar to most individuals are signs, barriers and shields, and 'communal features' such as neighbourhood playgrounds and public spaces. Each represents the decisions of individuals or groups to distinguish between spaces. There are also public and private landscapes that reflect safe and unsafe, secure and insecure feelings on the part of those inside and outside enclosed spaces. Sometimes the exact character of the spaces is ambiguous, as with enclosed shopping centres and malls, or industrial parks, security zones around financial institutions, and restricted areas for training military and security personnel or for weapons testing. These are spaces that are separated from most of the population by clearly marked signage as well as by security devices that are meant to discourage trespassing and unwelcome visitors. These landscape features are in essence subtle and overt symbols of power, power that is basically, it might be argued, protective or defensive in nature.

If we consider residential landscapes as specific examples, these 'landscapes of power' do not happen by accident. Rather, they are conscious decisions of governing or corporate leaders who wish to exercise power or wish to encourage and persuade others, through propaganda, advertising or even 'fear' to construct elements of a defence landscape. The power to zone and regulate the use of space in urbanised societies extends from a legitimation of that power in the interest of the health, safety and welfare of the public. The techniques and determination of that power, however, were historically predicated on the expertise of physicians, architects and corporate interests operating in the political process, often through expressed self-interest. Land-use zoning, urban services, police and fire protection, housing densities, speed limits, and traffic patterns are power decisions, in that they are based on the thoughts and plans of groups with major interests in how public and private spaces are to be used. Real estate developers, architects, and the construction, investment and banking industries make other decisions about human environments. Included in these landscape decisions would be plot size, dwelling styles, structures, materials and heights, and proximity to amenities or noxious land uses. Consumers are also party to many of the human landscape features relating to place identification and space use. Groups of consumers in neighbourhoods communicate various messages about themselves and to others through their landscape preferences and tastes for protection, privacy, security and preservation. This construction of landscaping for distinction and separation is achieved in various forms. These include forming separate political subdivisions or units, restrictive covenants, clearly demarcated landscape features such as walls, gates and security cameras, and erecting physical and psychological barriers that invite some and deter others.

The salient point made in the above examples is that social landscapes provide symbolic evidence of power and defence. This chapter focuses on the landscaping of power and defence, where 'landscaping' refers to *how* the physical and human landscapes are being built, designed or constructed for defence or defensive purposes. This landscaping may include modifying the existing land and water surface, or it may include constructing human settlement features or forms to convey power, protection and security. The built environment could include specially designed walled, contained or gated communities or televisions monitoring pedestrian and vehicular traffic in public or private spaces. These interfaces raise a number of questions about power, defence, security and landscaping that relate to protection and regulation by individuals, and non-state and state actors. The actors include architects, planners and other environmental engineers, the construction industries, investors, realtors, bankers, security forces (both public and private), advertisers and the media (as promoters of the meanings of defence and power in the human landscape), and technology companies producing security equipment.

In the discussion below we investigate the concept of 'landscapes of defence' in six sections. The first presents a theoretical context in which to study defence landscapes. This is followed by a discussion of the elements that comprise such landscapes and an analysis of the interest groups that are associated with these constructed landscapes. Next, we illustrate examples of defence communities that appear in the social and political landscapes. Fifth, we discuss the mapping of landscapes of power and defence. Finally, we present a typology of landscapes of defence that integrates our conceptual thinking and that will serve as a basis for continued thinking and research on the topic. We conclude by offering several promising areas for future research by those in the social science, public policy and design professions with interests in these landscapes and the processes that are responsible for them.

DEFENCE LANDSCAPING

The power expressed in landscapes of defence is predicated on the ability to arrange the materiality of those spaces. This power need not necessarily be legitimate in a legal sense, but the most common expressions of defensive landscapes extend from sovereign or legally sanctioned practices. Indeed, many defensive landscapes are designed to protect the very mechanisms of legitimation and control that enable their presence. Policy powers are granted to those whose organisational and institutional relationships can claim legitimate access to force. Similarly, the relation of private property establishes a parallel sphere of protected spatial practices, ownership of which has, until recently, implied near total control over space, objects, personal conduct, and even other humans.

However, this should not suggest that ownership alone is capable of rearranging materiality. Perhaps this is best expressed in the 1968 Paris expression 'Sous les pavés, la plage' (beneath the bricks, the beach). The power of this sentiment was its validation of the disassembly and, sometimes violent, displacement of building materials in the pursuit of an obtaining social legitimacy and its attendant imagined geography of urban space. Ephemeral practices evidenced in the barricades of 1968 Paris or the protests and graffiti of the 1999 WTO meeting in Seattle, then, are important rearrangements of both material and discursive space. The effects of momentary and spontaneous rearrangements are long-lasting and can, in some cases, reorganise the basis for legitimacy and legal sanction. Spatial rearrangement need not be manifested in simple brick and block, but towards altering access, usefulness and even the representation of a space (de Certeau 1984).

The arrangement of material and discursive landscapes for defence implies social reorganisation because alterations of the built environment and the creation of spaces of access and exclusion change social relations. Social positions are made evident in the restriction or availability of certain spaces – the symbolic register of exclusion and enclave, access and privilege denial from social categories of difference and privilege. When the practice of public transportation is altered by a strike or when the practice of walking on a beach is restricted by barbed wire, the possibilities and combinations of social interactions are changed. A person's identity is as much defined by their access, mobility and interpretation of space as it is by their race, ethnicity, nationality, gender or sexuality, because space and social position index each other through the social relations, or their prohibition, as expressed in the landscape.

In reading a landscape of defence we are presented with two competing narratives to understand its impact on society. Risk is often employed to explain how the technical advantage gained by defensive landscaping is applied to the reduction of risk. Yet risk must always be recognised as the erosion of a preferred condition – preferences, we must ask, defined by whom? Nuclear deterrence has been the preferred condition for superpower governments to reduce the risk of war, while other governments and many citizens' groups prefer nuclear disarmament to reduce the risks of nuclear war. However, the defence landscapes associated with nuclear deterrence represent more than risks of war. Ancillary risks associated with nuclear defence – reactors, refineries, processing plants, disposal sites – are increasingly the site of public health and environmental hazards, as witnessed by Rocky Flats, Colorado, and Paducah, Kentucky, among many others (see also Chapter 2). Defensive landscaping often successfully reduces risk by achieving preferred outcomes. For example, the risk to private property is reduced through anti-theft devices in stores or on car steering wheels. What is most useful about thinking through risk is that it is distributed unevenly, and its effect on human existence is not always well understood at first (Beck 1992). Geographers have an important role to play in studying both the intended and unintended landscapes of risk.

Another narrative form explains the social impact of defensive landscapes. In this approach, it is useful to consider the arrangement of material space, or the built environment, as a site of oppression. The racially segregated landscape – whether studied in its legal existence or in its current illegal maintenance – is a powerful example in the built environment. The defence of white privilege, for example, is a spatial strategy that builds social categories into the urban environment. Yet oppression in the built environment can also become something against which to landscape defensively. The clearance of bushes and installation of street lights has been used in an effort to reduce crime, particularly against women, in an effort to reduce risk to pedestrians. In approaching defensive landscapes as social practices that can increase or reduce oppression, we must also identify how that oppression is operative in rearranging social practices. Glenda Laws' work (1994) on responses to oppression in the landscape identifies violence, fear of violence, health and safety as possible axes of oppression, as well as exclusion, exploitation and immobility. She provides a typology to begin interrogating how different social positions and relations to a landscape make possible very different interpretations and reactions to that space.

ELEMENTS

Defence landscapes in the human environment can be analysed to distinguish their functional and formal elements. The spatial function of a defence landscape comprises the isolation of some human activity and the discrimination of who or what can pass between isolation and non-isolation. Isolation, however, does not imply total control, since the isolation of private freedoms in a gated community residence is of a very different quality to the isolation of solitary confinement in prison. The spatial forms of a defence landscape comprise the technologies used in arranging the space and the locational intensity, or spatial extent, of the technology. These two factors are not unrelated, since very few technologies are spatially flexible (Table 4.1). Gates, for example, are intended to be spatially discrete technologies, while minefields can extend quite far. In between are varying arrays of closed-circuit cameras or walls that might defend a boundary ranging from a few metres of perimeter to an international border.

The most apparent and observed elements in these landscapes are walls, fences and gates, which may enclose individual residences, apartment complexes or entire communities (Table 4.2). For example, an estimated six–eight million Americans live in gated communities. Blakely and Snyder (1997) classify them into three main categories, depending on residents' preferences: lifestyle, elite and security zone. Property and territorial markers would vary in design and construction from very simple to very elaborate. In some cultures, the walls and fences would be made of very crude materials, using wood, brush, rocks, piles of stones, poles, hedges, sticks, thorny plants, low stands of shrubbery, or single strands of wire. Alternatively, they may be elaborate and made of combinations of plastic, brick, stucco, concrete, topped with broken glass, iron or aluminium. Some very expensive walls may have ornaments and be overlooked by security cameras to connote the private and

Table 4.1   Elements of defence landscapes by form and function

Function	Form and extent		
	Localized	Medium-range	Long-range
Isolation	quarantine, prison	electric fences, minefields	economic embargo
Discriminant	metal detectors, locks, gated communities, public restrooms	military bases, transport corridors	US–Mexico border, Great Wall of China, airspace
Open access	in-store anti-theft gates	electronic collars, CCTV	EU borders, international waters

protected spaces of those on the inside. Bars in windows and glass or screened doors are additional devices to ensure security.

Walls can be constructed low to the ground, representing only a slight break from the land surface, or they can be constructed at eye level or higher to differentiate sharp landscape distinctions between those on one side of the wall and those on the other. Fences and gates are also ways in which individuals and communities exhibit how they perceive the need to protect personal or communal property, which may include crops, livestock, pets, factories, parks, gardens, public institutions, and memorialised and sacred spaces. These themselves can be simple or elaborate. In gated communities, for example, the entrance gate and possibly even a gatehouse or a guardhouse may have armed security guards who admit individuals only with approved coded vehicles and identification. The erection of these entrances, often very elaborate and expensive landscape features, is meant to signal symbols of power, control, wealth, privacy and protection. Outsiders and unwelcome visitors are, it is hoped, deterred by elaborate security systems, concrete barricades, walls

Table 4.2   Landscaping by individuals and communities

Past	Present	Future
Watch dogs	Alarm systems	Security-proof construction
Watch geese, large cats	Security cameras	Smart homes and offices
Fences – wire, wood, rock	Private security companies	Smart offices
Gates	Neighbourhood watches	GIS
Walls	Computer security	Coded vehicles
	Gated communities	
	Street lights	

within walls, the requirement to show security proof documents (personally coded cards or approved vehicles with acceptable signal devices), and walls or fences that from the outside are sharp social class and economic dividers.

For those who do not reside in gilded ghettos or exclusive and expensive sub-divisions, the privacy of one's space may be reflected in simple 'no trespassing' signage or the placement of outbuildings, flower pots and birdhouses, flower and vegetable gardens, low-growing or eye-level shrubbery and hedgerows (common in Europe), windbreaks (in the Great Plains), fruit trees, or even barking dogs. Open space itself may offer protection or the appearance of desired privacy by a resident or a community. Highly sensitive military and government intelligence spaces may have multiple entry checkpoints, voice-pattern and face-scanning identification, and approved authorisation by those in positions of power and command. Governments may demarcate official spaces and territorial boundaries by official signage, flags, monuments, walls and gardens. International borders will have border posts, customs houses and welcome stations. A state sends various messages to those wishing to enter its spaces by the architecture, colours, degree of openness, single or multiple checkpoints, and friendliness of its uniformed personnel, including whether they carry weapons.

A south Florida example illustrates the kinds of security system in gated communities. Embassy Lakes, in Cooper City, Florida, is a middle- to upper middle-income development for professionals, with approximately 1600 homes ranging in price from $100,000 for attached dwellings to over $400,000 for individual houses. Five developers built this walled development, and it contains seven neighbourhoods. These builders cooperated in obtaining the necessary zoning and constructing the walls and clubhouse. Embassy Lakes has three entrance gates with 24-hour security guards. Residents must have bar codes on their cars to enter. Visitors must show a picture ID and then the resident is phoned. Other complexes have a gate with a code, so if you visit someone there, you need to have them give the security code to enter. In yet other complexes, the visitor enters a code for a person on a numeric keypad. This code alerts their telephone. They then answer and press a button on their phone to lift the gate. Adjacent to Embassy Lakes is Rock Creek, another development that relies solely on the security code on the keypad system, and that applies only for some sub-developments. Perhaps as a result, some households have moved from Rock Creek to Embassy Lakes so that they can be inside walls.

Other features of the built defence landscape are roofs and rooftops, which may contain a variety of features to protect the residents of an individual domicile or those in an apartment complex. The roofs of embassies, airport buildings, research laboratories, banks and military headquarters often contain a number of antennae, satellite dishes, and hidden daytime and night-time motion cameras. Cameras may be focused on the streets below or on building entrances. These may be 24-hour cameras producing images observed by someone actually stationed in the entrance of a residence, factory, government office, museum or laboratory, or in the offices of a security company or public police force. Roofs may have heat-sensitive lights that are switched on when an individual or even an animal enters the security space, or sirens, shrieks, tiger growls or horns that sound when someone other than an approved resident enters the buildings or wanders into spaces without approval. Some buildings will also contain highly sensitive electronic devices to detect eavesdropping.

Signage is a key element in landscaping for defence. The words or symbols used on the signs may be subtle or visually apparent to the occupier, worker or resident of those spaces, or they may lead the person on the outside to wonder what transpires in the world beyond the walls and gates. Outdoor hoardings stating that the subdivision is a 'closed community' or 'protected community', or even advertising a 'gated community for a distinct group', would be likely to signal spaces of exclusiveness, security, privacy and wealth. In many high-income and exclusive communities, covenants restrict the size, shape and placement of signage on private or community property. At the opposite end of the class and income spectra, the signs of privacy and individuality may be as simple as 'no trespassing' or 'beware of the dog'. Signs constructed and erected by private corporations may inform the individual and literate walker, jogger or driver that 'this is a protected community' or 'this is a neighbourhood watch community' or 'this is a crime stopper community'. Political signage (on streets, highways and official buildings, and in dangerous sites) in multilingual states may be in only one language, namely the language of the majority, or in several depending on the multicultural character of the state. In some parts of rural Israel, for example, there are official signs only in Hebrew, even where Arab populations live nearby (see also Chapter 6). In other areas, such as the large cities, bilingual signage exists. In the USA, the growing Hispanisation of the Sun Belt has resulted in some states posting official signage on public buildings and infrastructure in Spanish.

CONSTRUCTING LANDSCAPES FOR DEFENCE

Individuals, communities and the state have constructed landscapes of defence throughout human history (see Nyström 1999 and Table 4.3). Hilltop settlements were important sites for castles and defensive installations by Mediterranean cultures. The thirteenth-century Cathar religious minority established high-walled defensive settlements on the crests of steep ridges where it could defend against political and religious persecution as well as watch for coming attacks. Walls were important in separating Chinese populations from steppe invaders. Moats, impregnable walls and huge castle doors were meant to deter enemies, as were open spaces around fortified settlements. Landscapes were also constructed for those with interests in preserving certain cultural features (such as churches, historic sites and battlefields) or in attracting consumers (e.g. believers, converts and those searching for heritage) to certain locations. In the latter group would be those on pilgrimages to holy sites, tourists to popular or isolated destinations, and retired people. Military and paramilitary forces also construct landscapes for offensive and defensive purposes. Similarly, corporations design landscapes that range in size and complexity from huge industrial parks to gigantic shopping centres. Landscapes for the defence of recreational enjoyment were built into theme parks such as Disneyworld by restricting guest access to approaches that highlighted particular features or imposed a visual

Table 4.3 Landscaping by the state

Past	Present	Future
hilltop settlements	military highways, rail lines	large-scale landscape modification
castles	planting forests	weather modification
impregnable walls	deforestation	satellites
moats	all-terrain vehicles	marine terrain analysis
huge gates	terrain modification	river diversion
drawbridges	secret maps	laser 'gates'
fences	secret cities and roads	virtual maps
caves	listening posts	terrorist-proof construction
mountain passes	radar	'smart' buildings
toll roads	land mines	night vision
clearing vegetation	lighted landscapes	new frequencies for communication
	cellular towers	multiple GIS uses
	tall buildings	'smart' transportation vehicles

distortion on aesthetic entrance effects. Disney continues to control the use of images taken in its parks, aside from personal collections, in order to defend the experience of its parks from commercial erosion.

A significant component of landscaping for defence is the variety of interest groups with a stake in the landscaping and in the defence of those spaces (see above: also Zukin 1991). Individual residences (apartments, town houses and condominiums) and subdivisions represent a wide variety of architectural designs and styles. They can be designed with specific concerns about defence or security, or they can be erected where security, privacy and protection are a lower priority. Newman (1972; 1996) refers to these as 'defensible spaces', a concept used to study public and private housing developments and map residential preferences, including residents' 'fear maps'. The architect plays a key role in designing not only human structures but also other parts of the built environment. The professional may work with the individual home owner or developer to prepare the actual site for the residential space. These may include decisions over what directions which rooms face, on what floor they are situated, and the nature of floor plans. They may also decide the uses of surrounding spaces – such as what appears in front of or behind the residence, whether it faces the street or adjacent buildings, whether or not it faces another residence, and the question about what one observes from one's kitchen, bedroom and living room windows. The architect may work with landscaping services to plant or preserve certain shrubs, trees, plants or landscape features in certain locations or to construct garden spaces, walls, and other territorial symbols and markers to designate the limits of one's territory. The 'visual' is part of that landscaping, but so are 'soundscapes'. Some residents will chose places that have little vehicular noise, that are not in the flight paths of regional airports but are on the flyways of migratory birds. Others will tolerate the sounds of heavy vehicular traffic and industrial plants and enjoy the sounds of children and outdoor neighbourly conversations.

The construction industries, and the building contractors working with them, have a major interest in the landscaping of power and defence. These industries not only construct civilian houses, apartment complexes and town houses but also shopping centres, malls and public facilities, such as sports stadiums, parks, playgrounds, gardens, museums and cemeteries. They are also major modifiers of the urban landscape through the construction of BIDs (business improvement districts); spaces whose developments are frequently subsidised by state or city bonds (Zukin 1995). Building contractors work with the providers and suppliers of various construction materials (concrete, aluminum, brick, glass, stucco, wood, plastic and special steels) to erect bridges, residences, parks, playgrounds, office buildings, streets, parking lots and other 'machine spaces' (Horvath 1974). Culturally acceptable standards may dictate what kind of materials are used for constructing walls, fences, gates and roofs and the preferred styles for roofs, heights of walls, gatehouse features, and publicly viewed signage.

Military or defence industries, contractors and consultants dot the economic landscape and are an important part of a state's economy (Brunn 1987). They construct airports, ports, military training facilities and even interstate highway systems – the US system, for instance, was constructed to move personnel and material easily around the country. Companies also produce jeeps, tanks, trucks, submarines, aircraft, all-terrain vehicles, and a wide variety of security and surveillance products ranging from communications satellites to night-vision equipment to terrain modification. During times of national emergencies and disasters, military and defence users may have priority over a state's communication and transportation systems. Vehicular traffic on two- or four-lane highways can be routed in one direction; streets can close; straight-line highways can be used to land aircraft (as is the case in coastal Israel); and basins for pleasure yachts can be converted to naval use. Security zones are established, and fortified landscapes include demilitarised zones, border posts and landscapes, and secret military testing facilities.

Landscape modification is a major component of designing landscapes for defence and power (Klein 1998). These modifications may be undertaken for military preparedness or actual military combat. States with regional and extra-regional military and security commitments train their land, air and sea forces in actual or simulated environments (including computer war games). Thus an understanding of the military geographies of sand, snow, ice, mud and tropical rainforests for national defence and security is critical (Winters 1998). Military construction teams design landscapes with these various environmental conditions in mind. Defensive landscaping may also include the draining of swamps, deforestation or chemical defoliation, the levelling of hills, the diversion of rivers, the construction of dams, rail lines, canals and airfields, even the alteration of coastal and ocean bottom landforms for military or security purposes. Weather modification and reporting may play a role in military operations: for example, cloud seeding, augmented pollution levels, false reporting on weather conditions, or preparing inaccurate weather forecasts to deceive the adversary. The modification can be extended to include camouflaged spaces of buildings, vehicles and surface land uses to avoid detection by the enemy or to confuse ground-level or satellite spying.

Protected spaces are also part of the state's landscapes of defence. These become important during times of conflict where there are calls for truces and temporary halts to enemy attack (Brunn 1991). Designation of these spaces as havens, sanctuaries, corridors or passages is testimony to the fact that states in conflict can agree to identify certain areas as being safe for children, the elderly, women and the disabled. The humanitarian spaces constructed for these peaceful purposes may be monitored and policed by the United Nations, or another regional peace-keeping force, or by non-governmental organisations such as the Red Cross, Red Crescent and Médecines sans Frontières. In Cyprus, Bosnia, Kashmir, Lebanon and elsewhere, there are UN peace-keeping forces; in East Timor, Sierra Leone and Liberia, regional forces seek to prevent the spread of conflict and assist returning refugees.

The method of border management and sovereign control over immigration is a constant issue in the USA and other Western states. Recent efforts illustrate the defensive landscapes put in place to maintain a US 'border that works – one that facilitates the flow of legal immigration and goods while preventing the illegal traffic of people and contraband' (INS 1999a). The US Operation Gatekeeper at the relatively short San Diego border with Mexico employs 2264 agents and 504 inspectors. It also has nearly 80 kilometres of elaborate fencing, 10 kilometres of permanent high-intensity lighting, fifty-nine infrared telescopes, 1214 underground sensors to detect footsteps and small vehicle movement, 1350 computers, 1765 vehicles and ten helicopters (INS 1998). The emphasis on 'force-multiplying technologies', allowing agents to survey the border remotely before committing to a field intervention, highlights their ability to remain invisible while using lights, cameras and motion detectors to increase the visibility of border crossings. At the gate, a new on-board communication system allows pre-approved drivers to pass quickly through the checkpoint – indicating the extremes in discriminating between legal and illegal movement through the same highly defended landscapes. Elsewhere, the INSPASS (INS Passenger Accelerated Service System) kiosk allows eligible businesspeople, 'diplomats, representatives of international organizations, or airline crews' from selected countries to bypass regular visa lines by inserting an ID card and placing their hand on a panel that confirms their identity through biometric hand geometry (INS 1999b).

In addition to the construction, landscaping and architectural groups interested in constructing landscapes for defence, there are other groups whose major concerns are in security and surveillance. These include the companies producing heat-sensitive lights, security day and night cameras, listening devices, electronic fences and gates, alarm systems, and security systems that can be turned on by voice command or from computers in one's office, car or boat, or from a police station or private security force office. Websites by firms promoting security and protection are being developed for home owners and businesses. Clearwater Landscapes Inc.

in Priest River, Idaho, a firm that has links to design, maintenance, installation and landscaping, offers what are termed 'hardscapes' (Eskelson 1999). Private Community Security Consultants (Breed 1999) states on its opening page: 'The public's perception of vulnerability has created a dramatic increase in the number of gated residential-resort communities. There is a recognized need for addressing those unique problems arising from the establishment and management of private community security operations.' Detailed information is provided for future or planned communities and those already existing, including community integrity risk assessment, vehicle access filtering mechanisms, security audits, and managing the human element in resort security. Ron Corbin (1999), who identifies himself as a crime prevention specialist for the Las Vegas Metropolitan Police Department, has a website entitled 'Reducing the Odds: Landscaping for Security'. His site offers suggestions for landscaping homes to deter burglary:

> Amid windows and narrow walkways, trim all bushes and ground plants to remain 3 feet or lower in height. For trees, cut branches so that the leaves and foliage are at least 7 feet or higher. What this provides is a 'clear space' for surveillance by you and your neighbours, and consequently less of a hiding place for criminal opportunities and activities.

He also adds this advice for those living in desert environments:

> [An] other practical crime prevention use for landscaping is to use a desert landscaping theme with cacti or other types of thorny plants along block walls and perimeter fencing. There are many different kinds of prickly plants that will provide a real deterrent to juveniles for climbing your walls or invading your property. A well qualified landscaper or plant nursery professional will be able to assist you in selecting the appropriate plant that will satisfy your needs for this prevention method.

Firms producing these high-tech security products play an important role in landscapes of defence and landscaping for defence. It might be argued that their business depends and thrives on not only convincing individuals, groups and communities of the need to adopt and purchase their products, but also on fear itself (Tuan 1979). Menzie (1998) gives several labels to this phenomenon, including 'defensive urbanism, paranoid architecture, and the architecture of fear.' The fear of the unknown, whether that is from fear of an outsider, intruder or innocent trespasser, plays a key role in how firms promote their products to potential consumers. Peddling fear and security is a component of their advertising on television and radio, and in newspapers, as well as large outdoor advertising. Their hope, it might be argued, is to instill or generate concern about one's private places or spaces and the need to protect those spaces and individuals through the purchase of certain security and surveillance products. The products themselves vary depending on the consumer's purchasing ability and wish to obtain protection and privacy. For some companies, their advertising might read 'a wish to retire in a community that has a private security force twenty-four hours a day.' For another, it might be 'residential security cameras installed that are connected to a local police force' or 'purchase the combination light and sound system to protect your children and possessions from unwanted outsiders.' Communities and companies may hire private security forces, perhaps in addition to public forces, to protect properties from criminals, potential criminals, or even terrorists.

Advertising and public relations firms also have an interest in the landscapes and landscaping of power and defence. Companies producing print or visual advertising can generate feelings of insecurity and also community distrust through the words and symbols that they use. Often children, mothers and the elderly are used in commercials and advertising on radio and television, and on outdoor billboards to generate the need to purchase a specific security or surveillance technological innovation. Fear, hostility and insecurity can be conveyed through the voice of a young child, a mother's or father's concern about neighbourhood safety, or an elderly resident's fear to shop at a neighbourhood store. Advertising companies working in tandem with building contractors can use powerful images of place (dimly lit streets *versus* well-lit streets, green spaces versus concrete spaces), select word usage (see examples above) and colours (of landscapes and people) to promote the exclusiveness of new subdivisions and reconstructed inner cities and to appeal to home owners and renters concerned about privacy, protection and security. There are also websites, as noted above, which promote their services to communities and individual consumers in several languages.

The local news media and government officials are two additional groups with interests in landscapes and security. The visual media are a potent force in influencing public opinion and consumer attitudes regarding privacy and security. Local television stations complement local newspapers and radio stations as important sources of public knowledge regarding places and what happens in those places. Reports about the frequency of burglaries, drug arrests, vandalism, shootings, and motor vehicle and industrial accidents, when seen, read about or heard, can influence how residents think about places in their community and surrounding communities. They are able through these reports to piece together places and environments that are dangerous and safe for walking, shopping, driving and working.

Interviewing residents on camera in places of crime and violence and public officials provides additional ammunition to those not wanting to venture into certain areas. Sometimes local evening newscasts are more 'crime reports' than accounts of civic and cultural events. The 'fear' and crime reports may also overtly or subtly play into the hands of those firms advertising and selling security and surveillance technologies, residential areas, or commercial spaces. Public advertising on English-speaking television stations may picture the stereotypes of non-native-speaking populations (voice and appearance) to instill fear. In short, those selling computer surveillance systems may benefit from news reported by the media and government law enforcement. It is not unknown for these firms even to sponsor the many police, law enforcement and other violent dramas seen in many homes during the course of a week and contribute heavily to the political campaigns of those seeking public office.

Finally, there are two other groups with specific interests in landscaping for power and defence and in the landscapes themselves: lawyers and politicians. Lawyers are charged with developing laws about definitions of property, the ownership and protection of property, and how official laws and regulations are to be enforced. Legal teams inform elected or selected members of local and state governments about a host of issues, including individual *versus* common property, the rights to ownership of land and places, zoning restrictions and regulations, and acceptable uses of designated land. Governments also seek legal advice in enacting building codes, in technologies to protect the rights of citizens, in approving and designating acceptable uses of public spaces for public events, and in the naming of streets, parks, hospitals, schools and memorials. The legal component of landscaping, which may find expression in both 'enclave and exclave zoning', is an important one that interfaces closely with the other interests groups discussed above. Elected politicians, or aspirants to public office, are also important actors in landscaping for power and defence. Some may seize on landscaping issues to identify campaign issues and thereby generate support for their candidacy. For example, a councillor or mayor may use the issue of distrust that some residents express for others to appeal to certain voters. The issues involved might be a real or perceived rise in violence or the lack of safety in public places to call for tighter security measures, including curfews, heavier patrols in certain neighbourhoods, and the need to install anti-burglary devices in homes, offices and cars.

MAPPING DEFENSIVE SPACES

Political maps contain many defensive spaces (Chaliand and Rageau 1985; Kidron and Smith 1991) that a state wants or wishes to protect, preserve and defend. Citizen groups, including non-governmental organisations, also designate spaces that they wish to protect, and they can also produce maps and atlases. Colours, symbols, languages and depictions are important features of such maps. State maps may show a variety of features, including military bases, military training facilities, military airports, ports, research laboratories, telecommunications centres, hospitals, storage depots and police stations (Brunn 1987). Aside from official protection, security and military land uses, maps may include the locations of private security forces, highways and airports, Interpol offices, and the sites of paramilitary and guerilla camps and held territories. Official maps may contain incomplete or misleading information about the placement of roads and towns and other surface features. For example, during the Cold War, the Soviet Union had several dozen secret cities not on any published maps. For its part, the USA has secret weapons-testing and storage areas in the Nevada desert (known as Area 51) that do not appear on official large-scale topographic maps. No roads or settlements appear on maps, just blank space. Official maps of many states are not likely to identify areas not under government control, disputed boundaries, or territories supervised by international peace-keeping forces.

Maps at another scale contain information important to individuals. For one farmer, they might be one's garden or cropland or livestock-raising area. Crudely constructed fences of brush or a grove or shelterbelt of mature trees may mark these spaces. For another farmer, they may be demarcated by rock fences or wooden fences or barbed wire. Alternatively, there may be no formal boundary separating one's person's rural property from their neighbour's, although the occupants know what spaces they own or over which they have rights. For a city or suburban resident the private spaces may be only where one resides, that is, there is no space outside one's apartment or town house complex that is private. Those with small amounts of private space may use a variety of symbols and structures to delimit their own personal property. It might be low or high-rising fences of brick, wire, aluminum or some combination, or hedges, the placement of fruit trees or outbuildings, or some ornamental symbols, or children's swing sets or gardens. It is very common to designate one's own space by the yard or grass that one mows. In these cases, there may be no formal protected spaces, but it is understood who owns what. In addition to these symbols and structures, individuals may install security and surveillance systems.

Many groups seek to protect and also thereby separate themselves from others. The 'others' may belong to different social and economic classes, or have different lifestyles, cultures and interests. This characteristic is part of the changing nature of cities that Smith (1996: 211) describes as the 'revanchist' city, which:

> expresses a race/class/gender felt by middle-class and ruling-class whites who are suddenly stuck in the place by a ravaged property market, the threat and reality of unemployment, the decimation of social services, and the emergence of minority and immigrant groups, as well as women, as powerful urban actors.

The maps of many medium-sized and large metropolitan areas experiencing these transitions contain spaces that insiders and long-time residents can identify as being for newcomers, outsiders, long-term residents, low-income occupants, the elderly or the homeless (Davis 1990). There are many ghettos that exhibit a high degree of similarity in age (e.g. elderly or college youth), religious affiliation, occupation, wealth, expatriate status or lifestyle. Enclaves and exclaves exist for the super-rich and 'first world' tourists. Many of these communities will have landscape features that have sharp transitions with surrounding spaces. Some may be walled and gated, others distinguished by street and advertising signage in different languages from the majority population, or institutions (e.g. religious, sporting, schools and entertainment) that are distinctive in the landscape. The 'defensive' nature of these ghetto spaces may be apparent to the outsider and insider by the street names, languages used, house types, distinctive place features, and also by distinctive entrances and signage in public and private spaces.

It merits mention that the separateness may be the result of groups who wish to be at a social distance from others; that is, they will voluntarily erect simple or elaborate gates, walls and symbolic features to reflect their wishes for group privacy, protection and security. By contrast, others social groups – including the poor, powerless, homeless, marginalised and ostracised, new diaspora populations (unwelcome guest workers from southeast Europe in European cities; Central Americans and East Asians in US cities) – are the victims of separateness (Smith 1996). They may reside in reservations or compounds that the rich, powerful and influential have designated for them (see Chapter 9). These groups may occupy places that themselves have offered some protection and privacy from others but are often labelled as 'being across the tracks, or the other side of town, or beyond the city limits.' Whatever the label, the fact is that involuntary segregation and separation exist and that those residing there may feel the need to erect some of their own symbols of protection. While they cannot afford high-priced security systems, home alarms and privacy fences, they may have barking and vicious-looking dogs, concrete or stucco fences with broken glass on top and barbed wire fences. There can also be an informal neighbourhood watch operation and casual surveillance systems that identify new occupants, passers-by and motor vehicles that 'don't belong'.

The map of protected or defensive spaces extends to other uses in a society as well. Their importance to the population and state will influence the degree to which they are protected. If a land use is designated as being of high value, it will most likely be worthy of protection and surveillance. These may include historic homes, government buildings, palaces, embassies, halls of justice, law enforcement, monuments and battlefields. Other state properties considered important to protect include military training bases, weapons-testing sites, nuclear weapons facilities, research laboratories (from genetic engineering to highway safety to planetary exploration) and centres for telecommunications, transportation and computing. Public housing, subways, interstate highway systems, toxic and hazardous waste disposal sites, and universities would similarly warrant protection, as might a long list of sites of cultural significance because of their heritage value. Some private properties also fall into the same category, since they contain very valuable properties that need surveillance systems. These may be large outdoor recreation areas, historic castles and walls, theme parks, precious mining sites, horse farms, shopping centres, banks, corporate offices, regional malls, churches, private hospitals, and environments for protected and threatened species. Some of the largest and most valuable horse farms in the Kentucky bluegrass region, for example, have security cameras on their gates, 24-hour security guards at entrances and television cameras in individual horse stalls. The state makes decisions on what is important to protect and preserve, as do business, educational and entertainment organisations and groups within the state. The map of protected, preserved or defended spaces is one that constantly changes, with a culture defining what should be and is worthy of being preserved, what needs to be protected, and how it can be preserved.

TYPOLOGY

A useful perspective on the geography of landscapes of defence can be obtained by integrating a number of the concepts and themes above into a matrix. This heuristic device contains constants and variables. The constants represent those elements that are important in studying these landscapes regardless of culture. They are the construction of these landscapes, designers, actors and decision makers, security providers, the built environment, and finally what we label the 'instillers of fear'. The variables are scale (local to global), sector (public versus private), culture, permanence (*ad hoc* or stable) and technology (low or high). As shown in this matrix (Table 4.4), one can consider, for example, the links between designers and sector or the links between security providers and scale and permanence.

Table 4.4 Typology of landscapes of defence

Constants \ Variants	Scale local – global	Sector public – private	Culture	Permanence ad hoc – stability	Technology low – high
Construction	walls, fences, security systems	public infrastructure, private construction	state construction, private construction, corporations	diurnal variations: open → closed, light → dark	signs, walls, gates, surveillance cameras, alarm systems
Designers	new communities, suburbs, cities	private consultants, government planners	architectural practices, planning styles	protecting all areas, privacy space, lights, open space	simple, crude, complicated, fashion
Actors/decision makers	zoning and re-zoning, allocation of police, laws and regulations	local, state officials, cartographers, real estate	zoning boards, planners, activist groups	planners, lawyers, utility companies, the state, mass media	individuals, bodyguards, police
Security providers	local police, Internet providers	public police and security, private security	firms, cameras, cellular phones	private security companies, public security	none (open) to extreme high technology
Consumers	individual, households, niche markets	all citizens, selected markets	families, elderly, poor, children, the vulnerable	children, women, disabled, 'fearful strangers'	all of selected niches
Built environment	fences, walls, gates, individual property, restricted areas	official constructs, public housing, private developers, gated communities	gated communities, hilltop settlements, castles, walls, moats	road blocks, lighted areas, humane spaces	signage, dogs, snakes, fences, alarms, security cameras
Instillers of fear	local media, consultants, terrorists threats, government reports	police actions, laws, new regulations, punishment, security firms	private TV, advertisements, outdoor advertising	sensitive journalists, publishing crime rates, violent acts, police visibility, security forces	word of mouth, news reports, weapons sites, nuclear accidents

CONCLUSION

The subject of landscaping for power and defence is one that benefits from the contributions of a multidisciplinary range of interests, including social and political geographers, landscape architects, engineers, environmental security consultants, and experts in public policy and gender relations. It would also benefit from research that employs archival as well as field investigations and examines rural, suburban and urban expressions of landscapes of defence in a range of different cultural and environmental settings.

We particularly identify five topics as being worthy of future study. First, we believe there need to be studies on specific industries and services that are defence-related. These would include the producers of walls, fences and gates, but also surveillance cameras, alarm systems and even cellular telephones. Their markets and perceived markets and niche marketing would merit scrutiny in different types of culture and society. Second, we would recommend investigating gender relations and landscaping for defence. These might include issues such as safe streets and home spaces, but also places of shopping and work. What kinds of product are being developed and marketed for elderly women, mothers and girls in different economies and societies? Third, we need more research on the kinds of training the professionals receive, that is, both those who design buildings, streets, gardens, gates and walls, and those who design more elaborate and expensive surveillance technologies. Are they sensitive to gender, age or class differences? Are their models for designing products and landscape features generic or are they culturally biased, especially towards Western experience? Fourth, the issue of lighting would benefit from additional investigations by geographers, architects and others. How has the lighting of streets and cities altered the landscapes of mobility, fear and identity? Are there cultural variants that operate in the designing of daytime and night-time landscapes in the lower and higher latitudes and in different seasons? Fifth, it would be useful to investigate the meaning of walls, fences and gates in different cultures. While these territorial markers exist in many societies, how are they perceived by residents, in particular, the new gates, walls and fences that become important elements in one's personal landscape and landscaping? Research into these and related areas will help social policy, and environmental scientists and landscape specialists in different fields to understand the fascinating worlds of defence landscapes and power that exist at all scales.

ACKNOWLEDGEMENTS

We want to thank Craig Campbell, Matt McCourt, Ira Sheskin, Jon Taylor and Jack Williams for providing us with insights into this topic from varying cultural perspectives.

CHAPTER 5

FUNDAMENTALIST LOYALISM: DISCOURSE, RESISTANCE AND IDENTITY POLITICS

Peter Shirlow

In many societies plagued by ethnic, racial and nationalist conflicts, the subordination and control of territory is symbolically tied to wider narratives of communal devotion and ideological interpretation (Thrift and Forbes 1983). Indeed, the actuality of place-centred resistance communities that actively oppose and repulse a racial, ethnic or ideological 'other' is commonplace in most arenas of intemperate and immoderate conflict. Such 'resistance' communities tend to be constructed by means of a communally devoted conception of place and the physical exclusion of those individuals and communities that do not accept a distinct place-centred narrative of socio-political congruity and cultural federation.

In simple terms, in arenas of violent conflict, the control of territory is usually achieved through dominating social contact, iconographies of devotion and political aspirations. In many instances, the words chosen to promote cultural devotion and resistance are also engaged in order to create a propaganda conditioning perspective of right or wrong, good or bad, that dominates the symbolic environment of landscape (Douglas and Shirlow 1998). Spradley and McCurdy (1987), for instance, argue that any interpretation of socio-cultural resistance should not rest upon the assumption that conflicting groups possess a rational linguistic expression directly reflected in the language of their essential interests. Instead, it should be acknowledged that communication within a morally defined community is based upon a particularistic translation of lived experiences and the promotion of an overtly restricted understanding of epistemology and belief. As a result, language and meaning come to constitute the 'tradition of discourse' that gives political dialogue its unique political identity (Finlayson 1997).

Furthermore, the ideological engineering of place as an instrument in the reproduction of resistance communities has provided many groups with the opportunity to engage wider populations in the struggles in which they have operated. Examples such as the Palestine Liberation Organisation's power bases in the refugee camps of the Middle East, the African National Congress's control of South Africa's townships and Euzkadi ta Azkaz's (ETA) support bases in the more rural Euzkadi-speaking areas of the Basque region indicate how the creation of resistance communities and the ideological and physical control of territory provided the infrastructure and support needed to wage long-term military and political campaigns. In Northern

Ireland, for example, the symbolically coded naming and quasi-political control of places such as 'west Belfast', 'south Armagh', 'the Bogside' and 'the Shankill' has been crucial in the reproduction of Irish republicanism and Ulster loyalism.

This chapter aims to comprehend the importance of creating such resistance communities, not only in relation to the defence of territory but also in the perpetuation of ethno-sectarian discourses in Northern Ireland. This is achieved through an examination of the ideological and organisational role played within Northern Irish politics by the Loyalist Volunteer Force (LVF) since its formation in 1996. In so doing, I aim to examine further the complexity of landscape readings in a society in which truth, fantasy and myth interlink to forge meaning, interpretation and, via the medium of power, violence (Baker 1992). Moreover, it is important to note that people's landscapes in conflictual societies are imbued with a power of intensity rarely witnessed in most Western societies. In particular, landscapes in Ireland can mean the very consciousness of being and believing.

However, the reading of landscapes is not merely motivated by a pre-modern tribalism, as many commentators on Northern Ireland assume. It is and always has been about recognising and reproducing identity in a social world. More importantly, landscapes and their readings are never inert, due to reinvention, removal and contestation. Operating at the conjunction of history, identity, belief and social reality, landscapes in Northern Ireland are the conception of intense passions, passions that manifest themselves in devotion, resistance, ghettoisation, legitimation and violent enactment. Unavoidably, landscapes possess an ideological context (Graham 1997) that provides an ordered and simplified interpretation of a social world in which the ideological reading of landscape exerts authority, signs and ethnic codes. In particular, the reading of landscapes in violent societies involves a landscape of symbolic violence, signification and the sanctity of place.

This chapter charts the evolution of the LVF's rudimentary reading of Northern Irish society through its promotion of a Protestant landscape, opposition to the loyalist ceasefires and the 'peace process'. This is achieved through acknowledging the manner in which the LVF adopted a culture of resistance to all agents and movements that were deemed to be opposed to what was self-identified as the 'Protestant and British' way of life. As such, the chapter aims to explain how political and cultural resistance came to dominate an organisation opposed to pluralism, parity of esteem, Catholicism, globalisation and the evolution of consensus building in Northern Ireland. In addition, it seeks to explain how and why the LVF has evolved a sense of politics, territorial control, landscape reading and resistance, where the imperatives of communal difference, segregation and exclusion have predominated over the politics of shared interests, integration, assimilation and consensus.

An examination of the language of oppression and the notion of resistance among members of the LVF is also presented through acknowledging the role that spatial factors play in the construction and reconstruction of this organisation's fundamentalist discourse. This understanding is achieved through examining how the symbolic order of the LVF is constituted by a diverse composition of discursive and material modes, each of which is assembled around a sphere of material reference. Furthermore, a central argument in this chapter is that the construction of this particular symbolic order is tied to modes of intentionality and technical practice. These are modes within which violence and 'reactive' resistance are 'real', even though such cognitive reasoning exceeds socially inclusive notions of conventional reality (Feldman 1991).

There is also a need to understand that the ideological divisions which exist in Northern Ireland, and which are played out through the medium of territorial affiliation and religious segregation, are not simply between the pro-Irish and pro-British sections of the population but, due to the complexity of political and cultural imaginings, can also occur within what are deemed to be homogeneous communal groups. In particular, attention is directed here towards interpreting the hostility between a more pluralistic 'new loyalism' and the spiritual convictions of the LVF's discourse.

THE FUNDAMENTAL 'RIGHT' TO RESIST?

In relation to the theme of territorial control and landscape reading, this chapter recognises many of the valid arguments made by those who have pursued the study of resistance and domination, and in so doing have recognised that the domination of groups, communities and individuals is neither static nor even (Law 1997; Routledge 1997a; Watts 1997). However, it should be acknowledged that much of the work undertaken within the arena of resistance and domination has been biased towards promoting an interpretation of groups and communities that advance the articulation of what are deemed to be 'radical' and 'counter-hegemonic' modes of political analysis and activity.

Somewhat alarmingly, these communities, which are deemed to promote sexist, racist, violent, homophobic or fundamentalist discourse, are often omitted or categorised as adopting an oppositional, yet largely unexamined, space in identity theory. In relation to this form of omission, Gallaher (1997) has correctly stressed that 'it is not enough to position the right as the opposition, while leaving its politics, often of exclusion and hate, unanalysed.' The general omission of communities that promote politically reactive forms of resistance impedes a diagnostic interpretation of the multiplicity of power relationships and their varied locations. The conspicuous exclusion of politicised identities that could be defined as influenced by moralistic, right-wing or inflexible fascist philosophies is not completely neglected by those who have studied theories of resistance and domination. Steve Pile (1997: 4), for example, acknowledges that: 'It is possible to recognise that resistance can involve resistance to any kind of change, to progressive and radical politics, and to social transformation.' However, his and other analyses are, on the one hand, capable of rejecting a significant and ideologically defective 'other' but, on the other, they may not provide significant evidence of how right-wing and fundamentalist discourses are articulated and reproduced geographically. In certain ways, the critiques of resistance and domination that have been provided may lack a much-needed scale in their modes of enquiry, especially given the rise of particular ideological forms that claim to be resisting ethnic, sexual and racial 'others' that they deem are not only defective but also physically and morally menacing.

Without doubt fundamentalist and right-wing groups produce identifiable geographies of socio-cultural domination and resistance in which power relationships are spatialised and imagined in distinct and observable ways. In particular, these reactive ideological forms are primarily concerned with the definition and defensive reaction to particular cultural and social forms that are construed as alien, hostile and unacceptable. Ultimately, the promotion and reproduction of fundamentalist spaces tends to echo strongly Keith and Pile's (1993: 222) assertion that:

> Politics is invariably about closure; it is about the moments at which boundaries become, symbolically, Berlin Walls. These politics hermetically seal these boundaries, creating spaces of closure; on the one side 'the goodies' and on the other 'the baddies'. An eternal struggle between Good and Evil.

Given the durability and production of spatial confinement and closure, therefore, it is important to interpret the predominance and durability of right-wing and fundamentalist discourses in the engendering and reproduction of various spatial arenas. Particularly crucial is the scrutiny of spatial and somatic ideologies among communities that have congealed their discourses into a social structure that has removed the immediacy of lived experience through a purposeful reproduction of modes of symbolic mediation. These modes of symbolic mediation are channelled through a framework of resistance founded upon chosen and exclusive representations of place.

REACTIVE RESISTANCE, IDENTITY MYTHS AND SPATIAL ENCLOSURE

The aim of adopting the term 'reactive resistance' is to acknowledge the complex nature of modes of resistance that are aligned to fundamentalist, ethnically supremacist, right-wing and ultra-nationalist modes of political practice. In general, reactive resistance should refer to forms of identity politics that are based upon an ideological framework which is itself conditioned and influenced through conflict and resource competition with a significant ideological 'other'.

The term 'reactive resistance' also refers to those communities in which identity is reproduced through unwavering communal devotion, the defence of what is construed as ethnic purity and kinship and a self-reflexive interpretative framing of power. This in turn implies those communities that generate a perspectivist illusion in which legitimation resides in the fabrication of a chimerical depth; a dimensionality of force that draws consciousness away from the concrete material investment in acts that reproduce domination in time and space (Bairner and Shirlow 1998). In socio-cultural terms, reactive resistance may, for example, refer to those Zionists (Kellerman 1996), Afrikaners (Cauthen 1997) and ultra-nationalists (Smith 1997) who promote a discourse of ethnic purity and racial superiority and whose resistance is based upon preserving a way of life that is categorised as morally uncontaminated. In relation to the reactive cultures of Catholic, Protestant, Zionist and Afrikaner fundamentalism, the perceived erosion of a valued cultural heritage is tied to opposing the eradication of a religious or ethnic culture that is being replaced by what is identified as an alien process of atheistic materialism, multiculturalism and pluralistic doctrine (Law 1994; Shotter 1993). Such fundamentalist discourses are, in many instances, articulated through the belief that the Bible is a practical guide for politics, governance, work, the family and the affairs of humankind.

In many instances, reactive forms of cultural opposition are tied to notions of cultural dissipation, besiegement and disintegration (Spradley and McCurdy 1987). Sectarian and racist discourses are in themselves cultural constructions that are reproduced and reworked through both time and space. It is in this sense, too, that the disquisition that constitutes ethno-sectarianism in Northern Ireland is reproduced through what are essentially 'lived experiences'. In particular, the defence of terrains of material reference and the perception that communally and symbolically cherished landscapes are, or could be, altered by the in-migration of an ethno-sectarian 'other' means that a reactive consciousness is not solely reproduced through ideology itself, but also in physical and spatial terms (Bell 1990). Clearly, when the defence of space becomes more than merely the ideological reproduction of kinship and location, then the production of violence begins a new phase in the reproduction of cultural practice and symbolic understanding (Feldman 1991; Massey 1995).

For fundamentalists such as the LVF, it is vital that perceived processes of cultural dissipation are not permitted through the sharing of 'loyalist' territory, as this would, it is believed, lead to a process of deterritorialisation and ultimately a weakening of social solidarity. Part of this process of promoting modes of reactive resistance is to tie the symbolic order into the political imagination of mythicised spaces that function as both objects and force (Anderson 1981; Sack 1980). Spatialisation, in such representations, ultimately seeks the creation of arenas that are confined through the communication and exclusion of cultural heterogeneity, on the one hand, and the promotion of illusionary homogeneity, on the other (Beck *et al.* 1994; Roseberry 1989). Indeed, the search for spatial enclosure and socio-spatial demarcation is clearly tied to Sack's (1997: 254) notion that the creation of illusionary homogeneous spaces produces 'boundaries which are virtually impermeable . . . [and which] isolate communities, create fear and hate of others, and push in the directions of inequality and justice.' In relation to spatial enclosure, it is important to recognise that language, its definition and use, creates chains of equivalence that envelop subjects within an entirely self-possessed and self-referential notion of identity, material practice, residence and communal devotion (Shirlow and McGovern 1998). For example, Foucault's (1973: 49) argument that 'practices . . . systematic- ally form the objects of which they speak' indicates how discourse can operate in constrained and constraining ways, especially in the process of socio-cultural entrapment between ideologically opposed communities. Moreover, practice and reproduction produce different imaginings of community. These allegories and mythic representations are themselves the outcome of discursively fabricated classifications of belonging. Undoubtedly, reactive resistance is centred on realising and promoting mythic traditions, temporal representations and increasingly redundant ideologies (Anderson 1981; Harvey 1989).

In sum, the physical reproduction of ethnically defined communities is achieved through the objectivisation of time and space. The narratives and reality of protecting and attacking are themselves interlinked devices in the whole enactment of violence and conflict (Franco 1985). A central part in the construction of identity in conflictual arenas is not merely the use of space to create 'sanctuaries' of ontological togetherness but to compact time and space through eulogising place-centred distinctiveness over topographical unity. Continuously being remade, the construction of territorial division encapsulates distinct, eulogised and communally devoted places, circumscribed by their very contrariness to the 'territorial other' (Jarman 1998). In highlighting the relevance of this identity dimension, in association with modes of socio-cultural and political activity, it can be argued that reactive and fundamentalist discourses aim to protect identities from their perceived 'powerlessness'. This is turn indicates the consequence of the reproduction of asymmetrical relations of power (Scott 1992; Ó Tuathail 1986; Varenne 1993). Undoubtedly, due to the significance of opposing an 'evil other' that lies at the heart of LVF discourse, a sense of the localised nature of politics of territorial control and resistance – where the imperatives of communal difference, segregation and exclusion have predominated over the politics of shared interests, integration, assimilation and consensus – is of concern.

THE LVF AND FUNDAMENTALIST SPACES

The LVF was formed in 1996 from disaffected members of the Ulster Volunteer Force (UVF). On 8 July 1996, the LVF was believed to have been responsible for the murder near Lurgan of a Catholic, Michael McGoldrick, during the Drumcree marching controversy of that time. Due to this murder the LVF's first leader, Billy Wright, known in the popular press as 'King Rat', was expelled from the Ulster Volunteer Force (UVF). Billy Wright was also sentenced to death by the Combined Loyalist Military Command (CLMC) due to his perceived involvement in this murder and his overt hostility to the Loyalist ceasefire and the 'Irish peace process'. Wright was subsequently shot and murdered by the Irish National Liberation Army inside the Maze Prison on 27 December 1997.

The LVF was responsible for a number of killings in 1997 and 1998. As an organisation it was vocal in its opposition to the present 'peace process', which it defines as a state-inspired betrayal of British identity and a vehicle of appeasement towards the Irish Republican Army (IRA). Between the summer and Christmas of 1997 the inmate population in the LVF wing of the Maze Prison grew from three to around thirty. The bulk of the members were those who had transferred from other loyalist wings of the Maze Prison. These individuals were a combination of ex-members of the mid-Ulster and other brigades of the UVF and politically estranged members of the Ulster Freedom Fighters (UFF). The LVF claims to have 800–900 members.

The organisation has supporters in Belfast, but the UVF in particular has mounted a concerted campaign to suppress its support in the city. This was made evident by the murder of John Mahood, a disaffected member of the UVF who it is alleged was trying to establish a support base for the LVF in West Belfast. As such, the LVF is seen as being divorced from a brand of 'new loyalism' that is currently practising and articulating cultural pluralism, socialism and constitutional politics. Due to its fundamentalist epistemology, the LVF derives the most acicular sense that everything around it is undergoing rudimentary and irrevocable change. Thus this is a form of cultural resistance that remains linked to a heritage of election, parental tradition, cultural preservation and armed resistance.

The depth and passion of the LVF's discourse of besiegement is based upon its particular brand of Protestant fundamentalism. In particular, it senses that Irish nationalism, republicanism and Catholicism are interlinked 'dark and satanic' forces ranged against it. Within this representation the oppressed innocents, who are central in the Christian gospel, become the Protestant people of Ulster (Morrow 1997).

In some ways, the LVF illustrates the model of a people conspicuously and self-consciously committed to following that which it sees as the will of God. Christian conviction has certainly been a defining factor in the construction of the LVF's identity. Many LVF members, for example, will constantly talk of 'a deep respect for God and Christ' as factors that motivate them (see also Dillon 1998). The LVF's interpretation of the link between Christianity and territorial sovereignty is underpinned by the notion that faith and belief in God is also, as stated by one LVF member: 'a belief in my country and the Protestant people.' As such, the loyalist legend of 'For God and Ulster' is, for the LVF, a very real place-centred theologically and covenantally designed relationship between the two, in which the latter is a landscape created by the former for a Protestant people. More importantly, the role of defending Northern Ireland is based upon a form of moral righteousness in which the maintenance of what is perceived as a 'God-given land' is both natural and conscientiously correct.

From this standpoint, the LVF claims to represent an elected people who have been anointed by God to execute a mission, and whose predetermination is spiritually inflexible. Collectively, it also possesses a divine warrant to subdue those who are seen as 'the heathens'. Furthermore, within an LVF analysis the rationale of the 'Protestant people' is established through the very will of God, and in relation to this it is inspired by a puissant sense of direction that surpasses and transcends more terrestrial considerations such as the social organisation of society. This cabbalistically sublime stewardship of place achieves an increasingly emotive dimension when the homeland is seen to have been protected through the sacrifice and blood of the 'elected' people. Furthermore, ethnic election demands a devotion and commitment to the defence of the stewarded homeland.

As such resistance is dedicated to defending a landscape within which the Protestant people are constructed as a singularly persecuted and innocent community – what Morrow identifies as a moral and spiritual battle between 'the saved and the damned, the sinners and the sinned against' (1997: 123). In terms of practice and in the face of violence and political turmoil, the LVF's epistemology is constantly driven back to the ontological security and comfort of righteousness, moral order and potential triumph. Threatened territorial domination, whether it be demographic change, territorial dissolution through reunification, IRA violence or British state duplicity with the Irish state will not be permitted to 'master' a community whose resistance calls for an ever greater physical effort to overpower the political and cultural capacity of the sinning 'other'. The influence of this fundamentalist epistemo-logy is that it accommodates self-assurance of certitude and cultural homogeneity. The primary importance of fundamentalist Protestantism is not merely doctrinal but its absolutist notion of a communal 'other' that is ultimately flagitious (*ibid.*).

Ultimately, the culturally hostile manner of reactive resistance and the desire to challenge pan-cultural contact leads in turn to what are essentially cultures of besiegement, which, somewhat depressingly, focus upon imaginings that distinguish the 'we' from the 'they'. Such a conception of peoples undoubtedly fortifies group togetherness, on the one hand, and provides a rationale for group action, on the other (Graham 1997; Shirlow and McGovern 1998). The biosphere of cultural opposition is immovably established upon the elementary binary opposition of the collective self (the elect) and the collective other, and upon the construction of a necessary relationship between the two. The mediating practice that delimits the necessary relationship between them is the notion of 'defence'. 'Communal devotion' in this sense is produced and reworked through animosity and identifiable defence strategies. However, given the cultural extremism of the LVF's discourse, the 'other' can, as detailed below, also include Protestants and even Ulster loyalists who do not subscribe to a fundamentalist and doctrinal interpretation of place, landscape, territory and communal devotion. 'Defence' and the protection of territory thus emerge as the primary discourse defining the mediating practice between the self and the other, the conceptual ordering of inter-communal and spiritual relations.

'RESISTANCE OR NON-EXISTENCE' AND THE MYTH OF ORIGIN

The defence of place and the promotion of a Protestant landscape are ultimately and purposefully articulated through not only an opposition to nationalist and re-publican communities but also through creating the imagined community of the Protestant and Catholic people. Casting the conflict in terms of a binary opposition between two 'peoples' is a central strategy in the search for ontological security, a search that is politically dangerous as it leads to a legend of 'the people' that may be unsubstantiated (Shirlow and McGovern 1998). Obviously, substantiating a claim of 'a people' requires a series of concepts that constitute a discourse of political legitimacy that is at the very heart of identity politics in Ireland (McGarry and O'Leary 1996). Such a process takes place when the LVF speaks of Ulster Protest-ants as an 'elect' community whose status is both distinct and inherited. In reality, communal devotion for the LVF emerges as a mode of domination in which shared class, gender and cultural experiences, which are inter-community in character, are concealed (Gellner 1964; 1983; Tonkin 1992). What has been left is a history of collective allegiance to the Crown, constitution and the Protestant people.

This process of historiographical incorporation and elimination has played a key role in establishing the ideas and beliefs of the LVF. The cycle of the past, as told in popular LVF histories, is traditionally conceived as a battle between Protestant civility and Irish barbarity. Historically, the civility–barbarity thesis not only legit-imises Protestant domicility in Ireland but also continues to suggest the need and right to defend such a position *vis-à-vis* the Catholic and nationalist community. What is clear is that a social memory and political landscape of conflict between two peoples, over the territory of Northern Ireland, has a perceived lineage and therefore a legitimacy in and of itself. Remembering, in such an arena, becomes an operative process, in which place and identity have to be worked on and utilised in order not only to maintain but also to explain the contemporary conflict and the continual need to resist the collective 'other' (Jarman 1998). To be established as a collective memory, there is a need to create images that are uncontaminated by other histories.

As such, the complex mosaic of socio-political and cultural heterogeneity within the Protestant community has been reduced to an ever-repeating cycle of threat, siege, resistance and deliverance. The siege mentality, which defines the nature of resistance, owes much to this conception of the past and therefore the production of a narrative of myths. Myths therefore simplify, dramatise and synthesise the charter for action and the right to be.

A particularly strong myth among the LVF is that of the Cruithin as the 'true' and indigenous people of Ulster. The Cruithin have been cast as the original stock of Ireland and Scotland. This origin myth, based mostly on the work of Adamson (1981), asserts that the evidence of a territorially distinct place in Ireland, that being Ulster, can be located within a justifiable and overtly succinct history. Moreover, the argument that the Celts drove the Cruithin out of Ulster and into Scotland during the Iron Age presents the idea that it is not the Ulster people who are the colonisers but the Gaels who are the oppressors of the truly indigenous and more importantly 'elect' people.

This classification of Ulster history allows an interpretation of the contemporary landscape of partition in Ireland as being legitimate in both contemporary and historical terms. It also permits the LVF to determine that the present conflict is based upon a direct lineage of conflict between two peoples (Jarman 1998). Serious academic analysis argues not only that the Cruithin are 'archaeologically invisible' but also that the argument 'that they are the original Neolithic population . . . up until the arrival of the Celts is quite remarkable' (Mallory and McNeill 1991: 177). Although an excessive and exaggerated history is presented, its ability to confirm the logic of territorial separation in Ireland and of a link with Scotland, and therefore Britain, is clearly part of the following comments made by a member of the LVF when interviewed in November 1998:

> The Pretani were . . . made up of the Ulster Cruithin, Scottish Picts, and the Welsh Britons. They were the first people to settle in Ulster and as such they are the true legitimate people of Ulster and even Ireland. They were driven out by the imperialist Celts, although they defended themselves well for a long time. They returned in the seventeenth century, from Scotland where they were exiled, as the planters to reclaim their ancient lands.
>
> The plantation was the return of the chosen people. The Cruithin have always been an Ulster people who defended their lands against the barbaric Celt.

Clearly, the plantation of Ulster in the seventeenth century is depicted as a rational reclaiming of an ancient elected homeland, what Graham (1997: 50) called 'the creation of an indigenous regional metanarrative, arrived from symbolic place icons and intermeshed with pseudo-history.' As argued by Graham, such origin myths adopt 'precisely the same foreshortening of time which . . . identifies as a primary characteristic of traditional Irish nationalism, the distant past being used in both representations to validate and legitimate contemporary social order.' What remains is a casting of a people who, like the Jews, are acknowledged as a chosen people whose epic narrative of election (although without direct biblical references), exodus, exile and ultimate redemption is sacrosanct. However, the origin myth is not only crucial in producing a mythic representation of the collective 'elect' but also underpins the LVF's belief that 'there are no nationalist places in Northern Ireland. Only places temporarily occupied by nationalists' (interview, January 1998). The central position adopted by the LVF is that, within the borders of Northern Ireland, the political and cultural desire for reunification with the Irish Republic is historically and temporally illegitimate and must be, as one respondent noted, 'physically removed' (interview, January 1998). Quite simply, as noted in a press release: 'the pan-nationalist front must . . . make a unilateral declaration that each recognises, without equivocation, Ulster as a legitimate political entity' (LVF press release 1998).

Even though this statement indicates the reality of territorial disputation in Northern Ireland, it also denies the right to promote any form of territorial governance that in any way dilutes the constitutional link with the UK. The LVF's position also makes it clear that under no circumstances can the future of Northern Ireland be resolved through any form of negotiation with parties such as the Social Democratic and Labour Party and Sinn Fein. This in turn means that the principle of consent, which is at the heart of any alteration in Northern Ireland's constitution, can only be articulated through the exclusion of any processes of cross-community-based consensus building. The territorial renegotiation of conflict – as outlined in the 'Good Friday Peace Agreement', which aims to create cross-border institutions, power sharing and the enshrinement of nationalist rights in the constitution of Northern Ireland – is inadmissible. The LVF positions itself around the supposition that Northern Ireland is a British and Protestant space that must be intolerant to any form of consensus building or cross-community linkage. As one LVF respondent noted (interview, November 1998):

> You can have Catholics living in Northern Ireland. That's fine. But you can't allow the cancer of nationalism or republicanism to exist. There can be no doubt about the fact that this is part of the UK. You cannot have anything but loyalty to that state and the majority who live there. The majority are the only people who can decide the future.

Resistance is in turn based upon pursuing ideological purity and resisting all discourses and modes of consensus building that fall outside this particular interpretation of habituation and landscape representation. Resistance, which is based upon the principle of subjugating compromise, can therefore be framed around direct resistance to and against republicans and nationalists. As stated by the LVF in a press release in January 1998:

A Protestant in a border area who refuses to leave in the face of nationalist intimidation is helping the struggle. A Protestant in a loyalist area of Belfast who refuses to leave in the face of nationalist intimidation is helping the struggle. A businessman who refuses to sell his business to a company from the Irish Republic or serve on a quango is helping the struggle.

However, resistance can also be directed against any person or community involved in cross-community work, cross-border trade or inter-party talks and negotiations. As stated (*ibid.*), the LVF:

recognises that key Protestant leaders in the church, politics, industry and commerce, and last, but not least, in the paramilitary world have succumbed to this blackmail and are presently colluding in a peace/surrender process designed to break the Union and establish the dynamic for Irish unity, within an all-Ireland Roman Catholic, Gaelic, Celtic state.

This type of ideological purity and devotion indicates the need for a communal set of images and landscape readings around which the LVF can represent itself to itself. It also shows how such mythic idealisations are inflexibly aligned around an invariable, anticipated and repetitive order of meanings in which those who are perceived to have dissented are intentionally excluded. This ideological framework, based upon a zero-sum analysis of territoriality, is fixed, rigid and ultimately denies the reality of a much more divergent political situation. For the LVF in particular, the process of ideological self-representation merely assumes the form of an allegorical reiteration of the origin myth. As such the demands, concerns and attitudes of the other, which are at the heart of the 'Irish peace process' and the negotiation of conflict, are invisible as they cannot be included in a discourse based upon the other's exclusion.

In this way, the place-centred narrative promoted by the LVF creates an identity-based determination of territory that precludes any permeation of the cultural boundaries that it imagines and upholds. This cultural and political reading of place echoes Sack's (1997: 254) contention that such place-centred discourses: 'imprison us. We cannot then see through our own culture and distance ourselves from it to see others. We do not notice that culture constrains, distorts, selects and obscures. We are unable to transcend its partiality.'

'NEW LOYALIST SPACES' AND HATING YOUR OWN

The exclusion of the 'other' through a specific discourse of election and exclusion means that history, time and space are condensed into a distinct abstraction of the past, an article assembled from categories of events in which faith and defence are mutually inclusive. The overall process is one in which history is not couched in a historical abstraction but is still being played out in the contemporary. As noted by Jarman (1998: 168): 'This past has not ended, but rather continues to structure the feelings, expectations and fears of those acting in the present. These pasts can be added to and extended with the commemoration of new local heroes.'

The spatialisation of fear, commemoration and suffering is clearly tied to the overall process of enclosing space through the instrument of religious and political segregation. However, the present division of loyalist territories and their control is now visibly divided between the LVF's power bases in Portadown and mid-Ulster and the domination of loyalist areas in Belfast and Derry by the UVF and UDA. The LVF is politically effective in these particular arenas for a range of reasons. First, the decline of the Protestant population in mid-Ulster and in particular south Armagh has intensified the sense of besiegement within this broad area. Second, loyalist areas of Belfast and Derry have been more receptive to cross-community and at times socialist politics, especially due to the strength and tradition of organised labour within these communities. Third, the fact that the 'Drumcree' dispute is located in Portadown has allowed the LVF at times to control the potency and intensity of this 'demonstration'. This sits in contrast to many redirected Orange parades in Belfast and Derry, where the UVF and UDA have also at times been actively involved in encouraging re-routing and the promotion of tolerance and inter-community dialogue. The fourth and most important factor is the legend of the LVF's first leader, Billy Wright.

The mythical stature of Billy Wright, even prior to his death, provided him with the status necessary to form the LVF as a counter to the UVF in the Portadown area. Wright's desire to leave the UVF was based upon his supposition that the UVF had betrayed loyalism by virtue of its engagement in the peace or what Wright called the 'surrender' process. As noted previously, Billy Wright's activities as leader of the LVF led to him being sentenced to death by the CLMC. This led to a wholly unprecedented event in the history of the contemporary conflict when over 10,000 loyalists attended a mass rally in Portadown calling on the CLMC to commute its death sentence against Wright. Members of this demonstration also called on the UVF and UDA to denounce the 'peace process' and their political representatives.

The combination of these four factors allowed the LVF to take control of paramilitary activity in loyalist communities in Portadown and Lurgan, but it has been unable to win noticeable support in other loyalist strongholds. This has been due to a range of factors. First, there have always been conflicts within loyalism because of the domination by loyalists of north, west and east Belfast of paramilitary activity. This has created tension in that loyalist paramilitaries from outside these areas have felt that their demands and opinions have been ignored. Second, the majority of loyalists in Belfast, especially those in positions of influence, detested Billy Wright (whom they sarcastically call 'Billy Wrong'). Several prominent loyalists stressed that he was insubordinate, a drug dealer and a pimp (an allegation made in the local, national and international press) and a person who was driven by, as one prominent member of the UVF stated, 'religious zealatory and blind bigotry' (interview, June 1999). Third, there is a strong anti-rural discourse among loyalists in Belfast, who tend to see those who do not reside in Belfast as politically backward. Finally, many loyalists in Belfast have been highly supportive of the 'peace process' due to the political leadership of their respective organisations. Indeed, among loyalists of influence there has been a concerted campaign aimed at the debunking of myths through the promotion of shared Protestant and Catholic histories and the promotion of cross-community dialogue.

In many ways there has been an attempt to remove the intemperate and sectarian histories that influenced loyalism in an attempt to reduce the nature and form of religious ideology within contemporary loyalism. It is evident that prominent loyalists are engaged in a process of rereading the Protestant landscape in order to alter the rationale of belief and understanding. This process of recasting the landscape through the adoption of histories that question the formation of Protestant identity means that it is now possible to speak of multiple Protestant landscape readings. These divisions over landscape readings and political direction has led to a particular divide between the LVF and its former colleagues in the UVF and their political spokespersons in the Progressive Unionist Party (PUP). The PUP has emerged largely from individuals who were linked either directly or indirectly to the UVF. Many of its spokespersons and elected representatives, such as David Ervine and Billy Hutchinson – both of whom have been imprisoned for terrorist activities – have engaged in the politics of conflict resolution. This has been achieved through the articulation of a quasi-socialist discourse, which is linked to the ideas of pluralism and cross-community dialogue. In relation to the loyalist ceasefire of 1994, the PUP has been instrumental in encouraging the UVF to engage in constitutional politics. Moreover, the PUP has participated in dialogue with Sinn Fein and the Social Democratic and Labour Party and has actively worked to support the 'Good Friday Agreement'.

Such activities are dismissed by the LVF as a process of duping the Protestant people and as evidence of covert republicanism. The PUP has been particularly vocal in its denunciation of fundamentalist Protestantism and has actively promoted ecumenism and pluralistic politics. Its agenda, as noted by Billy Hutchinson, has been 'to free the Protestant community from the grip of sectarian and fundamentalist politics' (interview, July 1998). For the PUP, the politics of identity and the recasting of landscapes of devotion are based upon what David Ervine has articulated as 'Freeing ourselves from spatial containment. A form of politics which takes us beyond traditional sectarian practice and myth' (interview, February 1998). Ervine and his colleagues continually emphasise the need for cross-community dialogue in order to deconstruct the myths of sectarianism that are perpetuated by the reality of religious segregation. Moreover, PUP spokespersons have been on record as sympathising with the victims of both loyalist and state terrorism. This particular form of political dialogue, based to some extent upon apology, clearly indicates the idea that loyalists have, at times, been involved in dangerously immoral and unacceptable behaviour. The desire to articulate a non-sectarian mode of Unionist politics is also based upon other forms of material cognition. As Billy Hutchinson commented when interviewed in July 1998:

> breaking out of containment actually means talking about the other realities that exist in loyalist communities. For example, we need to focus on economic and social deprivation, talk about women's rights and think genuinely about tolerant community politics. Breaking out means not simply accepting what we are told is tradition.

Hutchinson's and Ervine's renegotiation of loyalism clearly indicates the reconstruction of loyalist communities by means of less traditional sectarian discourses. In terms of removing the spatial containment of the loyalist 'sanctuary', the ideological goal of reconstruction is to develop what Sack would acknowledge as 'porous spaces': spaces that encourage interaction between the place-centred sanctuaries of religiously and politically defined communities. However, it is also clear that such a reconstruction process also aims to ensure that the interaction between the Protestant and Catholic sanctuaries does not dilute the desire to remain within the Union and to express what are seen as the values of a non-sectarian Protestant culture. Evidently, as noted by Hutchinson: 'There is nothing wrong in being a loyalist and to love one's country. But there is something wrong in articulating a mode of politics which is made blind by sectarianism' (interview, July 1998).

In spatial terms, given the legitimacy and power of the UVF, the PUP has been able to gain an electoral base in the working-class Protestant areas of west, east and north Belfast. In the active process of de-sectarianising working-class politics in those areas, the PUP has actively promoted the rights of ethnic minorities, gays and lesbians, and women. These secular pursuits have been actively denounced by the LVF as signs of a betrayal of Protestant doctrine. As one LVF respondent noted when interviewed in November 1997: 'The PUP are evil. They promote the rights of fruits [gays and lesbians] and Fenians [Catholics]. Fruits should be stoned to death not supported. It is against the will of God.'

For the LVF, the notion of breaking out of containment and adopting a mode of radical politics that challenges class, gender and sexual relations of power is problematic to the extent of being unacceptable. In terms of communal devotion and identity the LVF argues that loyalist spaces should be constructed around intolerance of what are perceived as non-traditional cultures. Effectively, the loyalist 'sanctuary' and its reading should be based upon a form of resistance not only to Irish nationalism but also towards secularisation and sexual equality.

Resistance, in this sense, implies reproducing and defending the cabbalistically sublime stewardship of loyalist places against all 'others' that are deemed to be ethnically or spiritually flawed. Clearly, the protection and reproduction of loyalist spaces demands a devotion and commitment to the antecedents of biblical scripture. Hence the division between upholding fundamentalism and the metanarrative of 'election', which is contested through the control of loyalist places, is set against an articulate, secular and non-fundamentalist loyalist and landscape-infused culture. As an LVF respondent noted in relation to the secularism of the PUP (interview, November 1998):

> Billy Hutchinson actually admitted that he was an agnostic. Can you believe he doesn't believe in God? How can you deny scripture and the word of God? These people have damned themselves and they must be stopped from damning others. It's a terrible thing to say but we hate some of our own!

Given these fundamentalist objections to the PUP, the LVF has actively been involved in the claiming and attempted control of loyalist territory. Given this struggle over the control and legitimacy of loyalist groups in certain areas, it has become apparent that what are deemed to be the 'true' loyalist communities have become immensely contested sites. This has involved not only physical intimidation and threat but also claims of loyalist legitimacy and power. As far as one LVF respondent was concerned (interview, November 1998):

> there are new and true Loyalist places, like Portadown. There is no atheist communism in Portadown, like you get on the Shankill Road. No talking to fenians. No cross-community work. Just the Red, White and 'True' Blue of Ulster.

The latter comment may seem churlish, but it does indicate the belief of many fundamentalist Protestants that evil and apostate loyalist communities have been created due to the machinations of the 'peace process', cross-community dialogue and the pursuit of radical politics. The emergence of these novel topographical constructs and the repositioning of loyalist identity reveals a distinct ideologicisation of loyalist places and their philosophical codes. What remains for the LVF is the need to control communities and to ensure their impermeability in the face of socio-cultural change and what is construed as secular contamination. As such, space is to be eulogised as pure, uncontaminated and contingent upon the reproduction of ethnic election.

Clearly, the articulation of secularism and the promotion of ecumenism and the adoption of nationalist–republican rights by Protestants is seen, by the LVF, as contradicting the rationale of religious and political belief. As one LVF respondent stated (interview, January 1998): 'Ecumenism is a serious breach of Protestant solidarity. It's really about "supping with the devil". It's depraved. You can't reject the word of God on these matters.'

In such a reductionist climate of cognition, it is clear that the LVF cannot comprehend politics as an arena within which socio-cultural antagonism can be resolved through normalised patterns of democratic practice and alternative forms of political governance. In simple terms, as one LVF respondent noted: 'there is no solution beyond the defeat of the pan-nationalist front, and the evil it promotes' (interview, June 1998). Any attempt to go beyond theological division through cross-community contact and constitutionally based 'power sharing' is seen as unwarranted. It is therefore not surprising that the LVF has threatened the lives of those individuals who try to dilute spatial confinement through the promotion of inter-community work and dialogue. For the LVF, any Protestant who engages in political forms contrary to those articulated by the LVF is categorised as a Protestant 'other'. Given this situation, Morrow (1997) has accurately argued that 'As such, Protestantism in Ireland may indeed be part of the glue of Ulster Unionist ideology but simultaneously it may be the source of the most profound critiques.'

Morrow's argument indicates that the significance of fundamentalist Protestantism as an ideology in Northern Irish politics lies in its relative durability, in that it sustains the fundamental division between 'good and bad'. In particular the rage and distrust of other members of the Unionist community becomes palpable, leading to the outward distrust of non-fundamentalist Protestants. Resistance is not merely about reproducing a particular identity but is also concerned with defending morally prescribed borders against traitordom, secularism or, as one LVF respondent noted (interview, January 1999): 'We have a God-given duty. We are ambassadors for decency and tradition. We take our stand against communism, homosexuality, ecumenism and Romanism.'

CONCLUSION

If anything, the realities of Protestant fundamentalism and the spatialisation of fear, commemoration, landscape readings and defence are a reminder of the significance of place in the reproduction of identity. The idea that globalisation is somehow creating cultural homogeneity and is continuing the process of modernisation is in many instances redundant. Without doubt, religion, as an instrument in the creation of landscapes and the performative enactment of violence, is crucial in terms of understanding the rise of militia groups in the USA, ethnic conflict and the growth of violent religious movements worldwide. In many cases, religion and its reading has re-entered world politics after centuries of enlightenment and social control. In particular, it is evident that as the world becomes more globalised it is also becoming increasingly fragmented. Fragmentation here stems from those who find loyalty and devotion in a recasting of the past, which is, in and of itself, a rejection of pluralism, globalisation and the threat to what was deemed to be a 'traditional' lifestyle.

In contrast, the evidence outlined above and the rise of fundamentalism on a global scale identify the need to question the nature of identity in a postmodern age in which the nation and its requital of inclusion and exclusion are no longer held to be a sufficient arbiter of belonging and devotion. If anything, the increasingly complex nature of image and identities reminds us that landscape readings are never unified but are increasingly fragmented and fractured, never common but multiply, fabricated across different and atavistic antagonistic discourses, beliefs and principles. Such conceptualisations of a multiple and inconsistent depiction of identity are clearly relevant in the elaboration of more diverse renditions of Northern Ireland and the deconstruction of monolithic nationalist and fundamentalist depictions of identity. Indeed, theorists often ignore the complex relationships between place and identity, which are central to any narration or understanding of communal devotion, collective action or socio-cultural modification (Roseberry 1989). In relegating these complex relationships, it is impossible to get beyond landscapes of defence that through their own definition and reproduction define the nature, volume and content of performative violence.

It is in this sense that there is an increasing need to focus upon the relative autonomy of ideology and collective consciousness as a determining factor in social action, the way in which material, political and cultural change is perceived within the context of pre-existing, if discursive, ideological frameworks (Finlayson 1997). Despite the representation of an increasingly postmodern world, it is vital to understand that multifarious ideologies still express a social group's need for a communal set of images and landscape renditions whereby people can represent themselves to themselves and to others. Most societies still invoke a tradition of mythic idealisations whereby communities are aligned within a fixed, foreseen and restated order of meanings. Indeed, this process of ideological self-representation will continually assume the form of a mythic reiteration of the act of community. Nevertheless, in Northern Ireland meaning goes beyond reiteration as it is forced back on to the plane of a landscape that must, for some, be defended.

CHAPTER 6

ISRAELI SECURITYSCAPES
Maoz Azaryahu

On the eve of the Jewish festival of Pesach (Passover), the following note appeared in the *Jerusalem Post* (31 March 1999):

> Security preparations under way for Pesach. Thousands of Israel Police, IDF and Civil Guard personnel have been posted in the nation's large cities and popular resorts in preparation for the Pesach holidays. While there have been no alerts regarding possible terrorist attacks, security will be intensified in crowded population centres and citizens are encouraged to stay vigilant and report any suspicious activities.

The prevailing sense of security alert was manifested in the presence of highly visible security personnel in crowded public spaces, most notably markets and shopping malls. Here security combined a mixture of deterrence and trust, in the form of young police cadets and newly recruited soldiers patrolling areas commonly associated with leisure and consumption. Their presence concretised the sense of threat and conveyed the message that precautions were indeed taken.

In contemporary Israel, the term 'security' is overwhelmingly, although not exclusively, evoked in the context of the Arab–Israeli conflict. 'Security' is addressed mainly in terms of national security. The concept of 'personal security', which became prominent in the beginning of the 1990s in the wake of terror attacks inside Israel, should also be understood in a national rather than individual context, since 'individual security' was a derivative aspect of 'national security' policies. 'Security' alludes to a shared concern among Israeli Jews. The term conjures patriotic commitment and resonates with a prevailing sense of a community under siege.

The prevalence of 'security' as a public theme is a feature of life in Israel. The term connotes many interrelated things: permanent alertness; a sense of a lurking threat and impending danger; awareness of vulnerability; and – last but surely not least – a strong measure of insecurity. According to common wisdom, the significance assigned to security issues in Israel is a prominent aspect of the Israeli condition. Those who object to its power as a political argument may disqualify 'security' as a mere 'social construct'. A fashionable academic catchword, it should however be noted that the term 'social construct' does not qualify something as 'true' or 'untrue', 'appropriate' or 'inappropriate'. Social phenomena are indeed social constructs, which is to state the obvious. In my view, the interesting question in this context is why individuals share certain 'constructs' as being relevant to their private and social life, while others are dismissed as irrelevant or 'academic', in the negative sense of the term.

The concern about, and quest for, security resonates with national survival, and therefore it transcends political divisions within Jewish-Israeli society. Beyond being a catchword, a political argument and a conceptual reference, security is also concretised in the landscape, where sights and situations, military installations and everyday experiences are combined in the formation of a variety of 'securityscapes'. Expressed in durable elements and ephemeral phenomena that pertain to a shared public concern, securityscapes pervade the Israeli experience of space in terms of impending threats, defence strategies and precautionary measures. As a refraction of projections, measures and needs, the variety of Israeli securityscapes both highlights and concretises 'security' as a tenet of collective faith and a symptom of the Israeli condition.

In a study that appeared in 1986, Arnon Soffer and Julian V. Minghi surveyed Israel's security landscapes, which, as the title of the article implied, constitute 'the impact of military considerations on land uses' (Soffer and Minghi 1986). Beyond drawing attention to the significance of 'security' as an aspect of contemporary Israeli landscape, this article was important in that it discussed a set of ostensibly separate issues in the contextual framework of 'security landscapes'. This chapter does not attempt to expand on all aspects of the 'Israeli security landscape'; the list is too long and repetitive. Details are therefore illustrative of the main argument of this chapter, namely, that securityscapes are variations on one and the same theme. Accordingly, it attempts to complement the discussion provided by Minghi and Soffer. Beyond updating their detailed analysis, I draw attention to the thematisation of security in terms of diverse securityscapes. The main concern of this chapter is to emphasise that securityscapes are not limited to specific geographical enclaves or pertain mainly to the military sphere or to the interface of the military and the civilian spheres. Instead, it is argued that these are also aspects of everyday experience. Accordingly, the analysis highlights issues and phenomena that from a more conservative perspective could be labelled as marginal. However, the emphasis on ordinary and even trivial examples is crucial for understanding the extent to which securityscapes indeed constitute an important aspect of ordinary life in contemporary Israel.

This chapter therefore highlights certain recurring motifs in the construction of Israeli securityscapes as a landscape genre. The investigation opens with a review of the semantic field in which the meanings and associations of the Hebrew term for security, *Bitakhon*, are produced. The main foci of the analysis are ordinary securityscapes, the notion of protected spaces at different scales, and a 'ritual of security' – the somewhat controversial security check at border crossings. With a reputation for rigour, this security measure amounts to a virtual 'rite of passage'. The perspective that I, as an Israeli citizen, assume in the analysis is that of a participant-observer; yet I also mention observations that represent the 'foreign gaze'. The 'foreign gaze' is not necessarily impartial, but it may offer insights where the 'local gaze' fails to do so; the 'local gaze' tends to take things for granted, as part of the 'natural order', and consequently, different approaches and interpretations are denied. The 'foreign gaze' is not better or worse than the 'local gaze', but it may be different. In this, it does not replace the 'local gaze' but complements it, for the benefit of the analysis.

SEMANTICS

The distinguished American geographer John Fraser Hart (1995) has expressed his dissatisfaction with the proliferation of the suffix '-scape', which he finds to be a linguistic corruption. He argued: 'Many users . . . have assumed, quite incorrectly, in my opinion, that the suffix -scape in landscape is related to the suffix -scope . . . and they have committed such awkward neologisms as cityscape, farmscape, roadscape, seascape, wildscape, windowscape and the like(scape).' To that list we could now add Netscape, a commercial neologism that seems to render the suffix familiar, at least to those wandering in cyberspace in search of appropriate websites. There is, it seems, no escape from 'scape', although the meaning is not always clear or consistent. In this chapter, we may note that the suffix '-scape' emphasises the visible, whereas the prefix highlights a special aspect of landscape as its main theme, thereby providing a focus and stressing a distinctive feature. In this sense, a 'securityscape' is the superposition of security on to the landscape to the effect that the landscape is permeated with the notion of security: 'securityscape' is a semantic aspect of the landscape, located at the interface of 'security' and 'landscape'.

Another issue is the meaning of the term 'security'. *Webster's Dictionary* defines 'security' as: '1. The quality or condition of being secure, specf.: a. freedom from exposure to danger; protection; safety or a place of safety. b. feeling of or assurance of safety or certainty; freedom from anxiety or doubt. 2. That which secures a means of protection, defence, etc.' The term 'security', then, conjures up the notions of safety, protection and defence. What the dictionary does not mention is that 'security' is coupled with control, which is the price paid for maintaining a sense of security.

The modern Hebrew term for security derives from the old Hebrew word *Bita'khon*. Reinvented in the course of the revival of Hebrew, it carries with it additional meanings and cultural associations that the English equivalent does not possess. *Bita'khon* is translated into English as 'security' and as 'safety' and defined in the Hebrew dictionary as 'Hope, expectation, belief in God and hope that he will fulfil wishes.' As mentioned in the Old Testament, *Bita'khon* in God means trust in God, which is reproduced in the famous 'In God we trust' printed on American coins and notes. The modern Hebrew word *Bitakhon* is used in a secular context. The irony is that in the process of cultural re-contextualisation, *Bitakhon* was transformed from a tenet of religious faith into a creed of Israel's civil religion.

Modern Jewish nationalism has been preoccupied with 'self-defence' as a prominent theme of the Jewish national revival ('Zionism'). Given that it was also a cultural revolution, Zionism sought to create a 'new Jew'. One of the features that distinguished the 'new Jew' from the 'old Jew' was his commitment to 'self-defence'. Significantly, the name of the main Jewish militia founded in Palestine during the British mandate was the *Hagana*' (in Hebrew, 'defence'). When the Israeli army was formally founded at the end of May 1948, shortly after the state of Israel was proclaimed, its official name was the 'Israel Defence Forces'. This name also established continuity with the pre-state militia. Significantly, the ministry in charge of the military is not the 'defence ministry' or 'war ministry' but Misrad Ha-Bita'khon, namely, the Ministry of Security. Significantly, the official English translation is 'Defence Ministry'.

The verbal association of the government agency in charge of the military with 'security' signalled a major semantic shift as it subordinated 'defence' to 'security' as the ultimate strategy of national survival. An Israeli commentator writing on the need to design a new 'security (grand-)strategy' for Israel pointed out that in 'progressive [*sic*] states *Misrad Ha-Bita'khon* (the Security Ministry) is called *Misrad Ha-Hagana* (the Defence Ministry), and it pertains not to semantics but rather to essence. It is clear to them that the term security is wider than the term defence' (*Maariv*, 21 May 1998). In this interpretation, the term 'security' – when applied to national security – also includes such ostensibly non-military factors as the economy, level of education and degree of technological development. However, by juxtaposing 'security' and 'defence' in terms of 'essence' as opposed to 'mere semantics', the author failed to notice that the Hebrew term *Bita'khon* suggested more than its English translation. The word *Bita'khon* therefore suggested more than additional terms of reference but also a multitude of connotative meanings that are essential for appraising the evocative power of the term for Hebrew speakers.

Whereas 'security' was traditionally aligned with the military and warfare, when terrorism later emerged as a prominent threat to public safety, 'security' was extended to cover spheres of public safety commonly associated with police jurisdiction and activity. Transferring responsibility for protecting the public from the threat of terrorist attacks had the effect of enhancing the notion that 'security' concerns the 'home front'. Semantically, this process culminated in renaming the former 'Police Ministry' the 'Internal Security Ministry'. The cognitive association of the activities of Israeli police with issues pertaining to national security is probably the reason for the (relative) popularity of the police with the Israeli public, despite the common impression that the police are incapable of fighting 'ordinary' crime (Yaar 1998). In this case, the popular notion of the police as a security agency contributed to its positive image. It is worth noting that many private companies that supply civilian 'security' services include the word '*Bitakhon*' in their names, thereby drawing on the positive connotations of the word in the public mind.

A THEME IN THE LANDSCAPE

The most notable securityscapes are military landscapes, especially the army camps that adorn the Israeli landscape. The architecture of military camps combines military functions with notions of discipline and authority. Signs on the perimeter fences indicate that these enclaves are subject to a set of rules different from the ordinary 'civilian' code. Fortified gates stress that these are restricted areas, inaccessible to unauthorised persons. Photography is prohibited, which enhances the sense of secrecy with which such secluded areas are shrouded.

Yet military landscapes are not limited to army camps. Military facilities of all kinds are instrumental in transforming an otherwise ordinary landscape into a securityscape. This is especially the case along the borders, where the military presence introduces and affirms the notion of a border as a 'danger zone'. Especially when the frontier zone is unpopulated, the visual presence of the army indicates the sensitive character of the border as an area that commands attention and caution. A case in point is a description of the area of Qumran, the archaeological site in the Judaean Desert on the northwestern edge of the Dead Sea, in the area where the Dead Sea Scrolls were found hidden in caves. Prior to the signing of the Israel–Jordan peace treaty in 1994, this area was a border zone (Baignet and Leigh 1991: 27):

> The Israeli Army is, needless to say, constantly in sight. This, after all, is the West Bank, and the Jordanians are only a few miles away, across the Dead Sea. Patrols run day and night, cruising at five miles per hour, scrutinising everything – small lorries, usually, with three heavy machine-guns on the back, soldiers upright behind them. These patrols will stop to check the cars and ascertain the precise whereabouts of anyone exploring the area, or excavating on the cliffs or in the caves. The visitor quickly learns to wave, to make sure the troops see him and acknowledge his presence. It is dangerous to come upon them too suddenly, or to act in any fashion that might strike them as furtive or suspicious.

The sight of the military in border areas conveys the special security character of these areas. A different thing is the visibility of the military, and especially that of soldiers in uniform carrying weapons in otherwise civilian contexts. Such recurrent sights make the military an aspect of everyday experience. Israelis are accustomed to it and do not pay it any special attention. Foreigners often find it anomalous. In a BBC World television *Time Out* feature dedicated to Eilat, the Israeli resort on the Red Sea, the reporter interviewed two British tourists sitting on the deck of a boat. Asked whether they were not afraid of the security situation in Israel, one tourist, identified as Joan, related her personal experience: 'First evening we went to a bar, a guy appeared with a rifle on his shoulder and his friend also had a rifle . . . that was a bit unusual.' Her friend described the experience as 'unnerving'. The reporter concluded that: 'apparently, when off-duty the army have to carry their guns, unloaded I am told.' Though expressed in terms of belief, the statement also articulated a tone of relief.

The travel writer Paul Theroux toured Israel in 1994. Despite his misunderstandings and hostile attitude, his references and commentaries demonstrate that the 'foreign gaze' is of much value in the evaluation of Israeli securityscapes. During a train ride from Haifa to Tel Aviv, the traveller was appalled by the omnipresence of soldiers carrying weapons:

> Something that bothered me greatly were the numerous people travelling armed with rifles, usually large and very lethal-looking ones, and pistols. Most of those people

were soldiers and one of the characteristics of Israeli soldiers on trains and buses was their fatigue. They always looked sleepy, overworked, and no sooner were they on a seat than they were asleep; I often found that their weapons bobbled in their arms were pointed directly at me. (Theroux 1995: 385–6)

The innocent – or perhaps hostile (e.g. contrast this with Theroux 1988: 274) – traveller was indeed terrified:

A soldier in the seat opposite made himself comfortable to sleep, put his feet up and propped the Uzi automatic rifle so that it was horizontal in his lap, and it slipped as it dozed, and was soon pointing into my face. I said, 'excuse me', because I was afraid to tap his arm and risk startling him and the weapon discharging. He did not wake, but after a while I raised my voice and asked him to take his rifle out of my face.

As it emerged, nothing lethal happened as the result of this unpleasant encounter. Yet the experience prompted the traveller to indulge in some further observations (*ibid.*):

Walking up and down the crowded coach, I counted the weapons: two in the next row, a frenzied man in a white shirt with a nickel-plated revolver, a soldier two rows back with a pistol and a rifle, ditto the soldier next to him, a woman in uniform lying across two seats with her big khaki buttocks in the air – a pistol on her belt, seven more armed passengers farther down.

Theroux (1995: 386) went on to contemplate: 'My feeling is that weapons are magnetic – they exert a distinct and polarising power, and nearly all attract power. The gun-carrier creed is: never display a weapon unless you plan to use it; never use unless you shoot to kill.' It did not occur to the traveller that for most passengers, the sight of armed soldiers on public transport induces a sense of security.

The notion of 'protected spaces' features prominently in the construction of Israeli securityscapes. Protected spaces entail demarcation and fences. The term 'fence', as Webster's relates, is related etymologically to 'defence'. A fence, also a measure of defence, also denotes an enclosure. To fence means to fend off danger, to protect. Fences include highly visible walls and stockades, yet they may also be invisible, a construct of the mind, or a metaphor.

The British mandate government had already implemented the idea of protected spaces when it faced Jewish terrorism in the late 1940s in the last chapter of British rule in Palestine. R.M. Graves (1949: 57), the nominated British mayor of Jerusalem between June 1947 and May 1948, related the situation of the British residents of the city in his memoirs:

After their women were banished, English residents were confined to zones girdled with barbed wire, with a strong guard and road block at all the gates . . . Between Zone A and the municipality one has to show one's pass three times. If one forgets it, one has to go back for it, no matter who one is and what one looks like. It is just too bad.

The urge to design 'protected spaces' dominates Israel's security thinking. Since threats assume different forms, the character and function of 'protected spaces' differ substantially. One kind of 'protected space' relates to the threat of missile attacks, an issue that became concrete during the Gulf War in 1991, when thirty-nine Iraqi missiles landed in Israel. On the micro-level, 'protected spaces' were constructed in the form of 'sealed rooms'. This was an improvised version of the 'security rooms' built in border areas. The latter had been introduced to provide each housing unit with a shelter – a room built of reinforced materials that can withstand even a direct hit by a Katyusha missile or a mortar. Later on, building regulations all over the country were updated to prescribe the building of 'protected rooms' in every new house and apartment building erected in Israel. These are the substitute for public shelters, of which there is a shortage.

The idea to construct a room sealed against chemical and biological weapons was devised a few weeks before the eruption of hostilities in the Persian Gulf in January 1991. The idea is attributed to Brigadier-General Ehud Barak, then Chief of Staff. According to the directives of civil defence, and fuelled by a sense of helplessness that bordered on hysteria, the public was asked to purchase sheets of plastic and to seal off a room from the outer world and the rest of the apartment. Supplemented with gas masks, the idea was to provide spaces that were hermetically sealed from the effects of chemical and biological materials. When it became evident that the missiles fired from Iraq towards the population centres in Israel carried conventional warheads, many commentators ridiculed this mode of protection. They pointed out that the improvised shelters were no more than 'make-believe', a sedative that concealed the lack of proper shelters for residents of the big cities. Nevertheless, in January 1998, when the possibility of a new war in the Gulf loomed, Israelis again rushed to purchase sheets of plastic.

At the macro-level, and as a consequence of the failure of Patriot missiles to intercept the Iraqi Scud missiles, an ambitious defence programme was launched with the aim of ensuring that anti-missile missiles would intercept long-range missiles directed at Israeli cities before entering Israeli airspace. The name given to the program is Wall, which connotes the Great Wall of China, Hadrian's Wall or even the Berlin Wall (it is worth remembering that the official East German name for the Berlin Wall between 1961 and 1989 was 'the antifascist defence wall'). Unlike these concrete edifices, the Wall system of anti-missile defence, when operative, will be defined by ranges, interception speeds and locations, not by permanent features in the landscape.

'Protected spaces' are marked and shaped by regular fences aimed at fending off infiltrators. The British mandate government in the late 1930s, in the period of the Arab Uprising in Palestine, first introduced the idea to seal off a border with a neighbouring country. In an attempt to prevent the smuggling of weapons and fighters from French-controlled Lebanon into British Palestine, the British government initiated the building of a fence along the border, augmented by fortified posts. In 1972, when the Israel–Lebanon border was peaceful, the border was described as 'two paved roads separated by a fence' (Judge 1972) – one road in Israel, the other in Lebanon. Harmless as it looked, the fence closed the border hermetically, and besides UN personnel crossing at Nakura (Rosh haNikra), no civilian movement across the border was permitted. The character of the border changed in 1974–75, when PLO squads crossed the fence on their way to attack Israeli settlements.

The ambiguity of the situation became apparent in 1976, when, following the civil war in Lebanon, a gate was opened near Metulla, on the eastern side of the border, and a controlled measure of civilian interaction between the two sides of the fence was re-established. In official Israeli language the fence was renamed 'the good fence', an indication of the humanitarian aspect of the partial opening of the fence. In 1978, Israel invaded southern Lebanon as a move to uproot the PLO presence there. When the operation was over, an Israeli-controlled 'security zone' in southern Lebanon was established. In 1982, Israel launched a full-scale invasion northwards. In the new situation, the fence seemed to lose its strategic significance. Following the partial withdrawal of Israeli troops from Lebanon in 1985, the fence was fortified and on the Lebanese side of the border a 'security zone' (in current BBC English: 'so-called security zone') was re-established as a buffer. Currently, Israeli politicians who advocate a unilateral withdrawal from southern Lebanon as the best way to enhance Israeli security in the region also recommend transforming the fence into a fully electronic, high-tech barrier.

A prominent feature of Israeli securityscapes is a fence that is much talked about but has failed to materialise in the landscape. The necessity of 'separation' between Israelis and Arab Palestinians appeals to many Israelis, irrespective of ideological and political persuasion. The idea of building a fence that would physically and mentally separate Palestinian Arabs and Israelis is, therefore, popular among many Israelis and has been repeatedly raised as a desirable option. The demands for a fence are especially loud on television discussion panels on days when bombs explode in Israeli cities. On such occasions, clichés such as 'high fences make good neighbours' abound and resonate well with audiences. Yet in spite of its popularity, it is only the Gaza Strip that has been surrounded by a border fence. As recurrent media reports testify, no fence is invincible: tunnels that connect areas controlled by the Palestinian Authority with Egypt are used to transfer people and weapons. Ladders are occasionally used by job seekers to enter Israel illegally. Army patrols alongside the fence have proved to be only partly effective.

The building of a fence separating the West Bank from Israel has never been seriously attempted. A stretch of wall 1.7 kilometres long and 2 metres high was erected to separate Jewish Kfar Saba from neighbouring Arab Qalqilya, its official *raison d'être* being the prevention of occasional shooting into Jewish residential areas. The fact that the wall has not yet been built is not only a result of the lack of financial means. The issue is also highly political, since erecting such a wall would also mean re-instituting the Green Line, the border between pre-1967 Israel and the Jordanian-controlled West Bank. On the other hand, the need to control the passage of people and stolen goods, especially cars, from Israel into the Palestinian-controlled area is constantly on the agenda. Significantly, responsibility for the 'seam' (as the frontier area is commonly referred to) is in the hands of the Internal Security Ministry and not the Defence Ministry, which emphasises that it is considered a police matter and not a military problem.

In a recent interview (*Makor Rishon*, 15 May 1998) an official of the Internal Security Ministry explained the current policy:

I intend to implement the building of a fence, which will actually be an obstacle for the transfer of stolen cars. There is no intention to separate between states or peoples. There is also no intention of making a 1998 version of the Green Line. The idea is to make it more difficult for car thieves. The fence we refer to will be 80 km long and 80 cm high. Actually it is not even a fence but a kind of steel grid that will serve as an obstacle.

In the meantime, the only visible measures of control are the military checkpoints on the roads connecting the West Bank and Israel. These are 'gates' in a non-existent wall, which explains why they are highly ineffective. As television features demonstrate, such checkpoints do not deter would-be infiltrators. Bypassing the control points is a common sight; all that it needs is to walk through the surrounding fields, as is done daily by thousands of Arab Palestinian job seekers who enter Israel without valid work permits. The separation fence features prominently in Israeli popular imagination; its power seems to be in that it suggests a simple solution to a set of security problems. Seemingly, the fact that the 'separation fence' has not been built only enhances its attractiveness as the ultimate solution to security problems. For now, the visibility of the fence is limited to the sphere of public debate.

A different kind of fence is represented by those built at the outer perimeters of Jewish settlements, especially those in border areas. The idea was introduced at the beginning of the century but became a settlement policy in 1936. Known as 'tower and stockade' settlements, wooden, prefabricated fortified fences and watch towers were part of the architecture of newly founded Jewish *kibbutzim* in areas prone to attack by hostile Arabs. After its successful establishment was secured, the settlement was expanded beyond the original wall. The idea of a settlement as a 'protected space' still prevails. The building of fences was reintroduced in the late 1960s, following a new wave of attacks on Israeli settlements along the Jordanian and later the Lebanese borders. This produced a typical picture: 'The kibbutz is surrounded by barbed wire and locked at night. An armed patrol is always on duty' (Baignet and Leigh 1991: 27).

Fences, with a guarded gate to control movements in and out of the settlement area, especially if they are small and isolated, also encircle Jewish settlements in the Occupied Territories. An equally important function of the fence is to mark land claims, and the building of a fence is often an occasion for a clash with Arab neighbours. The fence underlines the existence of the settlement as an enclave in a hostile environment. The resulting military-like landscape renders the simmering conflict and tensions between the Jewish settlers and the neighbouring Arab population visible. The ambiguity of fences, however, has not escaped the attention of residents. The head of the Alfei Menashe council, a Jewish settlement situated near the Green Line in the West Bank, explained his objections: 'We don't live in a ghetto here, and I don't have the slightest wish to let the residents here feel like animals in a cage' (*Maariv*, 21 May 1998). This was reinforced by a representative of the Sharon settlements, situated on the 'Israeli' side of the Green Line, when asked why the settlements are not fortified for their protection: 'we do not want to live in ghettos.'

Protected spaces are also to be found in the big cities. Impending threats of terror attacks are manifested in the presence of patrolling soldiers in the open markets of Tel Aviv and Jerusalem. A small sign notifies passengers in inter-city buses that the seats behind the driver are reserved for armed security personnel. After several attacks on buses, making buses into 'protected spaces' was high on the security agenda. For a while, in an attempt to quell public fears of suicide bombers, security personnel made random checks on buses. Their presence on the buses was intended to signal that 'everything is under control'. The transformation of public spaces into protected spaces was also initiated in the early 1970s after a series of bombs in movie theatres, supermarkets and shopping malls. Security personnel employed by the owners were positioned at the entrance to check the personal bags of people. Accustomed to this security ritual, Israelis open their bags automatically; no words are exchanged, and a slight nod of recognition and approval is enough to accomplish the check. Tourists, unaccustomed to local habits, have to be addressed verbally, adding to their confusion. In the wake of acts of terror in Israeli cities, the pressure exerted by parents forced the authorities to include schools in the category of spaces to be protected. Security personnel are positioned at entrances, and gates are closely controlled. Routine patrols along the fences complete the transformation of educational compounds into protected spaces, to the relief of concerned parents.

Small notes in buses warn passengers about 'abandoned objects', which may be anything from a suitcase to a handbag or a transistor radio. Abandoned objects look innocent and harmless. Yet they are rendered suspicious because they may be a 'ticking bomb'. The categorisation of an object as 'suspicious' is a common street scene in Tel Aviv and Jerusalem. The police transform the area into a 'sterile zone', which means that pedestrians are kept away and cars are barred. Car drivers, unaware of the reason for the congestion, honk to protest the delay. They regain their patience when they realise that the reason for the delay is security. An explosives expert checks the suspicious object and decides what further to do, while curious passers-by observe the scene from a safe distance. The common procedure is to detonate the suspicious object using a specially designed robot. After a few minutes, the disturbance is over and the street resumes its normal appearance. From an acoustic perspective, a 'suspicious object' generates a unique 'soundscape'. It is a sequence of police sirens and messages on loudspeakers that culminates in an explosion. A real bomb begins with the terrifying sound of explosion and is succeeded by a cacophony of sirens that disperses only after a few hours.

DISCUSSION

The effort to make public space into a protected space is sisyphean. Ever since the 1970s, in an attempt to make planes flying to and from Israel secure from bombs and hijacking, rigorous security checks have been part of the experience of the flight. Passengers to and from Israel are required to be at the airport three hours before the flight is scheduled. Security men and women, mostly young, serve as 'gate-keepers', and the checks are reputed to be thorough and uncompromising, to say the least. Some passengers are required to unpack their suitcases; when suspicion arises, they are likely to be questioned at length about personal details. Every Israeli who has travelled abroad is familiar with the questions 'Who packed your things?' followed by 'Has anyone given you anything to deliver?'

As a 'rite of passage' in the literal meaning of the anthropological term, the interviews can become an ordeal: 'Those who are familiar with the stringent security procedures there know that interviews can last up to an hour and culminate in a strip search' (*Jerusalem Post*, 15 March 1999). Guidebooks mention the security checks as a warning to prospective tourists. In its 1987 edition, *Fodor's Israel* (anon. 1986: 11) informed its users: 'Expect very thorough security checks at the airport before you leave, with bag searches and in-depth quizzing about your trip. You can also count on only slightly less of a grilling when you leave Israel.' In the 1992 edition of *Lonely Planet* (Tilbury 1992: 10), the issue was raised among such that are 'rarely mentioned in print but frequently discussed by travellers'. In a rather subtle manner, the author, reflecting a prevalent mood among tourists, cast doubt on 'the necessity for some of those countless questions you face from airport security and their response to certain answers you give them when trying to board a plane to/from Ben-Gurion Airport.'

Judging from the amount of printed space he dedicated in his travel report to the description of the security checks he underwent, Paul Theroux was both fascinated and repulsed by the intensity of the experience. He noted, 'at this point, faced by Israeli security and having questions barked at me, I was on the verge of asking whether this was a traditional way of greeting strangers: sharp questions and even sharper gun-muzzles in my face.' What he did not understand was that with a passport that testified to his recent visit to Syria, he appeared suspicious. To his relief, the woman in charge of the security control was familiar with his books and the questioning was over.

On 22 May 1993, a New York resident arrived at Kennedy International Airport to board an El Al (Israel's national carrier) flight to Tel Aviv (this episode is reported in *Makor Rishon*, 15 January 1999). After being asked the standard questions by the security personnel the woman passenger was assessed as 'high security risk'. She was taken to a separate room, where a security woman searched her body. Subsequently, the security personnel decided that she did not constitute a security risk, and the passenger was allowed to board the plane. After returning to the United States, the passenger decided to sue El Al for US$5 million since, as she maintained, she suffered emotional trauma and needed psychiatric treatment. In January 1999, the American court ruled by eight to one that the Warsaw Convention, to which the USA is a signatory, affirms the responsibility of airlines for international flights and forbids internal law suits. The legal aspect of the issue notwithstanding, this case demonstrated the extent to which the rigour of the security checks was a sensitive issue for all concerned.

The following letters sent to the editor of the *Jerusalem Post* in February 1998 exemplify the controversial character of these security checks, the fact that their evaluation differs substantially according to perspective. After arriving home, John

R. Lawrenson, from Suffolk in the United Kingdom, expressed his dismay (*Jerusalem Post*, 5 February 1998):

> The face of modern Israel is seen in the members of the security forces that greet every traveller . . . no intelligent individual questions the need for tight security. It is not what is done but how it is done that needs to be questioned. Is it really necessary for teenage girls to abuse travellers at Ben-Gurion Airport? And the word 'abuse' is carefully chosen.

According to him, the security check was 'humiliating and a never to be forgotten experience, something to warn any would-be tourist against.' A day later, Barbara Solomon-Brown from Ra'anana, Israel, reacted to Lawrenson's angry letter. In her letter (*Jerusalem Post*, 5 February 1998), she related how her mother, a Holocaust survivor and religious, 'was required to remove her wig in the presence of a youthful Israeli female security staff. They were looking for hidden weapons.' Her mother's assessment of the situation was different: 'I didn't like it, but thank God Israel takes such precautions to protect me and other passengers.' Ms Solomon-Brown concluded her letter by a direct appeal to Mr Lawrenson: 'It is nothing personal, just ask my mom.'

Some people may find security measures a nuisance and even a violation of their rights, whereas others may be satisfied that efforts are made to protect them, even if it entails inconvenience. The two versions of similar – though probably not identical – security check experiences and their opposite evaluation confirm that security spaces are not only 'what we see' but also 'how we see it'. Securityscapes, like landscapes, represent a point of view.

CHAPTER 7

FORTIFICATION, FRAGMENTATION AND THE THREAT OF TERRORISM IN THE CITY OF LONDON IN THE 1990S

Jon Coaffee

In the 1990s, a number of terrorist bombs were detonated by the Provisional Irish Republican Army (IRA) in the City of London, London's financial heart. These events were to have an enormous impact on the area's landscapes as the security agencies, the local authority (the Corporation of London), the business community, insurers and the government responded to the threat posed by terrorism. The resulting security strategies reflected a lengthy process of response that partly echoed the fortification of urban areas pioneered in Belfast in the 1970s and partly reflected new strategies aimed at reducing the risk of further terrorist attack. This chapter draws on an extensive series of interviews with individuals occupying key positions in the security industry to detail the development of landscapes of defence in and around the City. Before doing so, however, it is important to recognise that these strategies are also an expression of other, broader trends operating within Western cities.

URBAN FRAGMENTATION AND TERRORIST THREAT

In the last twenty years, Western cities have been increasingly restructured. On the one hand, technology has advanced tendencies towards globalisation, transforming our experiences of geographic space, reshaping and blurring traditional territorial patterns and boundaries (Anderson 1996). At the same time, paradoxically, heightened economic competition has led to increased localisation for those places that command the global economy (Robertson 1992; Lash and Urry 1994; Swyngedouw 1997). In such areas, 'the local is embedded within, and, superimposed upon the global, while global processes simultaneously appear to permeate all aspects of the local' (Brenner 1999: 438).

Today, the importance of these areas means that the groups that control them often employ defensive measures to protect the area and create new 'hardened' boundaries in the landscape, which in turn mirror their power and control over the spaces thus defended. Such spaces of financial supremacy can therefore be seen as new 'global enclaves' embracing inclusion in the globalisation process while at the same time excluding themselves from the rest of the city through their territorial boundedness. These manifestations of territoriality 'help protect and enforce the privileges of social elite areas, and areas of economic investment – the corporate office enclaves and new consumption spaces of the post-modern city' (Graham and Marvin 1996: 222). They also reflect two of the most important trends in contemporary urbanism. The first is the increased militarisation of urban space (Davis 1990), where 'form follows fear' (Ellin 1997). Here changes in the physical form of the landscape accompany increased perceptions of the fear of crime emanating from occupiers of a particular area. Such landscape changes range from simple removal of amenities to stop the homeless living on the street (Davis 1990; Smith 1996) to the creation of full-blown gated and heavily guarded residential and commercial enclaves. The second trend is the privatisation and control of space according to the preferences of the rich and powerful (Sorkin 1995). In recent years, this control has been increasingly asserted through physical and technological measures, with the explicit aim of excluding the sections of society deemed a threat to a particular way of life. As Flusty (1994: 67) noted:

> Traditional public spaces are increasingly supplanted by such privately produced (although often publicly subsidised) 'privately owned and administered spaces for public aggregation' such as shopping malls (and) corporate plazas ... In these new post-public spaces, access is predicated upon real or apparent ability to pay.

Flusty continued by indicating that such changes in the urban landscape are related to economic productivity: 'in such spaces, exclusivity is an inevitable by-product of the high levels of control necessary to insure that irregularity, unpredictability and inefficiency to not interfere with the orderly flow of commerce.'

It is possible to highlight three significant urban security strategies that arise as the result of these trends. The first is *management* of the landscape, where a series of spatial and temporal regulations are often socially enforced, or dictated by the forces of law and order. The second is *fortification*, which refers to the introduction of defensive measures such as walls, barriers and gates, which enhances the physical segregation of the landscape. The third is *surveillance*, referring to the control of space through the explicit visual presence of police and security guards, combined with electronic camera surveillance. Such approaches have commonly been used to address issues of crime, but in recent years these strategies have been employed in an attempt to prevent attacks on cities under particular threat from terrorists.

This threat grew rapidly in the early and mid-1990s as terrorist groups realised that attacks on major economic centres not only cause severe damage directly to valuable building structures but also guarantee maximum exposure for the groups concerned in the media. The bombing of the World Trade Center in New York in 1993, the bombing of central Bombay in 1993 and the Tokyo subway attack in March 1995 all achieved worldwide publicity. From a British perspective, the

Provisional IRA's attacks in the City of London (April 1992 and April 1993), the bomb under the Canary Wharf Tower in London Docklands (November 1992) and at South Quays station (February 1996) were the major events in a sustained campaign. As Timothy Hillier (1994) of the City of London Police stated:

> Massive explosions in London, New York and other major cities worldwide clearly demonstrate that important financial districts have become prestigious targets for terrorist organisations, regardless of their motives. In addition to causing significant loss of life, these bombs severely disrupt trade and economic transactions. Further, modern satellite communications broadcast grisly bomb scene images around the world within minutes adding to the lure of this type of target for groups seeking media publicity.

DEFENDING THE CITY

As a result of such attacks by the Provisional IRA (see Table 7.1), the City of London sought ways to improve its security against terrorism. Measures taken would eventually include the construction of roadblocks, the use of armed roving checkpoints, the development of a series of public and private closed-circuit television (CCTV) networks, a number of traffic restrictions, and increasingly visible policing. In addition, a series of private initiatives led to the fortification of individual buildings. This security operation was commonly referred to as constructing a 'ring of steel', a term that had first been used in the mid-1970s to refer to the high-level security measures, in particular the high steel gates, that were erected at all entrances into the centre of Belfast.

The 'ring of steel', however, did not emerge fully fledged but was the result of a lengthy period of development in which six distinct periods can be identified. These are:

- Apprehension (1990–April 1992)
- Containment (April 1992–April 1993)

Table 7.1   Provisional IRA attacks in the City in the 1990s

Date	Incident
20/7/1990	A bomb exploded in the Stock Exchange
29/2/1992	A device exploded at the Crown Prosecution Service in Furnival Street
10/4/1992	A large van bomb exploded outside the Baltic Exchange in St Mary Axe
25/6/1992	A device exploded under a car in Coleman Street
24/4/1993	A large vehicle bomb exploded in Bishopsgate
38/8/1993	A device was recovered from Wormwood Street near Bishopsgate

- Deterrence (April 1993–September 1994)
- Optimism (September 1994–February 1996)
- Reactivation (February 1996–February 1997)
- Extension (February 1997 onwards).

Apprehension

The Provisional IRA's bombing campaign in the 1990s, aimed at economic targets in London, sought to 'bring terror to the heart of London with a ferocity never before experienced in the capital' (Dillon 1996: 265). One of the first attacks occurred at the Stock Exchange in the City, in which a bomb exploded in the public gallery causing much damage to the visitor area. During 1991, perhaps aware of the state of the developing recession in Britain and the pressure on government finances, the Provisional IRA began to appreciate the value of inflicting massive economic damage on Britain, although this may have been more a question of stumbling on a strategy than long-term policy making. In late 1991 and early 1992 a number of large bombs had been detonated in Belfast city centre by the Provisional IRA, which noticed the economic cost of such bombs both in terms of actual damage caused and to Belfast's identity as a place of business.

In his annual report for 1991, the Commissioner of the City Police (Court of Common Council minutes, 1992) referred to the question of likely terrorist attacks by the Provisional IRA, given attacks elsewhere in London, most notably in Downing Street in February 1991. He noted that 'although we had no serious incidents in the City in 1991, the effects of terrorist attacks elsewhere in London had a significant effect in heightening the need for even greater security.' In particular, the potential for attacks on the City's transport infrastructure was noted, given the five main line railway terminals, the Thameslink line and ten underground stations in the Square Mile. As such, the police had been significantly enhancing security arrangements for major events in the City. Such high-profile policing had led to a 10 percent reduction in recorded crime (Corporation Police Committee minutes, 27 January 1993).

However, ideas for greater anti-terrorist security for the City began in February 1992, when a small terrorist bomb exploded in Furnival Street in the northeast of the City. Although this did not cause substantial damage, its impact was soon enhanced with the bombing of the Baltic Exchange in St Mary Axe, in the heart of the City, in April 1992.

Containment

The St Mary Axe bomb led to emergency plans being devised to prevent further attacks. As this was the first major bomb in the City, the Corporation of London took the view that an increased police presence was an appropriate response. As the Corporation's leader, Michael Cassidy, indicated in an interview: 'a ring of steel solution didn't arise then because it was the first such incident. I think people were taking the view that the policing was going to be very front-line and vigilant,' putting more officers on patrol. As a result, the police, stretching the Police and Criminal Evidence Act (PACE) to its limit, instigated a number of 'roving police checkpoints' at the major entrances into the City.

At this time the Corporation, in line with its unitary development plan, had been working on an environmental and movement policy called 'Key to the Future'. This planned to restrict access to some roads in the City and alter traffic signalling on others to improve traffic flow and reduce pollution. After the St Mary Axe bomb, with the support of the Commissioner of Police, some traffic management measures in line with these proposals were introduced on an experimental basis on three City roads. Furthermore, waiting restrictions on a total of fifteen roads in the vicinity of the Old Bailey, the Bank of England, the Stock Exchange and the Lloyds building were extended. This came as a result of a direct recommendation from the Commissioner of Police, who used his powers to suspend metered parking around these areas. Officially, these measures were done to ease emergency vehicle access, but they can be seen as an attempt to defend the most prestigious landmark buildings, considered to be those at greatest risk, from potential terrorist attack.

In addition, traffic-management CCTV was extended and adapted to focus on incoming traffic, and private businesses were encouraged by the police to install their own CCTV cameras. The City police at this time wanted to establish permanent vehicle checkpoints on all entrances to the City, but a combination of legal and financial restrictions and adverse public opinion made it impossible to do this at that time. There was a feeling that a permanent security cordon would be 'an over reaction, would make the city look like Belfast, [and] give a propaganda coup to the IRA' (Kelly 1994). As another senior police officer indicated in an interview: 'at this time we urged businesses to carry on as normal and avoid a fortress mentality.'

A minor explosion in June 1992 in Coleman Street further enhanced the need for improved anti-terrorism measures. This bomb, although only a small device, showed that security in the Square Mile could easily be breached. There were immediate attempts to raise awareness of the threats faced by City occupiers, but the security agencies had still failed to interest large sections of the business community in security matters. As the Corporation's senior security advisor noted when interviewed:

> St Mary Axe was a watershed as far as anti-terrorist measures were concerned – it focused the minds of people on what could be done. This advice was taken up by some but others held the view that lightning doesn't strike twice and failed to heed the warnings.

He continued by indicating that this view began to change after the 1993 bomb: 'Bishopsgate in 1993 was therefore much more significant than St Mary Axe, as it proved this wrong.'

Deterrence

The deficiencies of existing strategy were realised on 24 April 1993, when a Provisional IRA bomb exploded in Bishopsgate in the east of the City, killing one person, injuring ninety-four and causing damage initially estimated at £1 billion (but subsequently reduced to £550–600 million). With hindsight, although the road checks and other measures were seen as effective, the limitations of PACE meant that there could be no permanent checks on vehicles coming into the City. By using mobile communications and a scout ahead to tell them whether or not a police checkpoint was operating, terrorists could easily plant a bomb in the City.

After the Bishopsgate bomb, the media and sections of the business community began to suggest that drastic changes should be made to City security. An editorial in the *Sunday People* (25 April 1993), the day after the bomb explosion, captured the popular view that security must be enhanced in the City:

> If we are to wage effective war against the IRA, there must now be an urgent review of security at their most likely target. Since the IRA mortar-bombed Number 10 from a waiting van, nothing is allowed to park in Whitehall. IF IT CAN BE DONE FOR DOWNING STREET IT CAN BE DONE IN THE CITY.

This view was backed up by leading City figures cited in *The Times*, who indicated that a Belfast-style scheme should be implemented:

> The City should be turned into a medieval-style walled enclave to prevent terrorist attacks ... In private there is talk about a 'walled city' approach to security with access through a number of small 'gates' and controlled by security discs.

Initially, the leaders of the Corporation were sceptical about implementing such radical proposals. As Michael Cassidy, referring to the day after the bomb, noted: 'It's true that on that Sunday morning my reaction was that this is such a radical proposal that I couldn't immediately say it was one that we could embrace and support.'

However, in May 1993, given the heightened risk of further attack, the police confirmed that they were considering radical plans in the form of a security cordon to deter terrorists from the City. This was seen as the only way to enhance security significantly in the City. The construction of the scheme, which was considered on environmental grounds prior to the Bishopsgate bomb, began in June 1993 with the removal of all litter bins, and proposals to introduce security checkpoints to bar all non-essential traffic.

Such modifications were criticised by those who felt that it would cause traffic chaos at the boundaries of the City, as vehicles would be increasingly pushed into neighbouring boroughs. There were also fears that such a radical scheme could geographically displace the risk of terrorist attack to other areas. Kenneth Clarke, the Home Secretary at that time, summed up the situation facing the City. Indicating the delicate balancing act between security and business normality, he noted: 'There is a balance to be struck between having roadblocks which will frustrate what the terrorists can do, and creating enormous traffic jams which would disrupt the life out of the City' (*Evening Standard*, 29 April 1993).

Eventually, on the weekend of 3–4 July 1993, a full 'Belfast-style' ring of steel was set up in the City, securing all entrances. The main access restrictions imposed are seen in Figure 7.1, which shows that most routes were closed or made exit-only,

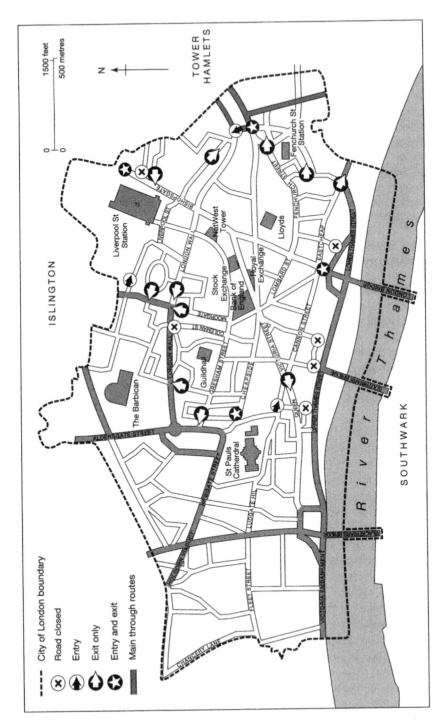

Figure 7.1   Access restrictions in the City of London (1993)

leaving seven routes (plus one bus route) by which the City could be entered. On these routes into the City, road checks manned by armed police were set up. Locally, the ring of steel was often referred to as the 'ring of plastic', as the temporary access restrictions were based primarily on the funnelling of traffic through rows of plastic traffic cones. Officially, it was called the Experimental Traffic Scheme in an attempt to remove references to terrorism.

The ring of steel did not provide full security coverage to the Square Mile, as much of the western side of the City remained outside the cordon (see Figure 7.2). This occurred for a number of reasons. Initially, the Police Commissioner wanted to make sure that all the key financial targets were included. Some businesses were unhappy about the exact placement of the cordon, as it left them outside the secure zone. As a senior police officer noted:

> Some people were discontented that they were excluded. The point is that we had to begin and end somewhere, and, like it or not, there is a part of the City that is more vulnerable and has a greater economic value to the nation and the international economy as opposed to just the City of London, which is why the inner cordon was focused on that particular area at the outset.

The ring of steel also had to be developed so that the traffic flow through the City remained the same. This meant that throwing a cordon around the entire City would have led to all traffic being diverted to neighbouring boroughs. Moreover, by setting up the cordon as they did the police could minimise the number of entry points into the secure zone. This aided security as well as minimising the personnel needed to run the scheme effectively.

In addition to access restrictions, the City began to enhance its electronic surveillance capabilities, which aimed to develop three separate, but interdependent, camera systems. In addition to traffic-management CCTV, security cameras were erected to monitor the entrances into the Square Mile at all times during the day so that every vehicle entering the security cordon was recorded. From November 1993, there were two cameras at each entry point – one read the number plate and the other scanned the front profile of the driver and passenger. By the summer of 1993, there were more than seventy police-controlled cameras covering the City, but there was still inadequate coverage of many public areas due to lack of private cameras. As a senior policeman noted: 'We conducted a survey [in 1993] to find out what the extent of the coverage was and much to everyone's amazement there were only something like 160 cameras that offered coverage of public areas.'

Subsequently a scheme called CameraWatch was launched in September 1993 by the police, which encouraged the police, the Corporation and City organisations to cooperate on camera surveillance, thereby creating an effective and highly visible camera network for the City. CameraWatch aimed to deter terrorism and other criminal activity by the knowledge that CCTV systems were operating and could provide the means of detection and evidence gathering should offences occur. The scheme was also seen as a potential catalyst for other security and community safety initiatives. However, the installation of cameras on City properties as part of

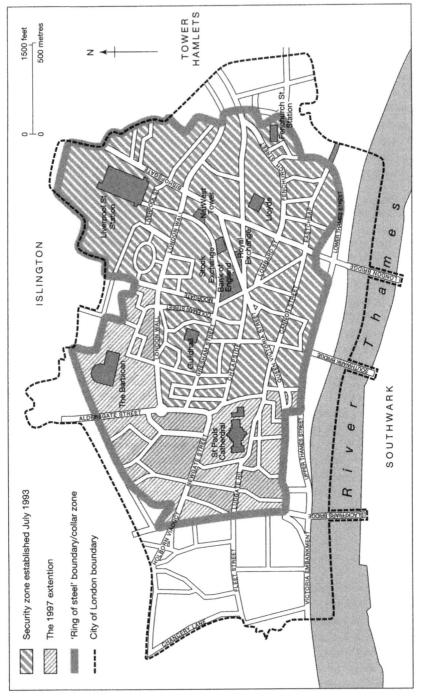

Figure 7.2 The City of London's extended ring of steel (1997)

this scheme was not always a feasible option, as an interview with the Corporation's security advisor indicated:

Many businesses are only tenants in the buildings and were reluctant to pay out for expensive surveillance equipment. Additionally there were objections from conservation groups about the aesthetic intrusion caused by cameras which have to be overtly visible. Therefore it was not just a matter of putting cameras up.

Nine months after CameraWatch was launched only 12.5 percent of buildings had camera systems, leaving a very large proportion of public areas without the security of constant CCTV coverage. Renewed effort on the part of the police and security advisors helped to boost the total to over 1000 private security cameras in 376 systems by 1996. It was also suggested that in the future it might be possible for these private systems to be linked into the police camera systems if the need arose (Kelly 1994).

The Corporation's response to terrorism after the Bishopsgate bomb was a result of severe pressure from the business community, especially from overseas institutions, to improve security. The Bishopsgate attack was widely seen as an attempt to undermine confidence in the reputation of the City as a financial centre. The Commissioner of Police, for example, noted: 'No one should be in any doubt that we are locked into a struggle with terrorists for the City of London and it is a struggle that we, the nation (not just the City of London), cannot afford to lose' (Corporation Police Committee minutes, 24 November 1993). He continued by highlighting the City's financial position, as well as countering the argument some had made that the security measures employed were a public relations exercise:

I know that I need not remind you that another massive bomb could make the City untenable as an international financial market place. Foreign investments and business would flee, perhaps never to return. The £18 billion a year earnings from City business could be lost and irreparable damage done to the country's economy.

Some ill-informed people think that all we are doing is protecting those 'fat cats' in the City. The reality is that if the City of London is brought down economically, perhaps never to be recovered, then all of us . . . will be the losers from the damage done to the nation's economy . . . It would be difficult to overstate the importance of securing the City against that threat. Of course, the terrorists too see the potential results of their activities and that makes the City their most desirable target.

As well as the centrally organised security response, many private businesses took additional precautions against terrorist attack in four distinct ways. First, they controlled access to their buildings by limiting the number of entrances and by security personnel carefully scrutinising visitors in the reception area. Second, they restricted access to areas under the building such as car parks and storage areas and created no-parking zones around the building to stop vehicle bombs being parked nearby. Third, there was an increase in surveillance through organised means via vigilant security guards and by CCTV. Finally, direct precautions were taken to reduce the effects of a bomb blast, including application of window film, installation of blast curtains and construction of shelter areas.

Optimism

Immediately after a Provisional IRA ceasefire was called on 31 August 1994, there were suggestions in the media that the ring of steel should be scaled down. This was subsequently done, with permanent armed guards being taken off most check-points and a less visible police presence on the street. This downgrading of security had a noticeable influence on recorded crime levels, which increased steadily during the ceasefire period. However, as a senior security advisor noted at that time: 'The ring of steel's function as a security initiative will be maintained. It will provide a framework in which to launch a security operation if needed.' Indeed, from August 1994, and throughout the ceasefire, permanent bollards began replacing temporary traffic cones.

As the ceasefire progressed, further moves to scale down security were suggested. A senior policeman indicated, 'the business community were also beginning to feel that the risk of further terrorist attack had gone away, making it more difficult for the police . . . to sustain a campaign of raising or maintaining public awareness of the need to be vigilant.' In particular, there was an attempt on behalf of a number of prominent organisations around Christmas 1995 to persuade the Corporation to disband the security cordon. The situation at the end of 1995 was one of optimism that the cessation of violence would continue, although the police remained concerned about potential terrorist threat, as the Corporation Police Committee minutes (31 January 1996) indicate:

> For a number of years, policing in the City [has been] significantly influenced by the necessity to respond effectively to the criminality of Irish Republican terrorists. During the past year, the cessation of this criminal terrorism has been heartening and optimism for a lasting settlement remains high. However, the measures put in place and being operated in the City are to deter all terrorist criminality. The City will always be a potential target for acts of terrorist criminality because of its high profile, high economic value and cosmopolitan business community.

As such, the police camera network had been continually upgraded to meet the perceived threat. In the early months of 1995, new high-resolution cameras for the traffic system were installed and a further thirteen cameras were added to monitor cars exiting the City. Prior to this date, security cameras focused only on cars entering the cordon; now the exit cameras offered the capability of tracking suspect vehicles across the City.

Reactivation and extension

The final two stages in the development of the City's security landscapes may be considered together, as one evolved from the other. Following the Docklands bomb in February 1996, the fortress mentality returned to the City. The full pre-ceasefire ring of steel was reactivated and operational within a number of hours as there were fears that the City would be attacked again. As reported to the Police Committee on 29 May 1996:

The resumption of IRA terrorist criminality earlier this year has once again highlighted the need for the City of London to be ever vigilant. The traffic control points are now equipped with entry and exit closed-circuit television cameras, and officers are deployed at the entry points, as judged necessary in response to the criminal threat.

Initially as a result there was a large increase in the police presence at entry points and on City streets in general. There was also an increased frequency of roving checkpoints.

After the Docklands bomb, further proposals to increase security in the Square Mile were made. Such suggestions centred on a proposed westward extension of the ring of steel, first suggested in February 1995, in order to reduce pollution, traffic and crime. The City Engineer emphasised in September 1996 that 'the scheme is operating efficiently and substantial environmental benefits have already resulted, with powerful enhancement of the City scene.' He believed that 'there was a strong case to build on our past experience and extend the environmental benefits further to the west by developing further complementary schemes currently in various stages of development.' In particular, it was felt that the extension would discourage through traffic from using local roads, reduce conflict between pedestrians and traffic on local roads, improve the environment in a larger part of the City, and give the Commissioner of Police improved security opportunities on City streets within the larger zone, which would cover nearly 75 percent of the Square Mile.

The Commissioner of the City's police strongly supported the proposals, despite downplaying their anti-terrorist importance:

I believe an extended zone will be of considerable benefit to the traffic and environmental conditions in the City . . . A by product of an enhanced traffic zone would be the opportunity to introduce security measures (as necessary) in a manner similar to that currently attaching to the present traffic zone. The threat to the City from criminal terrorism is high and whilst I judge that will change from time to time the source of the threat now and in the future is not one-dimensional and a strategic approach is essential.

The proposal for a western extension was implemented in January 1997, for an initial period of six months.

As well as the extension to the ring of steel, the police again extended their electronic surveillance capabilities. In particular, a CCTV system was completed at the entry points. This automated system, linked to police databases, allowed detection of vehicles entering the zone illegally. The digital number plate recording technology allowed the information to be processed and gave a warning to the operator within 4 seconds. The project to install this advanced system began in early 1995, with installation coinciding with the opening of the western extension to the ring of steel.

In April 1996, the ring of steel was given yet another layer of protection by new legislation relating to the Prevention of Terrorism Act, which allowed the police to search pedestrians randomly, as well as cars, in and around the Square Mile. Signs were initially put up in the City warning visitors that they were liable to be searched. This it was felt was an invaluable tool, as lessons from Belfast had indicated that when confronted with a security cordon the terrorists switch tactics and attempt to smuggle handheld explosive devices into the desired target area.

DISCUSSION

The security landscapes that have developed in the City were, as we have seen, the result of the actions of a broad coalition of interests. The Corporation of London provided the forum for discussions about security policy. Effectively, it was that agency that drove the process, as it needed to protect the general interests of the City as a financial centre and to reassure business, especially foreign institutions, that it was doing everything possible to prevent further bombings. The City police were the prime movers in creating a defensive landscape, most notably the ring of steel as a highly visible deterrent to set alongside private security approaches.

The construction of the ring of steel also focused attention on the effectiveness of such a radical security scheme, as well as the multiple meanings and values attached to it. Many occupants of the Square Mile saw it as an almost total guarantee of safety, despite the police publicly pointing out that there is no such thing as total security. Since the ring of steel's implementation there have, to date, been no further bombs in the City, and a number of other benefits emerged such as a reduction in recorded crime, pollution and traffic accidents. Overall, advocates of the ring of steel pointed out that a large proportion of businesses were in favour of the security measures, and that 'vetoing the fortress would be detrimental to traffic and business in the City' (*City Post*, 15 July 1993). Indeed, the landscape changes that have occurred in the City are highlighted as having improved the quality of life in place promotional material, in which the City is now portrayed as a safe area in which to conduct business. For example, the London Chamber of Commerce and Industry (1996), in an investment brochure for London, noted that the ring of steel is viewed positively:

> The City of London 'Square Mile' has benefited particularly from a cut in crime following the reduction in entry points . . . following a bomb in 1993 . . . The unplanned result of the initiative has been a cut in office burglaries in the Square Mile and a corresponding increase in business confidence.

However, other commentators and interviewees did not view the ring of steel so positively, with some suggesting that the ring of steel could have increased the likelihood of further terrorist attack, as media exposure would be beneficial to the Provisional IRA. For example, Dr Conor Cruise O'Brien indicated that in his opinion (cited in Dillon 1996: 292–3):

> The ring of steel increases the risk to the City in two ways. It increases the incentive to the IRA to strike, because of the propaganda value to be derived from penetrating that loudly trumpeted ring. The other way in which the charade increases the risk to the City is that it diminishes manpower available to counter the IRA threat. Fixed roadblocks need a lot of trained manpower.

Generally, two main criticisms were made about the construction of the ring of steel. The first was that it gave the Provisional IRA a propaganda coup. The police countered this criticism, indicating that publicity would have been magnified many times more if the Provisional IRA had bombed the City for a third time. The second criticism was that the ring of steel displaced the risk of terrorist attack to other areas. Lessons from Belfast indicted that, when faced with such a cordon, car bombers had targeted areas just outside the cordon or alternative 'softer' and less prestigious targets. In particular, within the City some businesses located on the edges of the Square Mile began to express concern that bombs could be planted at the edge of the newly constructed cordon. The City police and the Corporation were well aware of this Provisional IRA tactic and set up a so-called 'collar zone' around the ring of steel in an attempt to alleviate the fears of those businesses within this area (see Figure 7.2). As a result, extra police patrols, including roving checkpoints, were undertaken in this area.

Criticism about negative side effects also came from neighbouring boroughs. During the implementation of the ring of steel, they felt that they had not been properly consulted on what the City was planning to do. In particular, they became worried that there could be negative effects in their areas in terms of increased traffic and the risk from terrorism. For example, when the ring of steel was implemented Andrew Pharaoh, the director of 'Movement for London', indicated that severe disruption to normal traffic flow would occur, especially around the borders of the zone. On a similar note, the Islington council leader indicated that the implementation of the ring of steel had not been thought through properly and was 'completely ill conceived' (*Evening Standard*, 7 June 1993).

This sense of impending chaos was linked to the lack of official consultation between the boroughs surrounding the City and the Corporation of London. Initially the City police used traffic legislation, which gave them the power to restrict access to the area, to initiate the security cordon. Only the consent of the Corporation was needed to do this. Objections to that policy immediately came from neighbouring boroughs. Tower Hamlets and Southwark wrote to the Government and the Corporation over the complete lack of consultation, and Islington heard of the cordon by fax only the night before it was implemented. After the security cordon had been established, the local boroughs were consulted and objections were heard formally since a decision had yet to be made as to whether the ring of steel would become a permanent feature of the City landscape. Due to the limitations of the 1984 Road Traffic Act, the City based its legal case for 'renewal' of the ring of steel on transport and environmental grounds, given the reduction in atmospheric pollution and traffic accidents since the scheme had been implemented. In December 1993, consultation letters were sent out to the seven neighbouring authorities with regard to the question of making the ring of steel permanent. The responses of the authorities are shown in Table 7.2.

This table indicates that, despite initial criticism of the scheme, the majority of the neighbouring authorities were happy with the ring of steel becoming permanent. The holding objection that Islington had was immediately dropped, leaving Tower Hamlets as the sole objector from the neighbouring boroughs. This was a potentially

Table 7.2 Views of neighbouring authorities

Neighbouring authority	Consultation comments
City of Westminster	No objection to making scheme permanent
Islington	Holding objection, with concern expressed about 'geographical' areas that would subsequently be expanded
Camden	No objections, with the proviso that if any problems occurred ameliorative measures would be sought from the Corporation of London
Hackney	No objections, but would like to see better cycle access. Slight concern expressed about the effect of police activities on buses
Tower Hamlets	Objections on traffic management grounds
Lambeth	No objection to making scheme permanent
Southwark	No objections unless London Transport or the emergency services lodge complaints

Source: Adapted from a report by the City Engineer in February 1994.

serious matter for the City as, if this objection was not withdrawn, a public inquiry would have to be held at great cost. Subsequently, a series of meetings between Tower Hamlets and the Corporation allowed a 'memorandum of understanding' to be reached with a series of remedial measures planned to ensure better traffic control on the mutual boundary (Policy and Resources Committee, 2 June 1994). The objection was therefore withdrawn, allowing the ring of steel to be made permanent on 2 July 1994, a year after its initial implementation. As with the setting up of the 1993 cordon, the 1997 extension was criticised by the neighbouring boroughs, which again accused the Corporation of failing to consult with them. As a representative for Camden and Islington councils noted: 'The City is using Police powers to avoid going through the normal planning procedure. It is distributing its leaflets now, telling people it has already decided on this' (*Evening Standard*, 22 January 1997).

As well as complaints from neighbouring authorities there were complaints from civil libertarians, who had a number of concerns relating to the restriction of access to public spaces, the omnipresence of CCTV and the power of the police to stop and search. Civil liberties groups were critical of changes in the law that allowed people to be randomly searched. Just before the security cordon was constructed, John Wadham, director of Liberty, noted that 'we believe the balance as it is now is about right. We understand why the Police want more powers but do not think their power should be increased.' In particular, he believed that the new security arrangements would lead to harassment of the Irish community in London. Wadham further stated a year later in a letter to *The Times* (4 April 1994) that he believed the police were using the security cordon for measures not associated with terrorism:

> The statistical evidence for the ring of steel ... shows that Police have stopped a disproportionate number of black people, despite the fact that to date, no black people are believed to have been involved in the IRA activities in Britain.

CONCLUSION

As a result of terrorist attacks and the risk of further bombings, new security landscapes were constructed in the City of London between April 1992 and February 1997. They were based on radical developments of crime prevention measures commonly employed in other Western cities, the introduction of armed road checkpoints, the imposition of parking restrictions and the fortification of individual buildings. Additionally, three interrelated camera networks were established and continually updated. At the end of 1996, CameraWatch had 1250 private cameras, with the police in addition controlling eight permanent entry point cameras, thirteen exit cameras and forty-seven area traffic-control cameras. These numbers subsequently increased as the new extension was implemented.

Such changes have been constructed and activated as a result of a number of interrelated social, economic and political processes. In particular, the Corporation has attempted to control and regulate the space within the Square Mile, attempting to create a particular image for the City based on a balance between a flourishing business culture and safety and security. This is seen as vital if the area is to be competitive within the global economy. As such the Corporation, in liaison with its police force, used its considerable political power to enhance security with minimal consultation with other affected parties. These other groupings subsequently attributed different meanings to the landscape, viewing it as little more than a public relations exercise, an imposition on personal freedom and an exporter of traffic and, possibly, terrorist risk.

However, for the majority of occupiers in the City the new security landscapes were viewed positively – as an effective approach in relation to both crime and terrorism, as an effective traffic-management measure and as a beneficial environmental policy. The anti-terrorist measures employed in the City are now seen to have wider applications than just security and will remain a concrete part of the contemporary urban landscape in the City for the foreseeable future. For example, the chairman of the Corporation's Policy and Resources Committee summarised the effects of the City's security policies as:

> Totally positive in all areas. It's given people an extra comfort factor, the trend of banks coming here was not interrupted at all by the last two bombs, in fact the strength of the trend has built up ever since, so we now have even more foreign banks, even more headquarters. None left apart from for their own business reasons. None have left for security reasons, and, the typical City worker finds that the environment is better. From that point of view they wouldn't want to see it go. Not now.

CHAPTER 8

LANDSCAPES OF DEFENCE, EXCLUSIVITY AND LEISURE: RURAL PRIVATE COMMUNITIES IN NORTH CAROLINA

Martin Phillips

Mike Davis (1990: 223–4) has argued that much of urban social theory, while preoccupied with many aspects of postmodernism, has been strangely silent about at least one of its 'bad edges', namely the 'militarisation' of built environments. Davis argues that many people now live in: ' "fortress cities", composed of "fortified cells" of affluent society and "places of terror" where police battle the criminalised poor' (Davis 1990: 224; see also Davis 1992). Davis (1998: 364–5) has even sought to 'remap' the urban ecology of Burgess and the so-called 'Chicago School' of sociologists, an exercise premised on adding the social dynamics of fear to those of income, land value, class and race that underlay, albeit in a naturalised form, the earlier urban models. This emphasis on the militarisation of space has also been commented on by Blakely and Snyder (1995a,b; 1997), who have suggested that a growing 'fortress mentality' has emerged within America, a mentality that may have major social repercussions. As an illustration of these, they cited Calthorpe's (1993: 137) description of the social impacts of so-called gated communities:

> Physically ... [the gated community] denotes the separation, and sadly the fear, that has become the subtext of a country once founded on differences and tolerance. Politically it expresses the desire to privatise, cutting back the responsibilities of government to provide services for all ... Socially, the fortress represents a self-fulfilling prophecy. The more isolated people become and the less they share with others unlike themselves, the more they ... fear.

Davis also foresaw three similar social consequences stemming from the fortification of space. First, he suggested that it involves the creation of a new positional or status commodity whereby access to security becomes 'the decisive border between the merely well-off and the truly rich' (Davis 1990: 224). Second, he argued that the militarisation of space does not act to reduce feelings of insecurity but rather heightens people's levels of fear. Third, he claimed that militarisation acts to destroy public space, and particularly those spaces that allow social mixing. He also discussed a range of aspects of the militarisation of space. These included attempts to exclude groups such as the urban poor from the downtown areas of American cities, the creation of panoptic systems of surveillance in shopping malls and aerial surveillance of the streets, the incarceration of inmates in inner city penitentiaries, and the creation of what are often termed 'gated', 'walled' or 'private' communities.

In this chapter, I want to focus on this last aspect of the militarisation of space. The existence of such communities has been widely recognised in the United States, and they have a growing presence in many European countries and Australasia. Despite this, they have not been subjected to much research. As Blakely and Snyder (1995a: 1) commented, there are as yet 'no definite numbers on their extent'. Their best estimate is that although they were present in 'only a handful of places [in the USA] in 1985', they comprised some 20,000 communities housing between three and four million people by 1997 (Blakely and Snyder 1997: 3).

Private communities are defined as those that have 'restricted access such that normally public spaces have been privatised' (Blakely and Snyder 1995a: 1). Although they have also been termed 'gated' or 'walled' communities, the relationship between these terms is far from clear-cut. All these communities can be seen to be landscapes of defence in that they seek to construct some boundary between themselves and their surrounding environments. In the case of gated or walled communities, this creation of a boundary takes, at least in part, a physical form. Such communities have only one or two entrance and exit points, which often have barriers and a gatehouse. In the case of walled communities, a wall or fence often marks the boundaries of the community. In other cases, private communities make use of 'natural' physical boundaries, such as steep hillsides or water rather than walls or fences. For communities on the Outer Banks of North Carolina, sand dunes and the lack of made-up roads serve the same purpose. Many of these private communities do appear to have similar locational conditions to more traditional fortresses, such as being situated on islands or headlands.

While physical boundaries are present in many private communities, these are conjoined with symbolic, legislative and economic mechanisms of exclusion. There is a semiotics of defence and exclusion (Davis 1990: 224), with signs proclaiming that the communities are patrolled and watched by security guards and are not areas of public access, and that trespassers will be prosecuted. Security guards and closed-circuit television cameras create a regime of surveillance within the community area. Not surprisingly, property values and membership fees are high (Table 8.1). These symbolic, legislative and economic mechanisms of exclusion may indeed be present within communities that have no physical boundary markers. The category 'private community' should therefore be seen to subsume gated and walled communities but is not simply equivalent to these settlements.

Blakely and Snyder (1995a,b; 1997) further argued that there are three distinct types of private community beyond the previously mentioned distinction between private, walled and gated communities. Based on their assessments of the primary motivations of people moving into these communities, they suggested that it is possible to distinguish between 'lifestyle', 'elite' and 'security zone' communities. In the first, people move in to pursue particular activities and adopt particular lifestyles. In the second, people move into areas that symbolise distinction and prestige. In the third category, fear of, and perceived protection from, 'crime and

Table 8.1 Property values and membership fees in 'private communities' in North Carolina

Community	Champion Hills	Elk River[1]	Hound Ears	Jefferson Landing	Kenmure	Lake Toxaway	Lynks O' Tyron	Old North Club
Club membership (joining fee, resident)	$32,500	$35,000	$22,500			$15,700	$2,500	–
Club membership (annual fee, resident)	$4,164	$4,500	$3,000			$2,950	$1,650	–
Service fees (annual)	$1,060	$1,625	$1,280			$647	$825	$1,152
Properties (starting prices)								
Condominiums	–	–	$220,000	$135,000	$140,000	–	–	–
Homes (2 bedrooms)	–	$255,000	–	$149,000	$250,000	–	$160,000	–
Homes (3 bedrooms +)	–	$350,000	$117,000	$199,900	$300,000	–	–	$200,000
Land lots	$50,000 – $300,000	$75,000	–	–	–	$42,000	$22,000	$60,000

Note: Elk River figures for 1995, all other figures for 1997.

outsiders is the key motivation' (Blakely and Snyder 1995b: 3). Blakely and Snyder argued that these three motivations create distinct types of community, although they added that in many cases all three motivations may be significant within a community. They also claimed (Blakely and Snyder 1995a: 5–6) that there is a clear geography to these communities, with the greatest numbers of them appearing within areas with particular characteristics:

> metropolitan regions; areas with high levels of demographic change, especially large amounts of foreign immigration; areas with high median income levels; regions with extreme residential segregation patterns or without a clearly dominant white majority; areas with high crime rates and high levels of fear; and areas to which whites are moving, either for retirement or because of white flight.

Blakely and Snyder (*ibid.*: 5) suggested that private communities are most strongly present in the southwestern, southern and southeastern parts of the United States.

Potentially, the state of North Carolina would seem to fit many of the criteria identified by Blakely and Snyder. It is, for example, a growth area forming part of the US 'Sun Belt' (Debbage and Rees 1991; Rees and Debbage 1996), with much of the growth being stimulated by inward investment from multinationals (Furseth 1996) and the establishment of 'command and control' corporate and finance centres (Frey and Johnson 1998). The state has also been subject to considerable in-migration of retired people (Bennett 1993; Bohland and Rowles 1988; Serow and Haas 1992). North Carolina also still exhibits many of the patterns of social segregation characteristic of the 'Deep South', and there has recently been a series of highly publicised racial and other social conflicts, particularly within the urban area of Charlotte. As yet the existence of private, walled and gated communities in North Carolina has remained largely unrecognised and unexamined.

However, the existence of such communities in North Carolina is clearly discernible, not least in the 'place promotion' literature surrounding the development of this state. Towle (1995), for example, listed the existence of twenty 'exclusive communities' in North Carolina, and magazines such as *North Carolina* are replete with advertisements for 'exclusive', 'professional', 'private' and 'country club' communities. The use of this last phrase highlights how many of these communities are, at least symbolically, rural (see Phillips 2000). Furthermore, many of these communities are located at quite a distance from the metropolitan centres of North Carolina. This stands in some contrast to Blakely and Snyder's (1995a: 5) claim that private communities are predominantly a 'metropolitan' phenomenon. Many of the features identified by Blakely and Snyder as being causal factors in the emergence of private communities seem to parallel features identified as being significant in social change in rural communities in North Carolina and elsewhere. For example, studies by Bennett (1992; 1993) and Haas and Serow (Haas and Serow 1990; Serow and Haas 1992) have suggested that rural areas of North Carolina are being colonised by affluent in-migrants. These and other studies (e.g. Glasgow and Reeder 1990) have also highlighted how much of the retirement migration into the state has been rural in focus. The influences of 'fear of crime' and 'racism' in rural colonisation have also been documented across rural communities in America (e.g.

Krannich 1985; Frey 1995; 1996; Frey and Johnson 1998). Davis (1990: 224) also explicitly connected the militarisation of residential communities with the processes of counter-urbanisation.

For these reasons, private communities in rural North Carolina can be seen to warrant some attention. This chapter will draw on promotional materials, videos and Internet pages relating to these communities and upon interviews undertaken in 1996 with managers and property agents involved in these communities. It seeks to highlight the existence and social significance of these communities and to explore the extent to which they are constructed as spaces of defence, exclusivity and leisure, as suggested by Blakely and Snyder.

PRIVATE COMMUNITIES AS LANDSCAPES OF DEFENCE

As has already been noted, many private communities have locational and architectural features that are highly reminiscent of the defensive features of a fortress. There is a common perception of these communities as being places in which rather paranoid groups of people, most notably the ultra rich and the elderly, go to insulate themselves from what they see as the dangers of modern living, particularly crime but also '"anti-social behaviour" by "unsavoury" groups and individuals, and even crowds in general' (Davis 1990: 224). Blakely and Snyder (1997), in a survey of people involved in running private communities, found that over 70 percent considered that security was a very important factor in people's decision to live in gated communities. However, they argued that gates and walls actually do little to improve residents' security and often offer nothing more than symbolic defence. They further suggested that very little is made of security in the promotion of these communities, partly through fear of legal action: 'Even the most high tech security systems cannot guarantee a crime-free community, and developers are fearful of liability if they make such claims' (*ibid.*, 18).

In the promotion of these communities in North Carolina, there were many instances where defence figured highly, although often appearing seamlessly integrated into the promotion of other features of these communities. Here are two examples:

> Champion Hills ... offers all the advantages you expect of a premier residential community – full time roving security, tasteful architectural guidelines for custom homes, cottages and villas; and easy access to a number of fine medical facilities, the Asheville Regional Airport, the charm and activity of nearby Hendersonville, and a very special corner of western North Carolina. (Champion Hills, n.d.)

> Chestnut Hill is a community where elegant dining, recreation and peace of mind are all provided for your convenience and enjoyment. A community where safe-guarding you and your home is our prime concern. (Chestnut Hill, n.d.)

Sometimes the provision of gates and walls need not necessarily be interpreted as a defensive feature. In one promotional text, for example, the bricked boundary wall and gate was described purely as an aesthetic feature: 'The bricked wall entry creates an English Garden atmosphere' (Cornerstone Properties, n.d.). By contrast, other communities made very clear and direct reference to defensive landscape features in their promotional materials:

> THE CARRIAGE PARK GATEHOUSE. Elegant in its simplicity and functional in its purpose. The Gatehouse stands at the entry to this community and serves its purpose well. At Carriage Park, there is one way in and one way out. An exceptional deterrent to anyone who has mischief on their mind. The residents breathe a little easier and know in their heart of hearts when they are home, they are secure. Carriage Park is a gated community for all the right reasons. You decide your safety and we just added to it. (Carriage Park, n.d., b)

This promotion still places great stress on the provision of defence by individuals, perhaps partly an illustration of the caution highlighted by Blakely and Snyder but also pointing to the continued acceptance within these communities of individuals' rights to protect themselves and their households through a variety of means, including the carrying and use of firearms. Hence these communities were militarised internally as well as at the external boundaries.

Security at the Carriage Park community near Hendersonville did not simply involve the formation of an external boundary between the community and the outside world, or even the creation of this boundary and a series of more home-and body-centred zones of protection. Much was also made in this community, and indeed in many of the other private communities, of inter-subjective security. At Carriage Park (Hamlin 1997), for instance, much is made of the creation of 'systems of small villages':

> At CARRIAGE PARK, the 'physical' side of security is enhanced by being both a gated community with controlled ingress and egress and a system of 'small villages'. The 'small village' concept creates its own neighborhood watch program by the very nature of neighbor caring for neighbor. Restricted traffic flow and a parkway feeder road system with no individual driveways directly accessing further maximises security efforts.

This private community can be seen to incorporate a pre-existing strategy for creating 'defensible spaces' based on communal surveillance of territorialised public spaces, namely the cul-de-sac street layout such as is used extensively in suburban developments (see Blakely and Snyder 1997: 18; Southwood and Ben-Joseph 1997; see also Newman 1972; Coleman 1985). In private communities, however, this form of surveillance is taken to new heights through the more stark privatisation of who has access to the public space of the communal streets and, in some cases, also through the use of video camera systems to enable residents to keep watch on their neighbours (see Blakely and Snyder 1995a: 13; also this volume, Chapter 4).

For one private community manager, the level of internal communal surveillance was a clear motivation for people to move to these communities:

> Well, I think that one big reason [for people moving here] is that when you are in a community like this, you are in a controlled environment. The kids, once you drive through the gate, that gate closes behind you, the kids can roam around the community without, you know, getting into problems that are associated with the metro areas.

> And I think that is a big plus, and especially during these days. Most of the folks that I have sold property to last year, have decided that it is a factor in their purchase.

Notions of community in the sense of people watching over each other also figured quite highly in the promotion of another of these communities:

> At CARRIAGE PARK there is ... an inherent feeling of both security and comfort when accompanied by knowing your neighbors, knowing your street ... CARRIAGE PARK is a residential community whose beauty is enhanced by winding pathways, flower laden groves, mountain vistas, and neighbors ... friendly neighbors. (Hamlin 1997)

The promotion of community was frequently constructed as helpful in creating both social security and personal well-being. Many of the promotions can be seen to draw upon 'simulacra' of past social and personal experiences. Again the promotion of Carriage Park provides clear illustrations of both features:

> We believe the plan of Carriage Park is the perfect balance between intimacy and privacy, the best answer yet found to the personal isolation that has afflicted so many American communities in the last half of the 20th Century. Carriage Park's residents have confirmed our success in what we set out to do. (Carriage Park n.d., a)

> CARRIAGE PARK is, for its residents, a return to the wished for yesteryear of quiet walkways, an early morning sun grazing through mist shrouded trees, a wave from a friendly neighbor. It's that feeling of both security and peace that was felt on the day school let out and a long, never ending summer lay ahead. (Hamlin 1997)

While social and personal security was an element in promotion and in the descriptions of these communities by their managers and estate agencies, more materialistic constructions were also evident. Much of the security provisions, both formal and communal, were clearly geared to the protection of the residents' private property. Blakely and Snyder (1995a; 1997) argued that many private communities are composed of second-home owners and 'empty-nesters', that is people who are away from a residence for long periods of time, either for business or for pleasure. This was certainly evident in some North Carolinian communities, such as Elk River, where it was suggested that a large majority of the residents are from Florida and that these: 'Floridians tend to be members of an average in excess of four luxury golf clubs and they tend to be permanent residents up here in the Season and permanent residents of one place in Florida, and that seems to be for the rest of the year.'

For this group, protection of their property was an important issue, not least because the high proportion of second-home owners and empty-nesters in the development undermined the notion of widespread communal surveillance. As a result, great stress was placed here on the institutional provision of security, such as a 'stone security house' that was 'manned 24 hours a day, year round' and was 'located in a position to monitor and watch over' both the single entrance into the community and the community's own private airport and the private aircraft stationed on it (Elk River Development Corporation, n.d., b).

The defence of property was evident not only in the sense of security measures to combat crime and vandalism but also in the sense of a clear emphasis on the protection of property values. Sternleib (1990: 494) has attributed the growth of private communities to a middle-class fear of 'falling off the housing train' by buying properties that subsequently do not hold their value, perhaps as the result of a change in the physical or social make-up of the surrounding neighbourhood. In some of the private communities in North Carolina, there was a clear emphasis on the issue of property values, with one of the most explicit appearing in a promotional leaflet for Carriage Park:

> security wears more hats than physical well being and a sense of safety. Real security must also include having a sense of financial security. CARRIAGE PARK by the employment of exceptional architects, creative and inventive land planners, and experienced builders has created, in a literal sense, a haven of mountain living that, in marketing terms, is both desirable and sought after. Further, once residential construction is completed, there will be no more home building ... By limiting construction, the size of the market is also limited. Any home in CARRIAGE PARK is a select product in a very select market ... protecting the owner's investment and reinforcing their financial security. (Hamlin 1997)

Davis (1990) has also commented how these communities can confer economic advantage by allowing their residents to 'exit out' of tax burdens of social provision for other sectors of society. This is done either by 'incorporating' themselves to give themselves the power to set their own rates or remaining outside of city taxation regimes and yet providing urban levels of service provision for their own inhabitants. A striking feature of these communities is therefore how they fragment space physically, socially and economically. They also fragment space administratively through the establishment of autonomous systems of policing, social service provision and in many cases land-use planning. In many private communities an autonomous development control system has been set up, often under the title of a home owners' or a property owners' association or an architectural review board or committee. These bodies are often set up by the property developer and given powers to deliver services and to control house construction and land use via contracts made when individuals purchase individual plots of land from the developer to establish their own homes. Among the concerns of these boards or committees is the regulation of the size and number of buildings, and also their style of construction, as witnessed by the remarks of one community manager: 'We have an Architectural Board, we don't want any kind of extravagant kind of design, anything kind of ultra-modern, or Chinese Pagodas, or some type of extravagant design. We require natural building materials, stone, brick, wood.'

This concern to regulate building design can be interpreted as a key element in the economic segregation of these communities for the defence of property values: control over house design allows the reproduction of the desired 'select' or 'positional' commodity. This regulation may also contribute towards social segregation, attracting people with particular tastes (see Bourdieu 1984). As Blakely and Snyder (1997) commented, while many private communities place considerable stress on the creation of community they are communities where much of the social interaction is geared to the delivery of private ends. Most notable among these are the defence of private property and personal safety, and concern to be surrounded by people like themselves.

PRIVATE COMMUNITIES AS LIFESTYLES OF EXCLUSIVITY

Private communities are often seen as 'elite communities', inhabited by the rich and the powerful. It is certainly clear that many, perhaps the majority, of the promotions of these communities make a play of social positionality, making extensive use of such terms as 'elite', 'exclusive' and 'select'. Elk River, for example, is described as 'an exclusive residential community' with 'the finest social and recreational amenities' and being 'unmatched for privacy and exclusivity' (Elk River Development Corporation, n.d., a). The Hound Ears Club is praised as 'mountaintop living for the Special Few' (capitals in the original) full of 'testaments to the relaxed exclusivity favoured by people of taste' (Hound Ears Club, n.d.). The Old North State Club at Uwharrie is held to be 'uniquely rewarding', providing 'a myriad of activities and amenities' and 'perhaps best of all ... will always be safe, secluded, and available to the select few' (Old North State Club, n.d., emphasis added). Many of these constructions of exclusivity were 'social spatialisations' (Halfacree 1993; Shields 1991) that linked social relations and identities to particular geographical spaces, such as the countryside and the region (see also Phillips 2000). The promotion of Kenmure, for example, made a considerable play on historical associations with European aristocratic and Southern plantation society:

> A century and a half ago, Flat Rock began to be built up with large summer estates, after the English manner, by affluent Charlestonians, Europeans and prominent plantation owners of the low country ... Summers there became a round of Southern gaiety in antebellum days, Morning gatherings on latticed porches, picnics, teas, quadrilles danced under candled chandeliers – the essence of pleasant living, attracting many of the leading men of the era: families of four signers of The Declaration of Independence, the President of the Third Continental Congress, Revolutionary War Generals and later, Generals of the War between the States ... Flat Rock ... is the 20th Century seen against a backdrop of a distinguished past, a period when Southern leaders still had leisure for 'gracious living' ... Here, in Flat Rock, North Carolina, long the exclusive domain of a fortunate few, there is now an opportunity for you to own part of a mountain legacy ... The people now living here are only reaffirming what the wealthy Southern aristocracy knew over one hundred and fifty years ago. It is an exceptionally rare blend of the best there is. (Kenmure Enterprises Inc., n.d., a)

The construction of exclusivity is not purely symbolic. It is evident that the property prices, and additional costs associated with these communities – such as the membership fees of the communities' private social and recreational clubs and paying for the provision of security (see Table 8.1) – placed them beyond the reach of the majority of the population. A manager of one of these communities, for example, argued that: 'the price of the lots and the price of the condominiums and the average of the homes that are built here really reserve the place for the top, probably 1 percent, or fraction of 1 percent, of the economic strata in the US.'

This manager also clearly saw these communities as socially exclusionary and divisive:

> If you are in this upper class economically, you are going to live at a certain level of glibness to all this which is around you, regardless of where it is ... There's about 120 million dollars of property value [here] ... 13 years ago there was land here valued for tax purposes at ... one million dollars ... And, in that sense it has been a real boom to the people here. It doesn't create any sorts of skilled worker jobs but it does sort of provide a lot of construction related technical kinds of hourly opportunities ... [But] I think that there is a certain level of, what I suppose you would call natural resentment, a natural suspicion, of the newcomer ... The natural resentment, I think, comes from the fact ... the upper class economically develops a certain sense of arrogance or 'I'm getting what I deserve because I'm better, because ... I'm a superior person'. In reality, in many cases, they've just married the first grade money or whatever. But I think that, as a result of that, there's sometimes a snobbishness or an arrogance which gets conveyed, and the person who is say a security guard or someone cutting the grass or ... they pick up on that and so, that sort of resentment I think is more 'master–servant' kinds of resentment than any sort of fundamental conflict over policies or directions or anything of that nature. It is more like a class attitude, and it works both directions. And every time we have a member here who gets involved in a dispute with a local contractor, I just, I cringe, because it just reinforces, many times, the image of the wealthy outsider coming in and taking advantage of people who he perhaps thinks are part not up to his capabilities. I describe that because I think it is real. Fortunately it doesn't represent a major problem for us, it is not widespread, it is just one of your outlines. And then we've got some folks from the local community who have been known to vandalise fancy cars, just because they didn't like the people bringing them in. But that doesn't happen very often.

However, it is possible to overdraw the extent to which these communities are the preserve of the super-rich. A manager of another private community in North Carolina, for example, argued that there was, in the community he was responsible for, 'a broad spectrum' of people. He added: 'I don't think there is one specific profile that you can look at and say, look this is our profile for our owner. A lot of people choose to own here for a variety of reasons. I wouldn't say that we are an elitist or separatist type of (community).'

The community managed by this individual appeared to have quite a different market to that run by the other manager quoted above. For instance, while the previous place of residence for most of the residents of the first, highly elite, community was predominantly Florida, the residents of the second community were drawn largely from within North Carolina (see Figure 8.1). As shown, the property values and membership fees of private communities in rural North Carolina vary considerably. As Blakely and Snyder (1995a: 13) have argued, many of the private communities that make a play in their promotions of being socially exclusive are targeted at 'executive', middle-class households. They argue that these communities are designed so as to 'provide the cachet of exclusive living to those with non-exclusive incomes.' They further suggest that elements of a landscape of defence, such as the gatehouse, may indeed be part of this cachet: 'The service of gate guards and security patrols adds to the prestige of exclusivity; residents

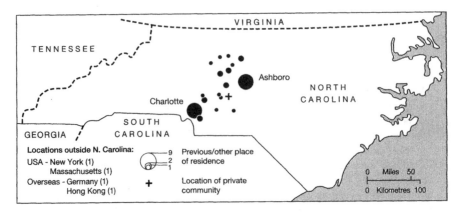

Figure 8.1   Previous or other place of residence for people buying property in one rural private community in North Carolina

value the simple presence of a security force more than any service they provide' (*ibid.*: 11).

As mentioned previously, Davis (1990: 246) made very similar claims, suggesting that access to security may have become a mark of socio-economic distinction and that security may be a self-producing commodity, with the militarisation of space acting to increase rather than reduce people's levels of fear and desire for security. He suggested that there may well be an emergent 'residential arms race' as an increasing array of people come to 'demand the kind of social insulation once enjoyed only by the rich' (*ibid.*). Both he and Blakely and Snyder highlighted the emergence of barricades within urban, low-income and racially marginalised inner city communities, and indeed in some low-income rural communities in North Carolina one can witness attempts to limit public access, albeit largely through symbolic barriers.

Blakely and Snyder (1995a: 13) suggested that even within many of the executive focused private communities, the defensive landscape may be more symbolic than material:

> Many have electronic gates, and they have guardhouses at the main entrance. Guards, however, are sometimes never hired by the homeowners' association because of the high ongoing cost, and the gatehouse stands solely as a psychological deterrent to outsiders. Individual home security systems are common.

It was certainly evident that there were wide variations in the level of fees that residents in these communities paid for their security provisions (see Table 8.1). Overall, it can be suggested that despite a similarity in their promotion of social positionality, in part through their landscapes of defence, there are significant differences in both the class composition of these communities and the nature of their security provision. The presence of significant differences is further highlighted when attention is paid to the forms of lifestyle and leisure that are also a prominent features of many of these communities.

PRIVATE COMMUNITIES AS LANDSCAPES OF LEISURE AND LIFESTYLE

To many people private communities are synonymous with elitism, and even Blakely and Snyder referred to 'prestige developments as "the purest form of gated community" composed simply of beautiful homes behind walls, guarded by gates and often sentries' (1997: 97). They note, however, that from the 1980s a range of 'lifestyle communities' has come to predominate in the development of private communities. These communities are, they suggest, distinct in the emphasis given to amenities, both for leisure and for particular consumption-oriented lifestyles.

Leisure and lifestyle certainly figure highly in the promotion of private communities in North Carolina:

> Champion Hills Lifestyle. Champion Hills is more than a place to live – it's a way of life. (Champion Hills, n.d.)

> Hound Ears Club has set the precedent for elegance and tradition in the mountains. Hound Ears has become one of America's finest four season, gated communities. Custom homes, luxury condos and spectacular homesites are available . . . Enjoy the finest facilities for golf, tennis, swimming and four-star dining. A very special lifestyle awaits you. (Hound Ears Club, n.d.)

The significance of leisure and lifestyle was echoed in interviews with the estate agents and managers of private communities. One, for example, exclaimed: 'the whole idea is more of a lifestyle thing which is being sold rather than just a piece of real estate.'

The notions of leisure and lifestyle were, however, constructed in various ways. Private communities in the United States generally, and in North Carolina more particularly, often started out as resort communities, places where people would spend their vacations. As part of this, many of these communities provided recreational facilities such as swimming pools, tennis courts and primarily golf courses. Golf courses in particular have become a mark of distinction, with communities vying with each other to sign up to be able to use the name of golfers and golf course designers in their marketing:

> Florida based architect Joe Lee has done a magnificent job in designing the course at Kenmure. (Kenmure Enterprises Inc., n.d., b)

> at the heart of the Old North State Club lies the championship golf course designed by Tom Fazio. The three-time National Golf Course Architect of the Year has created a course that's proven to be one of his finest. (Old North State Club, n.d.)

> The Champion Hills Club is a private, equity golf club featuring a 6,719 yard, par 71 mountain golf course designed by Tom Fazio. Champion Hills is Fazio's home course and club, a show case of his world-class talent. (Champion Hills, n.d.)

> For many years the name Elk River is synonymous with mountain golf at its finest. The Elk River Golf course was designed by Jack Nichlaus . . . and it remains his only mountain course in the Carolinas or Virginias. (Elk River Development Corporation, n.d., c)

According to one community manager, the presence of a golf course in a community has become almost ubiquitous:

> In the same sort of general economic strata is, there's Linville Ridge ... there's Grandfather Mountain Golf Club ... Then beyond ... is Linville Golf Club which has a golf course ... Those that don't have a golf course tend to be, maybe three or four tenths down the economic scale. For example, near here, between here and Blowing Rock, is a little community ... and it has no golf course, but it has wonderful tennis facilities, a swimming pool and great riding, equestrian, facilities. But it doesn't seem to attract the same economic backgrounds that are attracted to the use of golf courses.

However, as Blakely and Snyder (1997: 57) noted, although much is made of these golf courses in the promotions and many of the residents of the communities may well be keen golfers, 'many others simply value the open space and greenery golf courses provide.' This was explicitly recognised in some of the promotional material (see Phillips 2000). One community manager commented that he often saw 'a lot of people early in the morning hiking around the golf course' or else, 'in the course of the summer wading in the ... river out here, or just sitting on the log fishing.' In other words, beneath the common promotion a variety of lifestyle practices are undertaken.

Similar variety is also evident in lifestyles. In many instances the construction is quite directly positional: the lifestyle offered in the promotions is described simply in terms of its exclusivity. Nevertheless, lifestyle constructions frequently have a range of other elements incorporated into them. So, for example, particular geographical 'cultural textures' (Eder 1993; Cloke *et al.* 1995; 1998; Phillips 2000) were often intertwined with those of positionality and lifestyle. Hence one finds promotions offering:

> peak experiences ... mountaintop living for a Special Few. (Hound Ears Club, n.d.)

> an exclusive residential community that takes mountain living to new heights. (Elk River Development Corporation, n.d., a)

> exclusive and restricted country living at a 3,000 foot elevation. (Friendship Park, n.d.)

> A secret paradise that has attracted the most prominent families of America and Europe ... for its breathtaking beauty, exclusive country club and rich amenities. (Lake Toxaway Company, n.d.)

There were also lifestyle constructs drawing on more general social constructions such as those related to race, age, sexuality, family and sociability. Many of these communities are, for example, constructed as 'retirement communities' and are often promoted by drawing upon what Harper (1997: 191; see also Featherstone and Hepworth 1995) identified as the 'metaphorical body' of old age, that is people inhabiting 'an idyllic retirement' which is 'an extended plateau of active middle age typified ... as a period of youthfulness and active consumer lifestyles.' Hence the promotional materials are replete with images of '50 to 60 somethings', enjoying the good life of an active sport and socialising (Figure 8.2). There are also clear metaphorical and stereotypical constructions of race, family life, sexuality and childhood in the promotion of other communities (Figure 8.3). Once more one can

Figure 8.2 Youth and active retirement within a private community (Old North State Club at Uwharrie Point)

Figure 8.3 Images of community life (Old North State Club at Uwharrie Point)

see the notion of community being viewed in terms of a collection of 'the same' rather than encompassing social difference.

Indeed, as Davis (1990) has noted, many home owners' associations in the early twentieth century operated explicit social discrimination through deed restrictions that prohibited the sale of properties to, for example, non-Caucasians or non-Christians. Such discriminatory restrictions were ruled illegal in 1948 by the US Supreme Court, but Davis suggested that many luxury private communities continued to have explicitly racial restrictions until the mid-1970s. Implicit forms of social discrimination may well still function, for example, with membership of many of the social clubs that form a major amenity in these communities often being available only via personal recommendation from other members. Furthermore, there are clear economic and semiotic mechanisms of social exclusion. With reference to the former, mention has already been made of the financial costs of entry into these communities, and as Blakely and Snyder (1997: 153) noted, these communities are 'intentionally' segregated by wealth. However, given that 'race and class are closely correlated', economic segregation also leads to racial segregation. With regard to the semiotics exclusion, Blakely and Snyder (*ibid.*) argued that gates are a 'visible sign of exclusion' and may form 'an even stronger signal to those who already see themselves as excluded.' Davis (1990: 226) has likewise commented:

> the semiotics of so-called 'defensible space' are just about as subtle as the swaggering white cop . . . [They] are full of invisible signs warning off the underclass 'Other'. Although architectural critics are usually oblivious to how the built environment contributes to segregation, pariah groups – whether poor Latino families, young Black men, or elderly homeless white females – read the meaning immediately.

Although several of the managers and property agents whom I interviewed explicitly claimed that their communities were open to 'people of colour', they, and the promotional material they gave out, were exclusively 'white'.

CONCLUSION

This chapter has looked at private communities in rural North Carolina. If the existence of private communities in general has been neglected, then their presence in rural spaces has been almost completely ignored. This chapter has sought to demonstrate that they are a phenomenon of some significance and in particular has considered to what extent they may be seen as landscapes of defence, exclusivity and leisure and lifestyle. These notions have been derived from the work of Blakely and Snyder, who suggested that there are distinct types of community based on the extent to which security, exclusivity and leisure and lifestyle form the motivation for people to move into them. While this chapter has highlighted important differences in the private communities in rural North Carolina, notably with regard to the socio-economic character and migrational origins of their residents, it has been argued that the three factors outlined by Blakely and Snyder are all important in the symbolic and material construction of many of these communities. Furthermore, these factors may be seen as reinforcing one another. So, for example, landscape features such as the gatehouse may be as much an element in the construction of exclusivity as they are a feature of security, while leisure pursuits are also productive of exclusivity in part because of their privatisation, created through the landscapes of defence and exclusivity. Indeed, while Blakely and Snyder were somewhat dismissive of the defensive aspects of private communities, suggesting that defence is a major dynamic only within inner city and suburban areas, where groups of people seek to construct barricades and other defence features around pre-existing development, it can be suggested that defensive constructions are a key feature in the construction of symbolic and material landscapes of private communities.

It is important in saying this to recognise that defence and security can be taken to mean a range of different things. For Blakely and Snyder, the defensive aspects of private communities revolve around making provisions for the protection of the self from violence and protection of property from theft. However, these communities may also be seen as offering other forms of defence, such as protection of the self from the alienating experience of modern, urban life and the protection of financial investment. In all cases, defence in private communities involves, crucially but not exclusively, the creation of exclusion barriers. These barriers range from the physical and symbolic border boundaries encircling these communities, through the economic and semiotic constructions of exclusivity and into the performance of taste in leisure and lifestyle. While concurring with Blakely and Snyder that there are important, and often rather neglected, differences between private communities, there may well be a strong element of commonality in that they may be seen to create spaces that act to exclude an even greater degree of social differentiation than exists within them. Private communities may not represent completely secure defended landscapes but rather landscapes of social exclusion and segregation.

CHAPTER 9

LIVING ON THE EDGE: DEFENDING AMERICAN INDIAN RESERVATION LANDS AND CULTURE

Christina B. Kennedy and Alan A. Lew

I feel I am at home here. I am still within the bounds of the four sacred mountains.

This statement by the renowned Navajo artist and writer Shonto Begay (personal communication, 1999) conveys the power of the American Indian's attachment to place. An American Indian tribe's claim to a specific territory or physical place provides both a basis for nationhood and a framework within which cultural identity is claimed, nurtured and reinforced. Of all the minority groups in the USA, only American Indians have a traditional, collective land base, even though only about one-quarter of self-identified Native Americans actually live on reservation lands. Nonetheless, most American Indians are deeply attached to their homelands, which they see as critical in defining both their culture and identity. The land is indivisible from who they are as individuals and what they represent as a society. Defending, protecting and caring from their tribal lands is tantamount to caring for the future survival of Native American culture.

Native American reservations, while mostly in the western USA, contain extremely diverse landscapes and cultures. However, some common features and generalities can be identified. Compared with mainstream America, most reservation societies are similar to developing countries, with high levels of poverty. They also suffer from high rates of substance abuse and violent crime (Mihesuah 1996). As Rudzitis (1993: 579) cogently points out: 'Indian reservations, by most objective socio-economic criteria, are truly landscapes of despair, our shameful islands of poverty, perhaps more isolated and ignored because they are sovereign nations, cultures apart from the mainstream.' The crucial issues currently facing Native American tribes are the maintenance of a land base and political sovereignty, addressing the effects of the exploitation of reservation land past and present, and meeting the changing needs, attitudes and religious demands faced by tribes in the process of cultural change (Lewis 1995). Economic neo-colonialism, cultural imperialism, acculturation and assimilation are the global paradigms that shape American Indian struggles to defend their land and life.

These are the harsh realities of life for many tribes, but reservations have multiple layers of meaning. They are geographical and spiritual reference points for tribal members living off-reservation, seats of legend and memory extending back through countless generations, and 'the last outposts against a debilitating racism that shows few signs of abating' (Wilkinson 1992: 26). As Scott Canty, a Hopi lawyer, pointed out (personal communication, 1999), many Native Americans choose to live on reservations despite the isolation and poverty. Many return there after completing careers in the dominant society both because of their attachment to the land and community and because they are, at heart, separatists. Another common feature that many reservations share is a strong sense of place, engendered by the close attachment and relationship between tribal members and the land they occupy. Swentzell (1990: 64), in her descriptions of the connections between the people of the Santa Clara Tewa Pueblo and the land that they occupy, expressed this unity: 'These symbolic places remind the people of the vital, breathing earth. The plants, rocks, land, people are part of an entirety which is sacred because it breathes the creative energy of the universe.'

Despite the great diversity that exists among reservations in terms of their geography, history and culture, this sacred attachment to the land, probably more than anything, has enabled American Indians to maintain strong and vibrant cultures. Defending their land is tantamount to defending the living culture of their tribe. In the long run, rather than being a place of segregation and death, reservations have given American Indians the strength to survive.

In this chapter, we seek to categorise the defensive strategies used by tribes to maintain and strengthen both their culture and land base. The focus of our analysis is on the reservations of Arizona and New Mexico, although we recognise that the situation regarding reservations varies considerably from state to state. For example, unlike reservations in South Dakota or Oklahoma, where tribes from other parts of the country were often relocated, most of the tribal groups in Arizona and New Mexico occupy roughly the same locations today that they did when the Spanish first arrived in the sixteenth century. Furthermore, as Figure 9.1 shows, 26 percent of Arizona's land base and 9 percent of New Mexico's are occupied by reservation land (Lavender 1980).

The ensuing chapter has four main sections. The first briefly discusses the creation of reservations and the shifting relationship between tribes and the federal government. The next provides discussion of current issues that are being contested regarding the role and function of modern reservations, including the struggle to maintain sovereignty and self-governance, and defence against cultural appropriation. Specific examples of those strategies used by tribes in Arizona and New Mexico then follow. We conclude with a discussion of future directions.

INDIAN RESERVATIONS AND THE FEDERAL GOVERNMENT

There are just less than two million Native Americans in the United States. Most belong to tribes or cultural groups that occupy diverse environments and typically have their own languages and real or mythical lineages. The US government formally recognises over 330 American Indian tribal groups in the lower forty-eight states and 223 Alaska Native groups. Some of these have petitioned for recognition

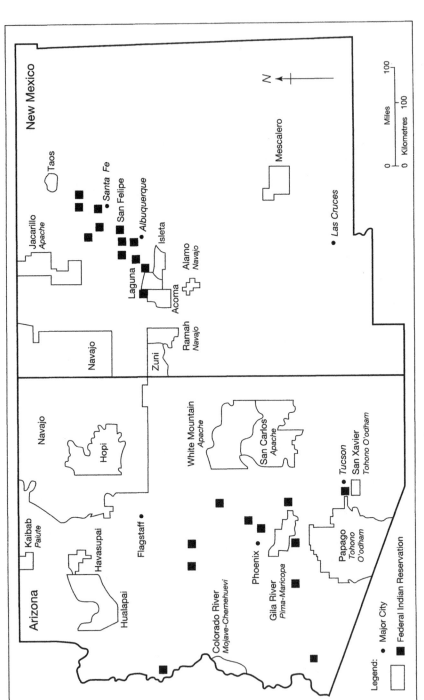

Figure 9.1 Map of Arizona and New Mexico showing the location of federal Indian reservations and major towns (map by Tina Kennedy)

(BIA 1999a), others for re-recognition (Wilkinson 1992). According to the US Census Bureau, in 1990 about one-quarter of all Indians lived on reservations, with another quarter living close enough to reservations to make frequent trips for family visits and ceremonies. Not all tribes have reservation lands, and some new reservations have been created in recent years. However, each reservation maintains a distinct and separate tribal government and administrative structure. Depending on the definition used, the federal government recognises around 275 to 300 reservations, rancherias and other land-based communities, while states recognise an additional twelve. These Indian lands total 55 million acres, with individual reservations ranging in size from less than 1 acre to more than 17 million acres in the case of the Navajo Reservation. Some tribes are adding to their reservation land base by purchasing non-reservation lands and placing them under federal government trust. Approximately 20 percent of the federally recognised reservation lands are privately owned by individual Indians, while the remainder is held in trust by the federal government and reserved for tribal use (BIA 1999b).

Reservations were established to protect non-Indians from the military threat of 'captive nations' (Snipp 1994) and are central to the federal government's policy towards Native Americans (Wilkinson 1992). The first true reservations were established by the state of California shortly after the discovery of gold brought a huge influx of non-Indians to the state (Findlay 1992). The apparent success of this Californian solution to the 'Indian problem' quickly spread throughout the west. In most instances, tribes were forced on to reservation lands through a form of forced dependency. Hunting tribes could no longer survive as the open ranges were subdivided and the US Army and other hunters killed nearly all of the buffalo around which their lives revolved (Hagan 1993). The Army also destroyed fields and orchards to force tribal groups on to reservation lands, where they could at least obtain government handouts of food and shelter. The policy was a conscious effort to respond to the American public's desire for more land and access to resources by concentrating, weakening or even eliminating the only potential threat to that desire, the American Indian.

Between 1853 and 1856 alone, the United States acquired more than 174 million acres of land from Native Americans through reservation treaties and presidential decrees. The General Allotment, or Dawes Act, of 1887 further sought to reduce tribal land holdings, increase Native American assimilation into the dominant culture and make more land available to non-Indians by converting communally owned tribal lands to individuals. Any remaining 'surplus' land would be open to non-Indian settlement. By 1933, when the allotment programme was stopped, reservation land areas had been reduced by 90 million acres (Tyler 1973; Wilkinson 1992). Although many of the Indians who accepted the offer for private land ownership were not able to compete successfully with non-Indian farmers and soon lost their land altogether, many tribes fought tenaciously against the allotment programme, recognising that the loss of communal land would mean the end of their culture.

In 1934 Congress passed the Indian Reorganization Act, which held that reservation lands would no longer belong to native peoples but would be held in trust by the US government. There are mixed views about its purpose. John Collier, the head of the BIA, thought it would strengthen tribes (Trosper, personal communication, 1999), but others saw it as another attempt to assimilate and weaken them. The Reorganization Act forced reservations to adopt American-style constitutions, with very limited separation powers, drafted for the tribes by the Department of the Interior. By 1940, and despite resistance, the BIA had imposed a tribal council form of governance on almost all Indian nations. The official goal was to wean Indian reservations from their heavy reliance on the US Treasury, which had become the new 'Indian problem' of the twentieth century. In some cases, the organisation of tribal councils enabled mining companies, particularly Standard Oil, to bypass traditional leaders and negotiate tribal contracts with younger and more 'malleable' secular leaders (Churchill 1994).

To further reduce reservation financial dependency on the federal government, Congress terminated 109 Indian nations between 1953 and 1958 in the belief that this would at long last make them true 'Americans'. Terminated tribes lost their official identities and their reservations (Tyler 1973). Termination meant that federal benefits were cut off, tax-exempt status removed and land taken. Terminations destroyed some reservations but threatened all. Furthermore, during this period the US Congress gave states more authority over reservation lands (Wilkinson 1992). In the 1950s and early 1960s, Congress cut funding for reservations while increasing support for urban Indians, thereby consciously encouraging a large exodus from rural reservations to urban centres.

Reservations were not originally viewed as a long-term solution to the 'Indian problem'. Instead, they were seen by the federal government as places where, through either benign neglect or forced assimilation, tribal culture would eventually die in the face of the dominant culture. It was believed that after about fifty years, the federal government's objectives would have been achieved and continued financial outlays for reservations would cease. This philosophy would help to explain the various efforts to eradicate reservations through most of the twentieth century. Yet, as attempts at cultural genocide throughout the history of the modern world have repeatedly demonstrated, deeply ingrained cultural traits are impossible to destroy. American Indians, as well as other Native Americans in Alaska and Hawaii, have survived the onslaught of more than one hundred years of the reservation system. The belief that reservations were transitory, however, has left the legacy of inadequate infrastructure and the major socio-economic problems that are observable on most reservations today (Wilkinson 1992).

ROLE AND FUNCTION OF RESERVATION LANDS

If cultural death was not the original intent of Indian reservations, their history clearly demonstrates purposeful control and subjugation by the federal government. These actions were justified through the dualism of a civilised 'us' doing what is best for the savage 'other'. Green (1995) argues that reservations were the epitome of nineteenth-century American imperialism and colonialism. In this way, subjugation, segregation, forced assimilation and even genocide of the 'savages' was rationalised.

Three ways in which this history continues to manifest its impacts on reservations today can be seen in struggles over culture, sovereignty and self-governance.

Culture

Possibly the one federal policy that has had the most devastating and bitter impact on American Indian culture was the forced assimilation imposed on large numbers of Indian children through the control of education by the Bureau of Indian Affairs (BIA). From the 1880s to the early 1960s, many Indian children were forcibly separated from their parents and taken to BIA boarding schools, many of which were in large cities off the reservations. Shonto Begay (personal communication, 1999) describes the experience:

> I was out herding the sheep and was picked up and put in the back of a sheep truck with a bunch of other little boys who were scared and crying. We were taken to school and had our long hair cut off, except for a long piece in front. That was called the BIA handle. They would grab us by that. We had to learn English. We weren't allowed to speak our language. There were pieces of paper tacked up all over. I didn't know what they said until I learned to read English. They said 'Tradition is the Enemy of Progress'. It made us feel we were wrong and inadequate. We were assigned religion. We were lined up shortest to tallest. The shortest group were made Catholics, the medium size Presbyterian, and the tallest Mormon. The next year we would be lined up again and the shortest would be Mormon. Our parents couldn't come see us. We only saw them for two weeks at Christmas and during the summer.

Since the late 1960s, federal intrusion into reservation life has been less invasive, though still ever-present. What may be an even greater threat to Indian land and culture is the overbearing presence of American mass culture – a phenomenon that is assaulting traditional societies throughout the world. The problem is particularly pernicious for American Indians, who have suffered from stereotyping and cultural misappropriation by European and Anglo-American society.

Film has been a particularly powerful force in creating and perpetuating the stereotypes of American Indians and forming the images that most people in the world hold of them today. Much has been written recently about problems in the portrayal of Native Americans in American western movies (e.g. Aleiss 1994; Leuthold 1995). Native American producer Phil Lucas (1979), director of the film *Images of Indians*, points out that westerns generally stereotyped Native Americans as cruel, vicious savages or as comic, drunken and valueless. Native American women were usually portrayed as 'Indian princesses' anxious for the attention of 'superior' white men or as unintelligent, subordinate 'squaws'. Lucas also noted a widespread failure to distinguish between different tribes and misrepresentations of Indian religions and rituals. Although there has been a significant change towards the portrayal of more sympathetic Indians in several films in the last fifteen years, the stereotypes gained from early westerns remain dominant among the majority of non-Indians.

Commercialisation of the stereotypical Indian image can be clearly seen along the former Route 66 highway, which passes through northern Arizona and New Mexico.

Figure 9.2   Wigwam Motel in Holbrook, Arizona (photo by Tina Kennedy)

A prime example is the Wigwam Hotel in Holbrook, Arizona (see Figure 9.2), which uses motel rooms shaped like teepees – a housing style of Great Plains Indians that is not used by southwestern tribes – to attract tourist dollars. Lew (1998) points out that this type of stereotyping and cultural misappropriation is common throughout the advertising industry and has even been used by states in their promotion of tourism.

Other forms of cultural appropriation are, however, more insidious from a cultural perspective. Traditional Indian arts and crafts had either utilitarian or religious significance. Many were transformed long ago into collectors' items and art forms, and major Indian artists are now recognised and appreciated for their talents (MacCannell 1994) – actual cultural appropriation. However, the popularity of Indian arts and crafts, and the high prices that authentic pieces demand, has created a market for inauthentic replicas produced in Asia and Latin America, or by non-Indians who are aware of the resale value of quality work. Some are labelled as being foreign-made, but others are sold as Indian-made – a practice that violates most tribal and some state laws. The New Age movement has created concern among Native Americans, as so-called 'wannabe' Indians appropriate, and often bastardise, Native American ceremonies and symbols for personal aggrandisement and financial gain (Mihesuah 1996; Whitt 1995).

While the American mass media promote cultural stereotyping and cultural appropriation, they tend to ignore American Indians most of the time. Yet the media continue to have a major impact on tribal culture and society. On even the most remote reservations it is common to see both traditional and modern homes with satellite dishes next to them or perched on their roofs. As West and Reid (1999: A16) point out, 'today's tribal children are growing up on the reservation to the sounds of MTV rather than traditional teachings.' Some tribal members see increased crime on reservations, the weakening of extended families and the erosion of valued cultural traditions as being related to mass media. Others, according to Canty (personal communication, 1999), welcome the exchange of ideas and information offered by the mass media. He points out that 'many Hopi see television as an opportunity to help their children see through the distractions of popular culture and to teach them basic traditional values.' Cultures change and evolve. We cannot expect them to remain static. At the same time, it may be argued that people within a culture need to have the choice of what aspects of their culture they wish to keep and what aspects of other cultures they do not want to adopt.

Sovereignty

If, as Wilkinson (1992: 28) claims, 'most powers of sovereignty are defined geographically', then to lose land is to lose sovereign power. Early reservations were established through treaties between the United States and Indian nations that were treated as equal sovereign states. This sovereignty was clearly defined by Supreme Court Justice Marshall in 1932 when he declared that tribes had governmental and judicial power within their boundaries – that they were sovereign and that the states could not intrude on them (Cohen 1998; Wilson 1996). Later reservations were established by presidential decree. When it comes to the issue of sovereignty, however, all tribes tend to speak with a single voice – they see it as a sacred promise given by the federal government and are concerned about federal and state court rulings that diminish it (Wilkinson 1992). However, Indian tribes are not sovereign to the same degree as member countries of the United Nations, since they are Fourth World nations, existing on reservations that are partially autonomous entities within a fully sovereign country. Indian nations do not have control over their international relations (Mihesuah 1996). Although some do issue passports that have been recognised by a handful of foreign countries, they cannot raise an army or enter into defence pacts with a foreign country. They are subject to the US constitution and must abide by laws passed and administered by the federal government. For the most part, they are not required to abide by state laws, although this is entirely dependent on the US Congress. Wilkinson (1992: 39) also points out that a majority of Indian rights are based on promises made in treaties, executive orders, agreements and federal statutes. A Supreme Court decision made in 1903 in the case of Lone Wolf *versus* Hitchcock showed that Congress can 'unilaterally break Indian treaties'. By that ruling, the Supreme Court effectively ceded plenary (absolute) power to Congress in Indian affairs.

Sovereignty remains a major issue. A large billboard on the interstate highway that passes through the Gila Indian Reservation between Phoenix and Tucson, Arizona, proclaims: 'Support Indian Sovereignty.' What it means by sovereignty is that tribal governments have 'control over reservation development and the power

Figure 9.3 Tribal police car at Acoma Pueblo, New Mexico (photo by Alan Lew)

to enter negotiations with non-Indians on behalf of the tribe' (Snipp 1994: 380). Probably the greatest fear is that the US Congress will give states increasing control over reservation activities. State governments have long sought this power from federal lawmakers, and some are of the opinion that the Indian Gaming Act may even allow states to tax gaming revenues on reservation lands, although this has yet to be tested (Rossi 1999). State and local governments often assist reservations, especially smaller ones, with law enforcement and social services, much as they would off-reservation communities, so there is some history of established working relationships between many tribes and non-federal government agencies.

The relationship between the US government, the state governments and the tribes may also be seen in law enforcement jurisdiction. According to McGuire, (1994: 15) a core element of sovereignty is 'judicial control over civil actions of tribal members. Tribal courts are empowered to hear such issues, based on their own rules of procedure and their own culturally specific rules of judgement.' In 1978, however, the Supreme Court ruled that tribal courts did not have jurisdiction over non-Indian criminal activities on reservations. This decision was a reminder to tribes that 'they are only "quasi-sovereign"' (*ibid.*). The legal jurisdiction of reservations reflects these political realities. Although many tribes have their own tribal police (Figure 9.3), felonies committed on reservations are under the jurisdiction of the Federal Bureau of Investigation (FBI). On the Navajo Reservation in Arizona, the Arizona Department of Public Safety is responsible for patrolling state highways crossing the reservation and for dealing with crimes committed by non-Navajos. They are also called in on other enforcement issues to work with the Navajo Police.

A non-Indian who is arrested on tribal lands is removed and tried elsewhere. Smaller reservations often do not have their own police forces and may be solely dependent upon county, state or federal officers.

Governance

Economic and social conditions on most reservations give cause for alarm (Russell 1993). Numerous efforts to address the problems of reservations have met with mixed success at best. Government administration on many reservations has been character-ised as complicated and requiring great expenditures of time, money and effort. Some of these problems in the past have been attributed to a lack of skilled labour and managerial experience, bureaucratic and cultural resistance to change, and complic-ated internal politics (UIPA 1977; USGAO 1987). Indeed, administrative models developed for non-reservation communities have seldom been effective on reserva-tions because they seldom consider the depth of the traditional culture and world view (Van Otten 1985). Reservation politics are themselves typically more complicated than off-reservation politics. This is because many tribes have both secular and traditional or religious leaderships – both of which are influential in the political decision-making process on reservations. Although today traditional forms of gov-erning still exist in some tribes, they do so side by side with elected tribal councils, which were established following the 1934 Indian Reorganization Act. The federal government formally recognises only those decisions made by the tribal councils.

Two examples where traditional governance can still be found are on the Hopi Reservation and the Navajo Reservation. The Hopi maintain a theocratic system that binds traditionally independent villages through common religious beliefs and culture. The influence of traditional leaders in this system varies from one village to the next; their influence can be felt reservation-wide, however, when they object to decisions made by the tribal council. The original Navajo governing system was based on 'family, clan, and group consensus' (Feher-Elston 1988: 57). This type of governing can still be found at the local or 'chapter' level on the reservation and, at times, the Navajo tribal council leans towards consensus rather than majority rule. Evidence of the older system is still apparent at the local level on the Navajo Reservation, where elected officials run chapter meetings, but all voting is inclus-ive. While consensus may not be achieved, everyone in attendance is given an open platform to talk at length on chapter issues. The process can seem tedious and inefficient from the outside, but it reflects and maintains longstanding traditional values and cultural beliefs about the integrity of tribal members and the importance of group decision making. Trosper (1998: 10) suggests that today, even within imposed governmental frameworks, many Native Americans have maintained, while others are returning to, a consensus mode in their own governance through 'a desire to be careful, to consider all viewpoints, to worry about minority views and the concerns they articulate.'

Another barrier to a greater sense of security over the future of tribal sovereignty and self-determination is the existence of the trust relationship between the federal government and the tribes. The fact that all reservation lands are owned by the federal government in trust makes them federal lands, which are generally outside the jurisdiction of state and local governments. Thomas Alcoze (personal communication, 1999) argues that the trust relationship is paternalistic and essentially maintains that 'Native Americans are incapable of taking care of their own resources and lives.' Through the trust relationship the federal government controls material and social wealth and 'established a reservation-management process that placed the federal government in the role of mediator and *benefactor* of Indian populations' (Henderson 1992: 115, emphasis added). The trust doctrine engenders a sense of dependency in tribes in that there is always a 'threat' that social services and federal money may be withheld (Trosper 1995).

The meaning of sovereignty and how to ensure it remain contentious issues, as are the questions of ownership of culture and the role of tradition on the modern reservation. What has changed in the 1990s is that the underlying debates are being played out primarily within tribes themselves, with less outside interference and manipulation. This does not make the debates any less intense or the solutions any easier. However, it may be argued that this is a trend that will help to strengthen the resolve and ability of tribes to defend their homelands.

DEFENDING RESERVATION LAND AND CULTURAL SPACE

Many tribes are taking active steps to counteract these attacks on their homelands and culture. On the Navajo Reservation, for example, children in the Headstart programme participate in a ceremony called 'The Children are Beautiful', in which they wear traditional dress (Figure 9.4). As mentioned earlier, tribes have their own laws and many have their own tribal police and courts. They increasingly seek to have more input into the management and protection of places that are sacred to them. Changes are occurring on reservations. Especially since the early 1970s there has been a resurgence of cultural pride, increased emphasis on tribal sovereignty, a desire for tribal control and use of resources, education more suited to Native American needs, and increased or at least more vocal concern over sacred places. These objectives are addressed through a variety of approaches that work to defend reservation lands and cultural space from excessive intrusion by the dominant society. These include limiting physical access to outsiders; reclaiming traditional and sacred lands; using education to enhance cultural preservation; combining traditional knowledge with modern science in a culturally sensitive manner; and establishing sustainable economic development. They are worth considering in greater detail.

Limiting access

Control over access and use of reservation land and resources by non-Indians or Indians of different tribes is essentially a legal issue (Canty, personal communication 1999). Signs are a visible means of place identity and help to control the behaviour of non-Indians on reservations (Lew 1999). They demarcate reservation boundaries

Figure 9.4   Navajo children preparing for 'The Children are Beautiful' celebration in Headstart, Navajo Nation, Arizona (photo by Tina Kennedy)

and may clearly point out that once on the reservation a person is subject to tribal laws and restrictions. They may also have more explicit purposes. For example, one at San Felipe Pueblo, New Mexico (Figure 9.5), clearly states that 'No picture taking or sketching allowed – Absolutely no liquor permitted. Solicitors and bill collectors must have governor's permission.' Signs like this are common on pueblo village reservations from the Rio Grande valley in New Mexico to the Hopi mesas in Arizona. Some will also indicate times when a pueblo is completely closed to outsiders when certain sacred ceremonies are held. Yet signs by themselves are not enough, and implementation or enforcement of rules is also necessary. McGuire (1994) points out that keeping non-Native Americans off the reservation or restricting their activities on the reservation is one way of counterbalancing the erosion of tribal sovereignty. Another means is through cross deputisation, wherein tribal law enforcement officers are deputised by state and county law enforcement agencies to give them authority to control non-Indian activities on the reservation and expel unwanted trespassers more easily.

Signs and rules are necessary because, while cultural tourism can provide considerable financial benefits, it can be intrusive. Taos Pueblo provides a good example of a people that has determined how to benefit from tourism yet defend their cultural space. This community, continuously occupied for 600 years, today uses signs with explicit rules to manage visitor behaviour. Signs indicate that visitors must register with the tribe and pay entry fees, as well as additional fees for such

Figure 9.5   Sign at San Felipe Pueblo, New Mexico (photo by Alan Lew)

activities as photography, painting or sketching. Visitors must also obey specific rules, as noted in a visitor pamphlet:

> Please respect the 'restricted area' signs as they protect the privacy of our residents and the sites of our native religious practices ... Do not enter doors that are not clearly marked as curio shops. Each home is privately owned and occupied by a family, and is not a museum display to be inspected with curiosity ... Please do not photograph members of our tribe without first asking permission ... Do not wade in our river – our sole source of drinking water. (Taos Pueblo, n.d.)

Acoma Pueblo in western New Mexico also receives large numbers of visitors to its Sky City pueblo. This village is located on top of a steep-sided mesa, allowing the tribe to require that all vehicles be parked off-site at a visitor cum arts-and-crafts centre. Tourists can only visit the pueblo on guided tours, which have a fee structure similar to that at Taos. Not all Pueblo reservations agree with the approaches used at Taos and Acoma. The Hopi, for example, use signage and the occasional closing of villages for ceremonies but do little else to manage tourists, in part because they do not want to create highly structured environments like those of Taos or Acoma.

Special, usually sacred or historic, places on reservations may be closed to tourists. Smith (1994) points out that the Hualapai tribe in northern Arizona has closed off a large part of its reservation to outsiders, who might interfere with or disrupt their traditional lifestyle. Even greater restrictions are imposed by the Havasupai, whose reservation is located in the Grand Canyon. Its main village of Supai is accessible only by helicopter, a raft on the Colorado River or a stiff 8-mile

Figure 9.6   Traditional Navajo landscape and hogan at Canyon de Chelly, Arizona
(photo by Alan Lew)

hike downhill. The village was devastated by floods in both 1990 and 1993, yet the
Havasupai chose to refuse help from non-Indians in favour of tribal members
making their own repairs – an action that perhaps reflects the sacred values attached
to their land and village.

A different but significant example of exerting control over special places is found
at Canyon de Chelly on the Navajo Reservation near the New Mexico border. The
canyon's magnificent scenery led to it being designated a National Monument by
presidential decree in 1931. It is the only National Park Service area in the continental USA that has a living culture as part of its management unit (Figure 9.6).
Until the mid-1980s, it was managed by a series of non-Native American National
Park Service superintendents and mostly Anglo park rangers. Today, the superintendent and the employees are predominantly Navajo or Native Americans from
other tribes. Some parts of the canyon floor that are used for traditional agriculture

and herding are closed to non-Navajos, and tours in other parts require a Navajo guide. The monument offers not only a place of scenic beauty but also an opportunity for people from other cultures to witness traditional Navajo culture.

Education and traditional knowledge

Control over the tribal education system is considered fundamental in ensuring cultural preservation and self-determination for reservations (Wells 1994). Efforts to achieve this have taken place at several levels. On the Navajo Reservation, there has been an increased effort to train and hire tribal members as teachers. Children in Navajo Headstart programmes normally speak in their native tongue and learn about their own culture. Although change is occurring, programmes such as Headstart are often the only means of introducing native content into school curricula (*ibid*.). The importance of education is emphasised by an official graduation ceremony where Headstart pre-kindergarteners wear academic regalia. Although the regalia are normal for US schoolchildren, the emphasis on ceremony conforms to traditional cultural frameworks. At the other end of the continuum, the Navajo Reservation's Dine Community College was the first tribal community college to be established in the USA. Its administration building and dormitories are shaped like traditional hogans: eight-sided with the main door facing east. The second floor of the administration building has an earth floor, its eight walls are covered with logs, and there are no divisions of its interior space. This floor is used for ceremonial and meeting purposes, in the same manner as a smaller conventional hogan that is also located in the campus grounds. On entering the room, movement is made in a clockwise direction in keeping with tradition. Courses at the college include those on Navajo language and culture as well as standard courses that help to prepare students for further or higher education in mainstream society. There has been a major increase in the number of reservation-based community colleges in the 1990s, as more tribes take the initiative and recognise the importance of education in their cultural preservation efforts.

Just as the Navajo nation has been a leader in offering a reservation-based higher education programme, they have recently instituted a medicine man apprenticeship programme. It may be, as Bill Donovan (1999) suggests, that this effort may be 'too little, too late', as many healing ceremonies may already have been lost. Currently, only twenty-nine ceremonies are still practised out of a repertoire that once consisted of more than 300. A further challenge is that training can take ten years and requires 'listening to chants over and over again until they are memorised' (*ibid*.: B1–2). Shonto Begay, whose father is a practising medicine man, suggested that because young people were forcibly removed from their families and placed in BIA schools, the passing of knowledge from one generation to the next was made nearly impossible and often left the younger people feeling incomplete (personal communication, 1999). Smith (1994) is more sanguine about the future of Navajo singers and traditional ceremonies. He suggests that as the Navajo tribe experiences economic development compatible with its culture, there will be increased incomes, more demand for ceremonies and, subsequently, an increase in income for singers, all of which can lead to a renewal of the profession of singer or medicine man.

However, there is no doubt that there has been a loss of knowledge from one generation to the next. There is also often a critical lack of understanding of Native American cultural backgrounds and world views on the part of educators and service workers in mainstream society. Both the Hopi and Navajo tribes, therefore, have developed cultural sensitivity programmes in an attempt to mitigate this problem. The programmes vary from living with a Hopi or Navajo family to more structured workshops where the people's creation stories are recounted, family or clan structure and obligations explained, and language and cultural characteristics that might lead to misinterpretation of Indian students by mainstream teachers are explored. Tribes are also attempting to improve recruitment and retention of American Indian students in mainstream academic institutions through grants and pledges of money. They do this because they recognise the importance of having skilled and educated tribal members so that they can reduce their dependency on non-tribal workers. Tribal workers would, understandably, be more sensitive to the perspectives and experiences of their own people. Universities in states with large American Indian populations, such as those in the southwest, also strive to increase Native American student retention by becoming more sensitive to their needs. Mentors and professors involved in American Indian programmes can assist students in balancing between the worlds of reservation and non-reservation. Some Indian professors even teach in styles they consider more in keeping with traditional ways of learning (Hunter 1999). Despite progress in these areas, a number of Native American staff in higher education wish to see more radical change. Professors and staff workers at Northern Arizona University (NAU), for example, have reported a strong need for a space on campus designated for Native Americans. They wish for a building of their own surrounded by grounds where important ceremonies and events could take place. They want the fact that they are 'different' recognised and seek to see that difference validated through the provision of what may be recognised as defensible space – a Native American centre. Some also argue that in order for Native American programmes to be effective at NAU, a critical mass of Native American professors is also needed (*ibid.*).

The growing presence of Native Americans in higher education in the USA and elsewhere has been accompanied by a rising interest in traditional Native American ecological knowledge. Kimmerer (1998: 16) points out that 'many tribal agencies are staffed with Native scientists and managers, who combine contemporary technical skills and traditional cultural values.' By teaching and sharing traditional ecological knowledge with non-Indians, tribes will gain greater respect and appreciation from these others, but perhaps more importantly they will strengthen their own self-respect and pass on this important cultural knowledge to future generations. An example of traditional ecological knowledge is the Hopi tradition of dryland farming (Figure 9.7), which makes productive use of an environment that would intimidate the most intrepid and optimistic Anglo-American farmer. At the same time, the Hopi have a sophisticated resource management programme that uses advanced technology, such as global satellite positioning and geographic information systems,

Figure 9.7 Hopi field where dryland farming is practised, Hopi Reservation, Arizona
(photo by Alan Lew)

to locate and map not only water sources, potential agricultural land, and mineral
and other resources, but also sacred sites and cultural artefacts.

Tribes and scientists often collaborate in order for the tribe to achieve specific
goals. For example, Pueblo people such as the Hopi and Zuni hire scientists to use
remote sensing (satellite images) and field work to find historic and prehistoric
trails, former agricultural sites and the ruins of greathouses. Trails are considered
traditional cultural properties, and the tribes do not want their locations or the
location of shrines, greathouses and other cultural artefacts to be known to the
public. Larger tribes have cultural advisory committees, the members of which are
usually highly placed in a tribe's religion. Members of the cultural advisory committees accompany teams of outside archaeologists and geographers on their 'field
truthing' expeditions. The work is sensitive on multiple levels, not only dealing
with relations between mainstream scientists and the tribes, but also the results of
the work may affect relationships between different tribes. Trails, for example, may
cut through contested reservation lands, such as a prehistoric Hopi trail passing
through part of the Navajo Reservation.

Land

When discussing the Indian Wars of the nineteenth century, Gribb (1992) points
out that 'in North America almost every conflict between the American Indian and

. . . advancing Europeans involved protection and preservation of traditional home-lands and sacred places.' Although today Indians avoid the Supreme Court if pos-sible due to the political views of the current Chief Justice (Trosper, personal communication, 1999), most conflicts are settled through legal action and legisla-tion. This was the case, for example, when the Taos Pueblo sought to regain control over the use and management of Blue Lake, a sacred place that, according to their belief system, is the source of origin of the Taos people and all else in the world (Gribb 1992). Regaining control of Blue Lake from the US government took 106 years, during which time they repeatedly refused offers of reparations, ending with a fifteen-year legal battle. The issue was resolved in Congress (Trosper, personal communication, 1999). In 1970, the sacred space of Blue Lake was transferred from the control of the US Forest Service to that of the Taos people, although, like other Indian lands, Blue Lake and the surrounding area is still held in trust by the US government. The struggle was not only to gain control of Blue Lake but also to 'preserve the ecosystem health of the drainage of the Rio Pueblo de Taos' (Trosper 1995: 85), which runs through Taos Pueblo. To gain control of activities affecting Blue Lake and the Rio Pueblo de Taos watershed, the Taos people had to accept a limitation on development in the area that 'restricts use of the watershed to wilder-ness' (*ibid.*). Similar efforts to regain former reservations lands managed by the US Forest Service or Bureau of Land Management have also been waged by other tribes, such as the Yavapai Apache tribe in north-central Arizona – although the process is time-consuming and is not always successful.

The Hopi people, who live on a reservation that is completely surrounded by Navajo land, have adopted a different strategy. The reservation's limited size and resources in the face of the needs of a growing population have led the Hopi to purchase four ranches that lie 75 miles south of their reservation and outside the boundaries of the Navajo reservation. The ranches were purchased fee simple, and the tribe pays state and county taxes on them. The tribe received transfer of the grazing leases on adjacent US Forest Service land, for which it pays associated fees, as well as rights to a checkerboard of surrounding lands leased from the Bureau of Land Management and the Arizona State Land Department. The tribe has also purchased a shopping and office complex in Flagstaff, Arizona. Known as Courtland Plaza, this development was purchased to generate investment income and is managed by a committee of Hopi Council members. Preliminary plans are to redesign the exterior of the buildings to make them look more like those in a pueblo. There is also a possibility of constructing a small cultural centre where information on the tribe could be provided and Hopi arts sold. Perhaps more import-antly, in addition to providing income Courtland Plaza offers a symbolic presence for the Hopi tribe in the regional population centre of northern Arizona.

Natural resource development

Allen (1994: 176) notes that: 'Natural resources development serves as both a prerequisite for and a manifestation of self-determination for some tribes' because the control of natural resource development allows greater 'consideration of tribal

Figure 9.8 Roadside business selling Navajo arts and crafts, Navajo Nation, Arizona (photo by Tina Kennedy)

lifestyles and financial needs.' This approach, which also incorporates the use of traditional ecosystem knowledge as discussed above, has close ties to models of sustainable development, especially those that incorporate cultural sustainability (Trosper 1995; see Figure 9.8). According to Smith (1994: 178):

> economic development is a means toward the end of sustaining tribal character; as such it is vital to formulate all development plans with an understanding of how they impact the overall societal makeup. Only when the individual tribe has control of its resources *and* can sustain its identity as a distinct civilisation does economic development make sense; otherwise, the tribe must choose between cultural integrity and economic development.

It is important to remember, however, that the federal government retains considerable authority over the management of the monetary and natural resources of tribes (McGuire 1994). Some researchers are deeply concerned about the potential negative consequences of resource development on reservations. American Indian reservations contain approximately 80 percent of the nation's uranium reserves, over 30 percent of its low-sulphur coal, and up to 10 percent of its oil and gas reserves (Snipp 1994). The 1934 Indian Reorganization Act was used in some instances to permit mineral extraction on reservation lands by off-reservation corporations. The value of these resources could continue to make some reservations 'internal colonies' and further stifle tribal sovereignty through economic domination by corporations and political domination by the federal government. Various strategies are therefore used by tribes in attempting to take greater control over the development of their resources. These include self-determination contracts allowing tribes to replace BIA employees with tribal employees (Trosper, personal communication, 1999); establishing tribal laws and codes to regulate business; imposing taxes on resource extraction; requiring hiring practices to give preference to Native Americans; providing and requiring job training programmes; and developing Native American expertise in business operations and conducting research (Page 1994). Tribes have also formed inter-tribal organisations such as the Council of Energy Resource Tribes (CERT) and the International Treaty Council to 'combat resource colonisation' (*ibid.*: 365).

Adopting sustainable development strategies can make it possible for tribes to adapt economic endeavours and resource management to their traditional values or world views. The traditional world view shared by many Indian communities is one of 'respect towards the world around us' (Trosper 1995: 67). Trosper sees this as having four components: community – requiring 'reciprocity in exchange'; connectedness, which describes the existing world; seventh generation – considera-tion for future generations; and humility – recognition of the power of nature. When each of these traditional values is given equal weight with the economic goals of development, then the option of 'non-development' in some situations becomes more apparent. Examples include decisions by Hopi and other Pueblo Indian tribes to ban tourists from some popular ceremonial dances, despite losing significant potential revenue; the Havasupai tribe's rejecting proposals for uranium mining on its land because it 'feel[s] that uranium mining desecrates the Mother' (Smith 1994: 184); and Taos Pueblo's agreement to a wilderness designation for Blue Lake (see above).

The rich natural resources that some reservations encompass certainly offer many opportunities for economic development, as shown by the pine forests on the White Mountain Apache Reservation. Originally, the Louisiana Pacific Timber Company harvested the forests on White Mountain Apache Reservation, but today the tribe manages its own vertically integrated timber harvesting programme. Tribal members manage the forest, harvest the trees and run the mills. The tribe also has a timber yard in the nearby off-reservation community of Show Low, where it sells its timber. Through this programme the White Mountain Apache receive all the proceeds and benefits from their resource and avoid dependence on off-reservation corporations and intermediaries. Another significant source of income for the tribe is fishing, and especially big game hunting. Fishing resources on the reservation have been less stressed than those in the surrounding national forests, and sports-men pay good money for Indian guides to take them to the best fishing locations (Figure 9.9). Even more money is paid for a tag for big game on the reservation, with elk alone costing from $10,000 to $12,000 a head, a fee that includes the ser-vices of an Apache guide. Tags are limited, and suitable hunting and grazing areas have been designated on the reservation to ensure the sustainability of this resource.

The White Mountain Apache also have an active tourism programme. They own, operate and employ tribal members in the Sunrise Ski Area and Resort, one of the more popular ski resorts in Arizona, as well as at the Hon-Dah Casino and

Figure 9.9 Signs on the White Mountain Apache Reservation indicating economic development (photo by Tina Kennedy)

Conference Center located near the popular resort of Pinetop-Lakeside, in an enclave just inside the northern boundary of the reservation. Casino gaming became an Indian reservation resource in 1988, when the US Congress passed the Indian Gaming Regulator Act (*cf.* Lew and Van Otten 1998). The first Hon-Dah Casino was built shortly after that and has since been replaced a couple of times by larger and more impressive facilities. Like any successful casino, Hon-Dah is brash and busy:

> On entering the casino my senses were overstimulated by flashing lights, bells, the sounds of slot machines. After adjusting to the noise and lights, I was drawn to the juxtaposition of stylish Apache cashiers in suits standing behind the iron bars protecting them and the casino's money from the gamblers. Behind them, on the wooden wall, hung historic portraits of tribal leaders such as Geronimo and portraits of traditional Apache women. (Kennedy 1999, field notes)

The juxtaposition of Indian icons and symbols with noisy slot machines may present a questionable example of cultural change, but equally it also suggests continuity with history and cultural values. Furthermore, the positioning of the complex in a peripheral enclave – though quite close to the nearest off-reservation population centre – makes it less intrusive into the lives of the Apache people living on the reservation. Nevertheless, because it is owned and run by the Apache, the economic benefits go to the tribe, not to outsiders.

CONCLUSION

In this chapter, we have shown how reservations have evolved from landscapes of defence originally created by the dominant culture to protect itself against the military threat of 'captive nations' (Snipp 1994) to cultural landscapes that Indian tribes are striving to defend against usurpation by mainstream culture. Reservations have become important cultural reference points – homelands to return to for those Indians who have left them – and well-defined cultural and sacred spaces. They have evolved from 'dying' places to potential wombs of renewal. There are significant social, economic and health problems on many reservations, the progeny of longstanding efforts to colonise or assimilate American Indian peoples. In recent decades, there has been a rebirth of hope, pride and independence. Although American Indian cultures are continually changing, the question need not be one of homogenisation of tribal diversity into a single 'Indian culture,' or one of complete assimilation into mainstream society. American Indians have resisted assimilation in the face of incredible pressures over the past centuries. In most cases, core values and world views have been maintained to a remarkable degree, as well as enough of the language and sacred ceremonies from which resurgence can be, and is being, successful. The extent to which this resurgence has occurred differs from tribe to tribe. Tribes that have had the most difficulty are those that were relocated to reservation lands far from their lands of origin and tribes that have totally lost their reservation lands through various federal programmes over the years. In the American southwest, however, where reservation lands frequently coincide at least partially with a tribe's original homeland, the reservations have in many ways helped to preserve important aspects of their culture. These are areas where conscious defence of the land has truly helped to achieve wider cultural and political goals.

ACKNOWLEDGEMENTS

Humanist geographers recognise the different perspectives or perceptions of landscape by 'insiders' and 'outsiders'. We are necessarily voices from the 'outside'. We offer what insights we might have with respect and humility. As such, we deeply appreciate the willingness of Tom Alcoze, Len Berlin, Ron Morgan, Scott Canty and Shonto Begay to share their insights. We extend a special thanks to Nancy Wilkinson and Ron Trosper for commenting on earlier drafts of this chapter and for their constructive criticism.

CHAPTER 10

CCTV SURVEILLANCE IN URBAN BRITAIN: BEYOND THE RHETORIC OF CRIME PREVENTION

Katherine S. Williams, Craig Johnstone and Mark Goodwin

Public-space closed-circuit television (CCTV) systems are a major presence in our towns and cities and are also critical to current crime prevention strategies. A number of authors have also positioned them as a central component of contemporary landscapes of defence (Davis 1990; 1992; Davies 1996; Fyfe and Bannister 1996; 1998; Norris and Armstrong 1997; 1998; 1999). Within such systems, CCTV cameras record the activities occurring in public spaces, and the pictures are relayed to a control room, where the images are recorded and may be watched. The images being relayed can then be used for a variety of purposes. In real time, they can inform police operational decisions as to how and whether to respond to situations being monitored. They can also be used by other agencies such as local authorities to meet objectives ranging from traffic monitoring to environmental control. The recorded data may also be used in the investigation of crimes and as evidence in court.

CCTV systems are usually justified, promoted and indeed accepted through recourse to an argument based heavily on the claim that such schemes reduce crime and fear of crime. This paper assesses this claim and suggests that most research is either inconclusive or concludes that these claims cannot be sustained. It therefore becomes difficult to view CCTV purely as a crime-control mechanism. Instead, if we are to understand fully the phenomenon of public-space CCTV, it has to be located in the context of four other processes: police, politics and financing; the changing role of the state in crime control; the politics of competition and entrepreneurialism; and popular support and political legitimation.

CCTV AS CRIME PREVENTION

In 1979, Margaret Thatcher's Conservative Party swept to power in Britain partly under a 'law and order' banner. Mrs Thatcher (quoted in Baker 1993: 32) maintained that:

> The first duty of government is to uphold the law, and if it tries to bob, weave and duck round that duty when it is inconvenient the governed will do exactly the same thing, and then nothing will be safe, not home, not liberty, not life itself.

Concern about issues involving crime lasted for the whole of the Conservatives' period in office from 1979 until 1997 and, as we shall see later, was subsequently adopted by the incoming Labour government. The recurrent rhetoric of Conservative ministers was that they would be tough on crime by providing more police, harsher sentences and more prison places. This rhetoric sat well with the electorate and with the more general theme of individual responsibility. In the area of crime it translated into the assertion that where one could blame, one could punish. It permitted a refusal either to acknowledge or to deal with any of the socio-economic factors that many researchers have associated with criminality (see Maguire *et al.* 1997; Lilly *et al.* 1995; Williams 1997). Having promised a tough approach to crime, it became politically expedient to produce results. Nevertheless, over the course of four successive Conservative administrations it became clear that more police and a tough sentencing policy were inadequate to cope with the crime problem or indeed with the public fear of crime that the law and order campaign had helped to feed. Crime rose from 2.4 million notifiable offences in 1979 to 5.5 million by 1992, while 'clear-up' rates declined from 41 percent to 29 percent over the same period.

Evidently, other tactics were necessary. Garland (1996) notes that politicians and police began to accept the normality of crime in everyday life and sought new ways of managing and learning to live with it. These included 'target hardening' through increased security, better management of personal risks, and efforts to reduce the fear of crime as a way of changing the perception that crime was a serious social problem. More specifically for our immediate purposes, two other shifts in approach were evident: namely, strategies that favoured crime prevention and inter-agency collaboration. With regard to the latter, the initial intention was that such collaboration should involve a range of public, private and voluntary sector organisations, but later there was an increasing expectation that the public would participate in local crime reduction strategies.

The emphasis placed by the government on crime prevention became more pronounced as its years in office passed, most notably after Mrs Thatcher's third election victory in 1987 (Brake and Hale 1992; Gamble 1994; Hay 1996; Jessop *et al.* 1988). The importance of prevention was first acknowledged by the Home Office in a few sentences in its 1985 annual report. By 1996, the equivalent annual report contained a full chapter on the subject (Koch 1998: 31). The continuing unwillingness of the Conservatives to accept that socio-economic conditions influenced crime rates meant that the form of prevention that found favour was situational – in other words, make something more difficult to attack or steal, largely by target hardening, or make it more likely that the offender will be apprehended if they choose criminality. From the late 1980s, CCTV increasingly became a central tool in this situational crime prevention programme.

The surveillance revolution began with private CCTV systems, using cameras to watch over banks, petrol stations, convenience stores and private car parks. As camera technology improved and as social acceptance and political backing grew, video surveillance was introduced in spaces of increasing significance – sports stadiums, shopping centres or precincts, and eventually the open streets of towns and cities. The first public-space surveillance scheme was established in Bournemouth in 1985, but the watershed for town centre surveillance came in 1994, when the Conservative government made the introduction of video surveillance an integral aspect of law and order policy. Importantly, political and ideological support was backed by a comprehensive financial package. Between 1994 and 1998, the Home Office ran four annual CCTV Challenge competitions, which together awarded £38 million to local authorities, parish councils, schools, hospitals and other bidders towards the capital costs of introducing video surveillance. These grants funded up to 50 percent of total set-up costs, the shortfall in each case to be made up by local partnerships, typically involving district and town councils and local businesses.

Allowing for this additional local funding and for schemes underwritten by the police or the Welsh and Scottish Offices, it would be reasonable to estimate that the total investment in public-space video surveillance hardware during this four-year period approached £120 million. In addition, each scheme incurs an annual running cost, which can be close to £250,000 for the most comprehensive multi-camera schemes. The extent of Home Office backing for, and reliance on, CCTV is indicated by the fact that by 1995 78 percent of the Home Office budget for crime prevention was being used to fund schemes to put CCTV in public places. By the end of the Conservative reign, 'crime prevention could be equated with CCTV' (Koch 1998: 173).

Today, cameras are a common fixture overlooking the central streets of cities, towns and villages in every corner of Britain (see also Chapter 1). Fyfe and Bannister (1996: 38) identified and mapped seventy-eight towns and cities that, by August 1994, had CCTV cameras watching over public spaces. Since then, the number of CCTV systems has increased dramatically. Using Home Office, Scottish Office, police and local government data we have calculated that in May 1999 there were 530 town centre CCTV schemes in operation or for which funding had been allocated (Figure 10.1). A comparison of the two data sets is instructive. In 1994, the schemes were mainly restricted to larger urban areas, but by 1999 CCTV had spread to encompass small and medium-sized towns. It is now the rule rather than the exception for any reasonably sized community to have CCTV surveillance of its public spaces, and Fyfe and Bannister (*ibid.*: 37) claim that CCTV cameras are now as familiar a sight in Britain's city streets 'as telephone boxes or traffic lights'.

In 1997–98 the Conservatives had intended there to be a further £15 million available for CCTV, but while the new Labour government honoured the commitment to a fourth round of the CCTV Challenge competition, it spent only £1 million on it. However, this was far from being an indication that it did not support CCTV. Rather, it showed that it wanted to see how it could integrate the technology into its wider crime-control policy. In 1999, a new CCTV initiative was announced with £50 million available in 2000 and a further £103 million expected in two future rounds. At the launch, Home Secretary Jack Straw said:

> CCTV is a very valuable weapon in the fight against crime. This £153 million programme represents the biggest ever investment in CCTV and will make a real difference to local communities . . . CCTV has a good track record and this joint initiative will help reduce the unacceptably high crime rates experienced in some areas.
>
> (Home Office 1999a)

This newest round differs from its predecessors in a number of important ways. First, whereas earlier rounds required local funds to match those made available by central government, the new round offers complete funding to successful bidders. Second, it is intended that much of this funding will support projects outside the commercial areas of towns and cities; in particular it will be targeted at crime-ridden housing estates. Third, the schemes must situate themselves within wider crime prevention and control strategies that are being undertaken locally. Yet in many important respects, the approaches of the two governments are similar. Funding was available only for CCTV. No other crime-prevention policy would receive such backing, even where it was felt that something different might be more advantageous. Moreover, partnerships were important to successful bids and in later rounds have been essential – the most important partners are the police and, to a lesser extent, the local authority. Crucially, CCTV was and is intended to reduce crime and the fear of crime. It is with this point that we now take issue.

THE CRIME REDUCTION FALLACY

The proliferation of CCTV is interesting since, in criminological terms, at least five different claims are made for it. First, it is argued that CCTV is a form of situational crime prevention; that by providing 24-hour surveillance it increases the chances of observing any criminal activity and detecting the culprits. Ideally, the resultant high risk of detection will deter any rational individual. This assumes that a rational decision is made before anyone participates in crime or disorder. Second, it is said that CCTV will reduce the 'fear of crime'. Popular belief that the cameras deter potential criminals will increase people's feelings of safety and lead them to visit the town centre more often. Third, it is claimed that heightened surveillance informs police awareness of criminal activity and helps to identify suspects. Fourth, since video footage can be used as an evidential tool, it is possible that more suspects will confess, thereby removing the need for lengthy trials. Finally, it is believed that cameras lead to more efficient use of police resources by indicating how, where and when police should be deployed.

Ministers have been particularly interested in encouraging belief in the first two – the deterrence effect and the reduction in the 'fear of crime'. Thus, in November 1994 Michael Howard, then Home Secretary, claimed, 'I am absolutely convinced that CCTV has a major part to play in helping to detect and reduce crimes and to convict criminals' (Home Office 1994b). That sentiment was reinforced five years later by Alun Michael, Secretary of State for Wales, when he said (Home Office 1999a): 'We are determined to help communities that have been hardest hit by crime and reduce the incidence and fear of crime. CCTV has a good track record in the prevention and detection of crime.' There have been numerous authoritative pronouncements about the success of CCTV. According to Peter Durham, Superintendent at Newcastle Policing Division: 'There is no doubt that the effects of closed circuit television had a massive bearing on reductions of crime and disorder in the area' (Home Office 1994a). Thus far these unqualified positive pronouncements on

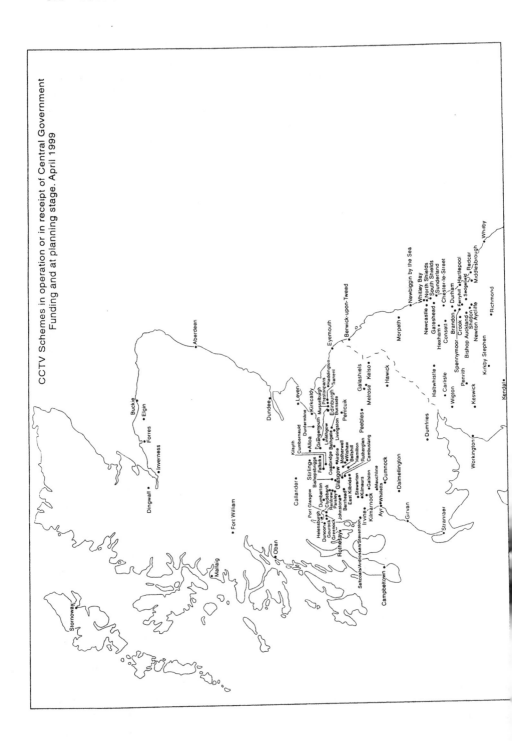

CCTV Schemes in operation or in receipt of Central Government Funding and at planning stage. April 1999

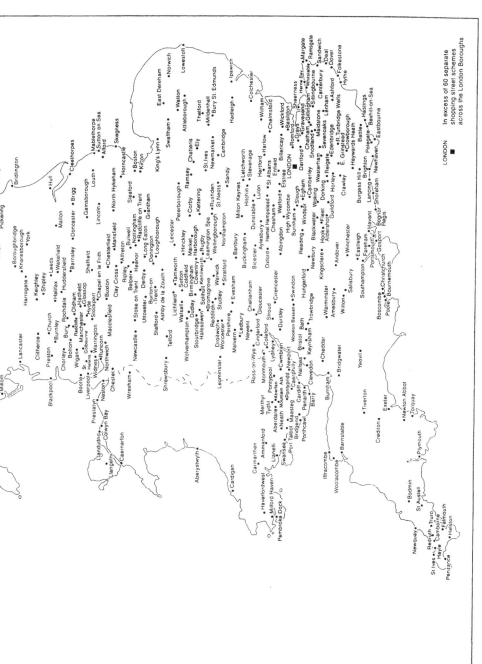

Figure 10.1 British towns and cities adopting public-space CCTV schemes, 1999

the effectiveness of CCTV are only expressions of faith, since they are not supported by research. The small amount of in-depth research available would disappoint CCTV's political supporters, since what exists is at best inconclusive about the benefits of CCTV and at worst it indicates that CCTV does not succeed.

Some of the earliest research did produce results that were mildly supportive of CCTV (Brown 1995; Tilley 1993), but this tended to study crime changes only over a very short period. As Brown (1995: 63) noted, the positive effects decrease over time and, in any case, one may see spatial displacement of crime. Similarly, Coleman and Sim (1998) referred to a displacement of crime from Liverpool city centre to Allerton, an adjacent shopping centre. A number of writers (see Brown 1995; Brown 1998; Nelson 1997; 1998) record that the preventive effects of CCTV arose only in property crime (vandalism and burglary) and had little effect on personal crime (such as assault or public order). In cases of assault its use was more to 'co-ordinate a quick response' and to direct the resulting police investigation. Short and Ditton (1996), working in Airdrie, also concluded that CCTV had a limited effect in reducing vandalism and burglary. More recently, Ditton and colleagues have evaluated Glasgow's thirty-two-camera CCTV system (Ditton *et al.* 1999). At the end of an extensive period of research involving the analysis of crime statistics and the testing of public opinion, they reached damning conclusions (*ibid.*) about the utility of CCTV as a crime prevention tool:

> Our main research finding was that, in the year following the introduction of CCTV in Glasgow [in November 1994] recorded crime rates rose effectively by 9%, and detections fell by 4%.

The use of CCTV systems in crime prevention therefore seems at the very least questionable, and the deterrent effects may be partly hampered by the fact that many people are unaware that cameras have been installed. In Ditton *et al.*'s (1999) Glasgow study, only 41 percent of those questioned fifteen months after installation were aware of the cameras.

With regard to their use to tackle the fear of crime, the evidence is similarly unconvincing. In 1996, Michael Howard could write: 'People want (CCTV) . . . because it makes them feel safer, reduces fear of crime and lets people use and enjoy their high streets again.' Nevertheless, the research findings are again unsupportive. Both Brown (1998) and Nelson (1997; 1998) question the use of CCTV in allaying the fears that women may have about personal safety. Brown (1998: 218) points out that much of women's feelings of insecurity are born out of their visibility to men, and CCTV merely heightens this visibility without necessarily adding security. Nelson, while acknowledging that some respondents felt that CCTV might be of some use in reducing some property crimes, concluded that: 'Both women and men felt the presence of CCTV would do little to reduce their fear of victimisation' and that 'people feel CCTV is as good as useless in reducing levels of violent crimes on the streets' (Nelson 1998: 17, 20). Ditton *et al.*'s (1999) Glasgow study included the collection of public opinion before and after the installation of surveillance cameras. They discovered that the number of people who would avoid the city centre actually increased after the installation of CCTV cameras – from 50 percent prior to installation to 65 percent fifteen months after installation. Both Brown and Nelson found that other solutions, such as enhanced lighting and putting more police officers on the beat, would be more likely to allay fears and persuade people to make more use of town centres. This is supported by the results of our own research, where CCTV did not gain strong support from the public as the preferred option of crime prevention (Johnstone *et al.* 1999). Only 20 percent of our respondents favoured it, and alternatives such as providing extra police officers on the beat and providing additional activities for teenagers were more attractive.

It would seem, therefore, that on the criteria most often used by government – deterrence, detection and reduction of fear – the cameras do not live up to the claims made for them. This is not to say that CCTV is without merit but that its claimed benefits, as presented to the public, are at least questionable. Even where successful in reducing crime, CCTV systems provide a very selective gaze, both spatially and socially. Unless their coverage is comprehensive, the cameras can by definition only watch some streets in some parts of some towns. Moreover, within these spaces of visibility, research (Norris and Armstrong 1997; 1998; 1999) has shown that specific types of people are singled out for observation – most notably those who are young, non-white and male and those whose appearance is considered somehow 'different'. Rather than being the neutral and democratic tool of its discursive construction, there is a risk that CCTV is being used as an instrument of social exclusion. This is especially problematic where CCTV is sited on housing estates, a practice set to increase under the Labour government, where the private lives of individuals will come under more scrutiny. In this case, the line between public and private space becomes difficult to draw and heightens concerns over invasion of privacy and other civil liberties issues. Finally, it is also worth questioning whether video surveillance is of any real utility in small towns with low crime rates, where, given this low level of criminal activity, its main role can only ever be as a tool to control public order. This function can in turn be seen as problematic, especially following Short and Ditton's (1998) finding that public order problems are often geographically displaced to back streets rather than prevented. From the point of view of its claimed benefits for reducing both crime and the fear of crime, there are clearly flaws in the arguments of those who promote CCTV, yet broad support for the technology continues. In the remainder of this chapter, it is our intention to explore why this is the case.

POLICE, POLITICS AND FINANCING

As indicated above, politicians make sweeping claims both about the state of crime and the powers of the state and police to deal with it. This political rhetoric has not been accompanied by successful policies – in the 1980s crime rates increased steadily and people commonly felt that the situation was getting worse. Yet, as Garland (1996) notes, the political claims of being able to tackle the crime problem successfully were anyway unrealistic – the police simply cannot prevent all crime (see also Bayley 1994; Johnston 1996). Recognition of this has given rise to changes in policing in order to attempt to deliver the political promises of safer communities through heightened crime control. These changes have included the embrace of partnerships with others and the use of alternative methods and technologies.

CCTV especially encompasses two of these changes in policing – partnership and the adoption of new technology. As such, its proliferation has proved important to the police but, as will be seen, it can also be argued that the police are also of critical importance in most CCTV schemes (Home Office 1999b; Norris and Armstrong 1997). During early rounds of Home Office funding for CCTV, bidders were encouraged to obtain the support of the police. The requirement for police support has slowly increased through successive CCTV Challenge competitions, and now police involvement both in the bid and as one of the partners in the scheme is a requirement (Home Office 1999b). Each of the competitions for funding has moved the CCTV schemes, and the partnerships that support them, closer to supporting the police. Hence, rather than necessarily serving the local community the systems have increasingly served the police. Assistance is provided in three key areas.

First, by literally watching criminal situations develop, CCTV can help to specify how and when officers should be deployed. In theory, by allowing more accurate estimates to be made of the number of officers needed at each incident, police should be released to patrol those areas not covered by CCTV, so ensuring greater protection for all. Should a system be used for social control and suppressing minor infringements, however, it could mean that the police are called more frequently to the area under the gaze of CCTV, leaving other spaces even less protected. This attention to minor infringements and to social control clearly happens. In our case study work we have found evidence that the police use the schemes to control non-criminal activity (Johnstone *et al.* 1999). As a local police inspector put it when discussing the use of CCTV, often people 'are not committing any offence, but their demeanour may appear intimidating. So we move them on' (*Cambrian News*, 7 October 1999). In King's Lynn, a town with one of the most extensive CCTV schemes in the country, a council official recognised that the cameras are used to deal with 'anti-social behaviour', including littering, urinating in public, traffic violations, fighting, obstruction, drunkenness and evading payment at parking meters; they are also used to intervene in undesirable behaviour such as under-age smoking and a variety of public order transgressions (Davies 1996: 176). Furthermore, the Home Office (1994a: 12) itself states that CCTV can be used to police such problems as vandalism, drug use, drunkenness, racial harassment, sexual harassment, loitering and disorderly behaviour.

Second, cameras may help the police considerably by immediately identifying key offenders in complex public order situations, so dispensing with the need for lengthy interviews and the cross-checking of statements. Third, and related, the likelihood of a guilty plea is enhanced, so saving time in court (Home Office 1994a). In King's Lynn, 98 percent of those arrested whose activities are caught on camera admit the offence, some of whom may well have a valid defence but fail to obtain legal advice before they make a statement (see also Dawson 1994). Thus even if the utility of the cameras can be doubted in terms of crime prevention and the reduction of fear of crime, much of the evidence suggests that they are effective in helping the police in both their operational and administrative functions.

Drawing on this evidence, it seems that the positive role that the cameras can play for the police is critical in accounting for the introduction and use of public-space CCTV. The police play a crucial role in supporting the scheme from the planning to the operational phase. They will advise on and choose camera locations, and others involved in the scheme will usually, without question, bow to their knowledge of the local crime situation in discussions over where the cameras are to be sited. This means the acceptance of very traditional ideas of crime and safety (see Brown 1998: 47). The police will also advise on monitoring procedures and their advice over the operation of the cameras is taken more seriously than any other information. In our research we found that public opinion strongly indicated that the main purpose of CCTV should be to enhance personal safety. Over 80 percent of those questioned wanted the systems to be manned between 10 o'clock at night and 2am, or at least in the evening. The majority also favoured the siting of cameras in quieter non-commercial locations, which they perceived to be dangerous. By contrast, the police wanted the same scheme to detect crime, mostly property-oriented, and so preferred camera locations in the commercial areas, which would be monitored in the daytime. It was impossible to meet both preferences given that resources were limited. In the event, the police argument prevailed and the cameras were sited and monitored in line with their preferences. Since the scheme has opened, however, the police advice and the decision to follow it has been seriously questioned in the light of experience. Crimes that have occurred when the cameras are not monitored have been missed, and there have been calls from business proprietors and the local council for the cameras to be operated manually, at least between 10pm and midnight every night.

The police will offer training to operators, and the Police Scientific Development Branch has produced a number of publications on the training of such staff. Often they will also offer the physical space for a CCTV control room to be located in the police station. However, actually siting the screens in the police station can incorporate CCTV into police culture, and conversely incorporate police culture into the operation of the CCTV system since the operating staff are likely to be influenced, and even directed, by the police. Although usually not police employees, such staff will watch and report on those whom they believe to be acting suspiciously – a definition that may amount to no more than appearing on the CCTV screens in the wrong place or at an unexpected time (Norris and Armstrong 1997; 1998). These processes are exacerbated if training is also placed under police management. In our research, both CCTV screens and camera operators were housed in police stations, while local authorities, the owners of the schemes and employers of the operating staff, left all operator training to the police. Thus, any understanding of the politics of CCTV is incomplete without an understanding of the often hidden role of the police in promoting, manipulating, controlling and sustaining each scheme. It also needs to be noted that the behaviour and culture of groups such as the police are difficult to control by law (McConville *et al.* 1991;

Mansfield 1993). Although voluntary codes of conduct may give some protection, the only real control can come from the other stakeholders involved in a scheme. All too often this is not forthcoming. The partnerships that are responsible for promoting and controlling most CCTV initiatives could certainly take on a useful role in ensuring the protection of the whole community. This would need active control of the system, care to protect citizens' rights and protection of privacy. These would be facilitated by careful operational guidelines, tied into both the police and local authority complaints systems (through a code of conduct) and through rigorous care in controlling the copyright in recorded tapes to prevent abuse. Further control might be maintained by the removal of CCTV screens from the police station, by the operators being employed and directed by the local authority, and through independent training. Police would then obtain access to tapes only on a need-to-see basis, and many CCTV systems do indeed have these arrangements.

From the above, it is clear that the police have perhaps most to gain from the introduction of CCTV. Its presumed ability to define precisely the who, what and where of city centre crimes can save considerably on operational and administrative police time and police resources. However, police involvement rarely extends to contributing directly to the cost of the scheme. Somewhat paradoxically the police, the principal beneficiaries, perhaps pay the least. Operation and monitoring are usually paid for and staffed by local authorities, sometimes with private funding, and the hardware is installed with the aid of a government grant and local authority or private funding. In this climate, it is unsurprising that the police are operating assiduously to ensure development and acceptance of the schemes and, in an era of declining resources, they view CCTV as essential in maintaining effective policing in many towns and cities.

THE CHANGING ROLE OF THE STATE IN CRIME CONTROL

From the inception of CCTV funding, central government has made it clear that bids to its grant schemes must come from local partnerships involving, at a minimum, a local authority and the police, but preferably other agencies drawn from both the public and private sectors. This redistribution of crime reduction responsibilities from the central state to other actors has evolved rapidly during the last two decades. Indeed, this is only the latest development in a relationship between the state, crime control and the police that has become increasingly complex. Throughout most of the twentieth century, it was widely accepted that the central state was responsible for the control of crime and for protecting its citizens from criminal activities. This responsibility was willingly accepted, and until the late 1960s governments of all political colours were confident that they would win the 'war against crime' (Garland 1996). The state carried out this function through what Garland (*ibid.*) called the 'penal-welfare complex', comprising the police, criminal courts, the prison service and the welfare state. Since the mid-1980s, this line of state responsibility has been eroded or, at the very least, complicated.

Faced with the evident failure of increasingly costly law and order policies, the Conservative governments of the 1980s accepted that a shift to a more affordable and realistic strategy was desirable. Consequently, criminal justice priorities were refocused. Ministerial speeches and government publications made it clear that the state could not and would not accept sole responsibility for crime prevention. This process began as early as 1983, when Circular 114/1983 from the Home Office (1983) informed police forces and authorities:

> The potential demands on the police are such that they could not be met solely by increases in police strength, even if substantial additional resources were available. It is necessary to work in co-operation with other public services, voluntary bodies and the public themselves.

This was developed the following year in another circular, which stated that 'every individual citizen and all those agencies whose policies and practices can influence the extent of crime should make their contribution' (Home Office 1984a). Both statements show a desire to widen responsibility and action with regard to the management of crime. Later, more stress was laid on the concepts of community partnerships and active citizens (see Brake and Hale 1992). The partnerships that resulted meant that the blame for any future rise in crime could be placed on a range of agencies and individuals and would not have to be shouldered by the government alone.

As the main providers of local services, local authorities were obvious candidates to be included in any crime-prevention partnerships (Bright 1991), but there were considerable tensions at the time between local and central government (Duncan and Goodwin 1987). Often, especially in Labour-controlled metropolitan areas, other partners were found or created – from the private sector (often through city centre partnerships as in Liverpool), the probation service or the voluntary sector. Important local partnerships emerged from the Safer Cities Campaign (Tilley 1992; Sutton 1996) and from urban development corporations (Duncan and Goodwin 1987), both being used partly to bypass local authorities in politically sensitive inner urban areas. Hand in hand with inter-agency partnerships, the police engaged with responsible citizens through schemes such as Neighbourhood Watch (Husain 1988; Evans 1992; Pease 1992), mediation and conferencing (Wright 1995), and special constables.

The system of special constables is an old idea that gained added impetus following its incorporation into Michael Howard's 1994 Partners Against Crime campaign. In 1996, Howard launched the ill-conceived Street Watch, which encouraged people to 'walk with a purpose' around their neighbourhood. Hughes (1997) and other critics expressed concern that vigilant patrolling could degenerate into patrols by vigilantes. Koch (1998: 38) notes that there was a distinct change in the Conservative Party's discursive construction of partnership in the early 1990s. Between 1984 and 1992, partnerships were to be multi-agency and inclusive (see Morgan Committee 1991). By 1994, the definition had narrowed to individuals helping the police, with the role of agencies being downplayed. The election of the 1997 Labour government saw the meaning and structure of crime-control partnerships shift again. The Crime and Disorder Act 1998 requires local government, the police, the probation service and health trusts to engage in crime and disorder partnerships and also place a duty on local authorities to audit and manage community safety (Home Office 1998).

It may be argued that the reliance on other agencies in the fight against crime is part of a larger transition. Pavlich (1999: 103–4), for example, argues that 'the rise of community crime prevention is not just a slight adjustment to political technique; rather, the very notions of who is governed, who governs and what governance entails are in the process of significant revision.' This revision is not unique to criminal justice, and authors across the social sciences (Garland 1996; Crawford 1997; Smandych 1999) have identified an increasing shift from government to governance over a wide range of policy areas. According to Stoker (1996: 2–3), this shift is underpinned by an increase in 'governing mechanisms which do not rest on recourse to the authority and sanctions of government.' Instead, we are witnessing 'the development of governing styles in which boundaries between and within public and private sectors have become blurred' (*ibid.*).

Thus, the state guides policy, while others are responsible for delivering it. As Kelsey (1993: 78–9) notes, it entails 'the government retaining power over essential resource and policy decisions, while delegating delivery to the voluntary or private sector.' Osborne and Gaebler (1992) liken the concept of governance to that of a boat, where central government acts as the cox 'steering the boat' (through accounting, financial and management procedures), while others row (i.e. deliver the services). While superficially illuminating, this analogy is rather simplistic, and it overlooks a number of factors. First, there are a range of different interests actually on board the boat, and individual 'rowers' may well be concerned only with those of their own organisations rather than with those interests that might collectively be described as the 'general good'. Second, and arising from this, the government may well be dealing with a number of different 'service providers' (or rowing partners) in the delivery of any one service, such as crime control, to any particular area. Not all of these will necessarily be pulling in the same direction. The government must try to resolve these tensions through its overall steering and control. For Hill (1996), the role of the state in a governance system is to give leadership; build or help to build partnerships; and promote opportunities for citizen participation and for the involvement of other partners in service delivery. In contemporary Britain, even more complexities need to be considered. Devolution in Wales and Scotland means that the steering control and therefore the policy targets set for each of the organisations in any given partnership may be different, widening the likelihood of policy tensions. The result may be a plethora of 'service providers', each playing to different policy requirements. In fact the complex web of actors and partners that results renders it nearly impossible to define who is responsible for errors and policy failure (Goodwin 1998).

Despite variations in the composition of local partnerships, the rhetoric widening responsibility for crime control, as in many other policy arenas, is that of 'empowering' local communities. Such rhetoric is not confined to Britain. In the case of New Zealand, for example, Pavlich (1999: 120) commented:

Here, a dubious managerial 'freedom' is granted to both governors and the governed; they are invited to locate themselves in 'communities' and to choose crime prevention programmes they deem appropriate. This freedom is practically 'managed' through devolution, or decentralizing crime prevention initiatives to a fragmented 'community' space, whilst retaining control over important funding decisions and monitoring the entire process through protocols, accountancy and evaluation.

In its widest application, this might involve empowering local communities to make all their policing and crime-control decisions within funding controls and monitoring guidelines set by central government (Brogden and Shearing 1993).

In Britain, the process has not been as 'free' as this, although claims are regularly made that the empowerment of communities is taking place in a wholly healthy manner (Pavlich 1999; Morgan Committee 1991). In his introduction to the Home Office Guidance Notes on the Crime and Disorder Act, for instance, Home Secretary Jack Straw stated that the Act was 'the culmination of a long held ambition to empower local people to take control of the fight against crime' and that local people should be 'invited to participate actively in the process of tackling local problems, not just passively consulted about them.' Yet there are many problems that have not yet really been resolved, including, at a fairly fundamental level, what is meant by the term 'community'. The language of the Crime and Disorder Act 1998 seems to imply a geographical concept, set by local authority boundaries, but geographical demarcation does not necessarily correspond to, or even represent, other ideas of community that rely more on sociological concerns of inclusion and cohesion (Currie 1988; Gardiner 1995; Crawford 1995; 1997; Williams 1997: 484–5; Gilling 1997). Moreover, the establishment of partnerships does not guarantee democratic participation. In many instances, the groups who become stakeholders in a partnership, and those individuals who represent these groups, will not come from those sections of society most likely to be the target of criminal justice policy. There is thus the potential for greater rather than less social exclusion (Norris and Armstrong 1997; 1998). Moreover, the partnerships that have been formed in the past have often been more rhetorical than real (Garland 1998). In our research on CCTV, we found that the partnerships that manage the schemes do so in a very informal manner (Johnstone *et al.* 1999). The local authority provides money and personnel, while the police provide premises for monitoring the cameras and largely use the technology to fulfil their own policing strategies. Other agencies, such as those from the voluntary sector, and more local levels of government, such as town or community councils, are largely excluded. Both of the principal partners are fulfilling the formal requirements placed on them, but neither is really working in a true partnership.

In spite of claims that there has been a move away from central state responsibility for the control of crime and for protecting its citizens from criminal activities and despite the existence of some new strategies, the move from government to governance in this area of policy has been neither clean nor clear. The relationships emerging on the ground are very complex and the lines of responsibility and action increasingly blurred (Garland 1996; 1997; 1998; 1999; Stoker 1996; 1997; Crawford 1997; Wilson and Charlton 1997; Osbourne and Gaebler 1992; Rhodes 1996; Johnston 1996). Certainly in the case of crime control, the partnerships are not firm and stakeholders do not always operate as true partners (Garland 1998; Crawford 1997). Through its control of both policy and resources, central government has retained rather than lost power, and although it may have handed over the responsibility for delivering some services, it is still responsible for the overall 'government of governance' (Garland 1998; Crawford 1997; Goodwin 1998). There is still the need to ensure that clearly defined roles are set out for both local and central agencies in terms of responsibility and accountability for crime control. This will apply to the implementation and control of CCTV schemes as much as to any other crime-control strategy, and it will mean that we cannot ignore issues of governance and state restructuring when considering the ways in which such schemes work.

THE POLITICS OF COMPETITION AND ENTREPRENEURIALISM

The move from government to governance has a wider context. The increasingly global nature of both trade and employment has led to fierce competition for the location of manufacturing, service and consumption activities. As Mitchell (1997: 303) observed: 'capital is able to behave ... like a plague of locusts circling the globe, touching down hither and yon, devouring whole places as it seeks even better comparative advantage.' This has intensified competition between places at both international and national scale and, in such a climate, towns and cities have to compete at a local and regional level for investment in production and consumption facilities (Gold and Ward 1994). An especially vivid illustration of these trends is found in the world of retailing. Increasingly, town and city centres have become locked in battles for trade with each other, with out-of-town shopping centres, and recently with regional shopping malls to maintain their levels of retail activity. Faced with intense competition for consumers, those responsible for planning and developing city centres have moved to create an environment that is attractive to consumers and businesses alike. Their goal, Mitchell argued, is to create 'a legitimate stay against the insecurity of flexible capital accumulation' (1997: 316).

It is increasingly argued that a necessary part of such strategies involves bringing the perceived safety of the enclosed shopping mall into the realm of the public street. CCTV has been a critical element in this process. It is used as a key part of a package of investments designed to attract and sustain commercial investment. Initiatives are pursued that are felt to contribute to a reduced sense of risk or danger, and CCTV is increasingly seen as central to what might be termed the politics of urban revival and competition. Here CCTV is viewed as a technology capable of rendering a space 'safe' both for individuals *and* for investors. Writing with regard to Glasgow, for instance, Ditton *et al.* (1999) note:

> As well as eliminating crime and the fear of crime, the cameras were also supposed to increase annual inward investment to the city by £43 million a year, generate 1500 new jobs, and bring an additional 225,000 visitors each year.

We would argue that the proliferation of CCTV has been partly fuelled by the desire of one town or city to emulate a rival. This constant competition for investment and consumer confidence leads to an escalation in both the nature and the number of CCTV schemes as places argue that they require a better and more effective system than their neighbours. In this environment the spread of CCTV is almost impossible to halt, for crime control may become a secondary consideration, or at least one that has to be viewed as part of a wider strategy of place competition (Taylor 1996; Bannister *et al.* 1998). Even when the areas to be rendered 'safe' are housing estates rather than city centres, as is the case with Labour's CCTV initiative, the link drawn between CCTV and regeneration is still very strong. As the Home Secretary (Straw 1998) stated:

> Crime and the fear of crime are a blight on some of our most deprived communities. Tackling crime is central to the Government's wider strategy of regenerating run-down areas. This scheme will improve safety and security for the people who live in these estates.

In such strategies, the fact that CCTV does not deliver crime reduction or ease the fear of crime does not necessarily matter, since it is being used more as a tool to stimulate economic investment.

As we have noted, part of the reason for using CCTV in this manner is to emulate the perceived safety of the mall, or private shopping area, on public streets (Goodwin *et al.* 1998a,b). The mall is a private space into which all appear to be welcome but in which there is 'an overt security presence that reassures preferred customers that the unseemly and seamy side of the real world will be excluded' (Goss 1993: 27). This necessitates the exclusion of any individuals who do not fit in with the commercial ideal themselves or whose presence may discourage others from feeling secure. This especially applies to those non-consumers who do not have the privileges associated with 'credit-card citizenship' such as teenagers (especially those belonging to ethnic minorities), drunks, beggars, street traders and the homeless (Judd 1995; Jackson 1998; Norris and Armstrong 1997; 1998; Bannister *et al.* 1998). Those who are made unwelcome will be become keenly aware that they are watched, even if nothing else is done. It is hoped that they will thereby feel suitably uncomfortable and eventually choose to exclude themselves from these increasingly 'purified' retail spaces (Jackson 1998). This decision to avoid certain spaces may technically be their own, but in practice it has been imposed by the operation of a supposed crime-control measure. Such processes of exclusion have led some urban commentators to speak of the 'fortress city' (Christopherson 1994; Davis 1990; 1992). In such cities the white middle-class citizen enjoys unfettered access to the homogenised spaces of consumption and leisure, which, according to Davis (1990), is directed with 'behaviourist ferocity'. Those less fortunate are increasingly forced to live and shop elsewhere. The presence of CCTV plays a key role in enforcing these divisions, and those who are deemed to be different are increasingly ghettoised. This in turn has led to an emergent geography of purified and impure spaces, resulting in what Christopherson (1994: 421) has labelled as 'the piecemeal spatial mosaic of the safe and the unsafe.'

The critical point is that safety may not have actually increased, but danger is less evident. The desirability and possible use of CCTV to achieve these ends has been furthered by the policy of the Home Office (1994a: 12–14), which sees CCTV as a legitimate means of dealing with the 'problems' of public order such as groups loitering, disorderly behaviour, drunkenness, or those whose behaviour is merely suspicious. In many towns and cities, those who do not fit the perceived notion of acceptable consumers, often the homeless, are identified by CCTV and then moved on, or arrested for minor transgressions such as begging. In turn, the careful use of bail conditions can result in these individuals being excluded from the town centre in future. The demands of the business sector often necessitate the siting of cameras in the main commercial streets of towns and cities so as to facilitate this process. Alternative sites, such as on children's playgrounds or near schools, are less often considered as suitable for the gaze of the cameras, although it is here that the local community will often want the cameras in order to feel 'safer'. Moreover, the cameras will be largely monitored during retail opening hours, rather than in the evening when most people feel afraid, and in this way they will be used more to control order and 'improve' the feel and look of a place rather than to control crime. Through these uses of CCTV, the public spaces of our town centres are increasingly preserved for the consumer citizen, while those whose spending power is low, or who *look* as if their spending power is low, are effectively excluded. Thus to view the technology of CCTV solely in terms of crime control would be to miss its role as a central component of local development strategies that have at their core the creation of 'safe' and purified retail environments.

POPULAR SUPPORT AND POLITICAL LEGITIMATION

There is also a further political context within which CCTV needs to be seen. In the immediate postwar period, there was a political consensus in Britain that a penal welfare approach to the crime problem could and would deliver results. In 1970 Edward Heath, then Prime Minister, brought this to an abrupt end when he turned law and order into a party political issue. Partly through carefully constructed speeches and partly through press articles, his Conservative government success- fully constructed a series of 'folk devils', such as criminals, scroungers, trades unions and marauding youths. These, it was argued, needed to be controlled by laws notionally designed to deliver an ordered and crime-free society. Due to other political imperatives in the early 1970s, little was done to convert this law and order rhetoric into legislation, but as Hall *et al.* (1978: 78) noted:

> the law and order campaign of 1970 had the overwhelming single consequence of legitimating the recourse to the law, to constraint and statutory power, as the main, indeed the only means of defending hegemony in conditions of severe crisis.

By the end of the decade the newly elected Prime Minister, Margaret Thatcher, was ready to use such legitimation in order to unravel the postwar consensus. Her governments' policies simultaneously aimed to permit the free market to flourish in economic terms while using a strong state and draconian laws to preserve the 'traditional British way of life', to keep the streets safe, to control the 'dangerous' classes and to protect property (Gamble 1994). The media were increasingly used to persuade the public that not only did the 'folk devils' still exist but also that their ranks were expanding almost daily to include terrorists, 'subversives', single mothers and the homeless (McRobbie 1994), so legitimising increased recourse to authoritarian state action. In playing on the electorate's fears and by offering common-sense solutions – more police, harsher punishments – the government gained popular support for its hard-line law and order strategy, generating what Hall (1980) termed an 'authoritarian populism'. The press was central to the maintenance of consent. Koch (1998: 175), for example, concludes that Home Office policy decisions were 'political and irrational' and 'rarely based on research but on what [was] thought by the individual minister to be popular with *The Sun* readers.' The media construction of public opinion was then often used by those in power as 'impartial evidence of what the public believes and wants' (Hall *et al.* 1978: 63). Hall (1980: 4) adds:

> By . . . first informing public opinion; then, disingenuously, consulting it, the tendency to 'reach for the Law', above, is complemented by a popular demand to be governed more strictly, from below. Thereby, the drift towards law and order, above, secures a degree of popular support and legitimacy amongst the powerless, who see no other alternative.

The press welcomed its central role: crimes (particularly violent crimes) are sellers of newspapers. Readers easily understand political messages attacking crime. These stories feed on and are fed by fear and unease. Law and order politics and media presentation of crime issues have therefore gone hand in hand, and policies have had to become more and more stringent to answer the call of press and public for something to be done.

In 1993 Michael Howard, from the right wing of the Conservative Party, became Home Secretary. His appointment was made at a time when the government was deeply divided, was behind in opinion polls and perceived to be lacking in direction. He used the political opportunity to implement a number of new authoritarian initiatives, which quickly gained media attention and brought law and order issues back to centre stage. According to *The Guardian* (23 October 1993):

> the country and the party needed a new fix from the Conservatives that couldn't be Europe – or the economy. Law and order fitted the available space because there is a genuine problem (or multiplicity of problems) which unites and enthuses the party . . . So although the primacy of law and order was partly a response to some real incidents, trends and issues, it was also essentially a contrivance.

Howard's policies were populist in the extreme (for example, American-style 'boot camps' for young offenders) and can be seen as an attempt to show the Conservatives to be the genuine party of law and order, and to generate political legitimation at a time of crisis. To avoid being cast as a party soft on criminals, Labour could not afford to attack the government's seemingly common-sense policies. Throughout 1995 and 1996, Howard and Jack Straw, then the opposition spokesman on home affairs, continually raised the stakes with regard to who could produce the most authoritarian rhetoric:

> Mr. Straw's main ambition – sometimes, it seems, his only intention – is to prove he
> wields a thicker baseball bat than Michael Howard . . . Whenever Howard pops up
> with a 'crackdown', Mr. Straw's standard response it to say that it was his idea first
> and, what's more, his crackdown would be much harder. (*The Guardian*, 2 June 1996)

It is within this context that CCTV funding began. From the outset ministerial rhetoric presented video surveillance as a common-sense solution to the crime problem,
one that everyone should want, and indeed welcome, for their town. Any criticism,
regardless of its foundation, was attacked. In 1996, the Home Office minister David
Maclean vehemently attacked early CCTV evaluations, which had suggested that
cameras might not reduce crime (Home Office 1996). He stated that: 'recent misleading reports try to criticise CCTV. They're rubbish. CCTV is working – the
police and everyone who lives and works in areas with CCTV knows it reduces
crime.'

   This forceful appeal to some form of common knowledge – 'everyone knows it
reduces crime' – rather than to evidence-based research was maintained across a
front that included government ministers, the opposition, the police, local authorities
and the media. When coupled with the public's fear of crime, and their intolerance
towards difference, it formed a very powerful ideological legitimation of CCTV.
Again the acceptance and proliferation of CCTV should be viewed as one part of a
much broader process of political and ideological legitimation, which helped to
underpin the rise and development of an 'authoritarian populism'. Such a strategy
has proved to be very strong, enabling policies such as CCTV to survive the change
of government in 1997. Certainly in terms of law and order, Tony Blair's famed
promise when Shadow Home Secretary to be 'tough on crime and tough on the
causes of crime' has been realised in his premiership. This in turn has led to a
convergence rather than a divergence between the main political parties in terms of
law and order policy. CCTV is now being even more extensively funded under a
Labour government than it was under the Conservatives.

CONCLUSION

The central thrust of this chapter has been to consider critically the grounds on
which CCTV has been so extensively introduced in towns and cities across Britain.
The justification for such proliferation is almost always in terms of crime prevention
and the reduction of fear of crime. In assessing this justification, it is notable that
existing research lends little support to the claims that CCTV contributes significantly to either crime reduction or to reducing fear of crime among the public. This
is not necessarily an argument against CCTV, but it does suggest a need for much
closer scrutiny of official pronouncements, and also for a recognition that other
factors have been serving to drive the extension of public surveillance.

   The paper identifies four such influences. The first is that the police have promoted and supported CCTV for reasons other than crime reduction. They have
benefited both financially and operationally from CCTV, and we would claim that
these advantages should be recognised as drivers of police support just as much as

the claim that CCTV reduces crime. Indeed, the police have also taken the opportunity offered by the rise of CCTV partnerships to abdicate responsibility for both civil liberties violations and failures in crime control that may result from CCTV use. Responsibility for these faults is passed on to the operators, who are not normally employed by the police, or directly on to other partners. Thus the police have been keen supporters of CCTV and must be included in any explanation for its proliferation – but not necessarily for the stated reason that it reduces crime. The hidden benefits enjoyed by the police are just as significant in accounting for their support.

A second influence has been the fact that with the move from government to governance, responsibility for crime control and crime prevention is no longer the sole prerogative of the police. Other agencies, at both local and central levels, now have a stronger voice in crime-control issues, and this voice has increasingly spoken on behalf of CCTV rather than any alternative strategy.

Third, the rise of CCTV can only be understood against the background of increasing competition and entrepreneurialism at the local level. Cities and towns now incorporate CCTV into their development packages, and its role in supposedly securing 'safe' environments to live, work and shop is seen as critical.

Fourth, all of these influences need to be seen in the context of the 'great moving right show' (Hall and Jacques 1983), in which increasingly authoritarian policies were not only legitimised but also accepted. Britain stands out in this respect from every other Western democracy, and public-space CCTV has gained a greater level of acceptance here than anywhere else. Indeed, the proliferation of CCTV in other countries has been checked by public concern, especially over issues of civil liberties. Not only has Britain seen little outright opposition to CCTV, but it has also witnessed very high levels of positive support. We would contend that this can only be understood in the context of three decades of political and ideological support for popular authoritarianism.

Hence, while CCTV may not reduce either crime or the fear of crime in the manner claimed by its proponents, we would argue that there are other, very good reasons for expecting it to continue to be a major part of our 'landscapes of defence' well into the foreseeable future.

CHAPTER 11

URBAN DESIGN APPROACHES TO SAFER CITY CENTRES: THE FORTRESS, THE PANOPTIC, THE REGULATORY AND THE ANIMATED

Taner Oc and Steven Tiesdell

This chapter focuses on urban design approaches to creating city centres in which people feel safer. Contemporary urban design has broader concerns than simply the physical design of public space, the physical public realm. It is also concerned with what has been termed the socio-cultural public realm – the activities that occur within that space (Oc and Tiesdell 1997). It is therefore intimately concerned with the design and management of public space and the various processes that create city centre landscapes. City centres represent possibly the last significant concentrations of universally accessible, urban public space where people of different classes, races and cultures can meet (Tiesdell and Oc 1998). They are places of social exchange, transactions and interaction between people. Feelings of personal safety are prerequisites for a vital and viable city centre; if a city centre is not perceived to be safe, those with choice will decide not to use it, making it less safe for those with fewer choices. Hence, there is an important social justice dimension to efforts towards making city centres safer. In Britain, the recent Crime and Disorder Act (Home Office 1998) has placed a statutory duty on local authorities and the police to prepare local crime and disorder reduction strategies. Such strategies will inevitably have urban design dimensions (see also previous chapter).

This chapter identifies and discusses four urban design approaches to making city centres safer. Before presenting them, some background is necessary in order to place the approaches within a crime and safety context. The first section of this chapter therefore discusses crime and safety and measures to tackle crime and fear of victimisation. The next sections discuss the four approaches, while the final section discusses the regulatory and animated approaches within the context of the city centre of Leeds, in the north of England.

CRIME AND SAFETY

Considerations of safety are related to, but distinct from, concerns about crime. Crime is about offenders and offences, while safety is about victims and the fear of victimisation. Although both crime and fear of victimisation manifest themselves in a variety of deleterious ways, the overarching effect is a reduction in the quality of life for a city's inhabitants. In discussing crime and safety, two distinctions must be made.

The first is the distinction between crime and incivilities. The general definition of a crime is a transgression of a formally constituted law and thereby prosecutable as an offence; some crimes are injurious to people and some to property. Nevertheless, much of the conduct and behaviour deterring many groups from using city centres is not technically a crime. These are the host of offences, usually termed disorder or incivilities, that provoke anxiety and apprehension. They are sometimes referred to as 'quality of life' crimes. Jane Jacobs (1961: 39), for example, aptly referred to them as 'street barbarism'. A distinction should also be made between *social* incivilities and *physical* incivilities. The presence of vagrants and down-and-outs, for example, is a physical incivility, but 'aggressive begging' is a social incivility (it may also be a crime). A further distinction between crimes and incivilities is that for the latter there may not be an active offender. Women are often disproportionately victims of incivilities and suffer greater distress as a result of them (see Trench *et al.* 1992). They are also more sensitised to environmental cues and more likely to take appropriate precautions (Riger *et al.* 1982). Wekerle and Whitzman (1995: 4), for example, observe that many women spontaneously identify what criminologists call 'hot spots' of predatory crime and fear.

A second distinction is between *fear* and *risk* of victimisation, in other words between 'feeling safe' and actually 'being safe'. The subjective assessment of actual risk is often distorted and exaggerated. Thus, fear of victimisation may also be out of proportion to the risk of victimisation, but fear of victimisation should not be pathologised as 'neurotic' or 'irrational'. It is an inherently complex phenomenon that has a number of different dimensions. These include, for example, vulnerability and perceptions of personal risk; reputations attributed to particular places; confidence in the police and other public guardians; and particular environmental cues and conditions, including the presence of certain people (Box *et al.*1988; Oc and Tiesdell 1997: 33–8). Perceptions can also come from many sources: the reporting of crime in the local and popular press, for instance, can often be responsible for the persistence of mis- or half-truths in the popular imagination. In general, women are more fearful of victimisation than men. While men never become as fearful as women, as men grow older the gender–fear gap narrows (Box *et al.* 1988).

In terms of the impact of crime on urban life and urban living, the fear of crime is often more important than the actual statistical risk. Those most at risk statistically in the UK are young males, but those who exhibit most fear of victimisation are women, the elderly and ethnic minorities. Arguably, one explanation for this is that vulnerable people tend to be more cautious and risk-averse, and therefore take precautionary measures, including placing curfews on themselves. Hence, as a direct result of not putting themselves at risk, they are less well represented in terms of victimisation. In response to fears of victimisation, many people take precautionary actions either to avoid the risk or, where risk avoidance is not possible or desirable, to reduce their exposure through risk management (Oc and Tiesdell 1997: 38–40).

Another strategy is to practise denial. As many people, especially women, limit their activities due to fear of victimisation, fear is as great a problem as crime itself. The *Home Office Standing Conference Report on the Fear of Crime* (Home Office 1989), for example, emphasised that fear of crime can be as instrumental as actual victimisation in lowering the quality of people's lives. This report defined a four-stage spectrum of feelings about victimisation – complacency, awareness, worry and terror. The first level, *complacency*, might entail a failure to take appropriate precautions. The second is a healthy *concern* and *awareness*, entailing the adoption of realistic preventive and precautionary measures whereby the management of crime risk is successfully integrated into daily life. The third is *worry* and *fear*, where there is a preoccupation with harm and danger, leading individuals to adopt perhaps unnecessary measures to protect themselves, which can affect their quality and richness of life. The fourth is *terror*, leading to an obsession with crime and a total disruption of life.

Avoidance of city centres through fear of victimisation is a consequence not only of fears of certain incidents but also of certain environments. Physical and social disorder or incivilities, such as graffiti, litter, broken windows, vandalised public property, vomit and urine in shop doorways, drunks and beggars, for example, signal an environment that is out of control, unpredictable and menacing (Painter 1996: 52). Many people are apprehensive about or fearful of certain parts of city centres, such as pedestrian subways, dark alleys, and areas that are deserted or crowded with the 'wrong kind of people'. Many are also disturbed by situations that restrict choice or offer no alternatives: for example, subways as the only means of crossing busy roads, narrow pavements or constricted entrances, particularly if these are obstructed by 'people who create anxiety' (winos, beggars, indigents, rowdy or drunken youths).

TACKLING CRIME AND FEAR OF VICTIMISATION

Crime-prevention measures tend to focus on the actuality of crime rather than on fear of victimisation and on crimes rather than incivilities. Measures to address crime can focus on any or all of the three 'minimal elements' of all direct-contact victim crimes – a motivated offender, an opportunity, and a target or victim. Removing any one of these elements can prevent a crime. Beyond their taking reasonable and appropriate precautionary measures, it is undesirable that measures focus on the victim, since such actions diminish their quality of life. Hence, measures tend to focus, first, on reducing the offender's motivation or disposition for crime and, second, on the opportunities for crime to occur. There is a continuing debate regarding which is the more effective (see McLaughlin and Muncie 1996). In theory, reducing the offender's motivation to offend is innately superior to opportunity reduction but, because it is often difficult to do, opportunity-reduction measures can be justified on practical grounds. The best-known opportunity-reduction approach is the 'situational approach' (Clarke and Mayhew 1980; Clarke 1992). Initially codified by Hough *et al.* (1980), it was revised and extended by Clarke (1992). As shown in Table 11.1, the main thrust of the approach is that *once* the offender has made the

Table 11.1 Opportunity-reduction measures

More difficult
- *target hardening* i.e., increasing the difficulty of the offence
- *access control* i.e., admitting only those with legitimate purpose
- *deflecting offenders* i.e., channelling potentially harmful behaviour in more acceptable directions
- *controlling facilitators* i.e., removing the means to commit crime

Increase risk
- *entry/exit screening* i.e., increasing the risk of detecting someone who does not have authority to be where they are
- *formal surveillance* i.e., by staff whose primary responsibility is surveillance and control
- *employee surveillance* i.e., by staff who work in the public realm
- *natural surveillance* i.e., by people going about their normal activities

Reduce reward
- *target removal* i.e., removing the target from the potential crime scene
- *identifying property* i.e., reducing the resale value of property
- *removing inducements* i.e., reducing the material gain or 'thrill' of the criminal act
- *rule setting* i.e., reducing the ambiguity between acceptable and unacceptable behaviour

Source: Clarke 1992.

initial decision to offend (i.e. has become motivated), then the techniques make the commission of *that* crime in that *particular place* more difficult. The opportunity-reduction methods make the offence more difficult, riskier for the offender and/or less rewarding for the offender.

Opportunity-reduction approaches have been criticised on a number of grounds. First, they raise concerns about the image presented by such measures and the ambience of the resulting environment. Second, they raise concerns about repressive social control and the infringement of civil liberties. Third, and perhaps most challenging to claims about their effectiveness, it is argued that rather than reducing crime, such measures simply displace it. Although these criticisms do not necessarily invalidate the approaches, they expose some of their limitations and must be borne in mind. While the first two criticisms are discussed below in the review of the fortress and panoptic approaches, it is necessary to discuss displacement here.

Opportunity-reduction measures inevitably pose the possibility of displacement. Commentators who emphasise motivation as the most important element of the crime event tend to argue that restricting the opportunities in one location merely redistributes crime. The degree of displacement is, however, likely to correlate with the availability of alternative targets and with the offender's strength of motivation. Arguably, 'opportunistic' criminals are more likely to be deterred, but 'professional' or instrumental criminals may be spatially displaced or resort to other means to achieve their aims. As displacement may take place in different ways, many now accept that conclusive demonstrations of the absence of displacement are extremely elusive and even impossible to demonstrate. Nevertheless, an inability to detect displacement does not mean that it is not present, and the possibility of displacement can never be precluded by research. However, it is probable that displacement or deflection will serve to dissipate at least some criminal energies and motivation.

Displacement is not a compelling argument against opportunity-reduction measures, particularly if, as planners, urban designers, landscape architects or town centre managers, we are concerned about particular places. Barr and Pease (1992), for example, have usefully distinguished between 'benign' and 'malign' displacement. Benign displacement involves a less serious offence being committed instead of a more serious one, an act of similar seriousness being moved to a target or victim for whom the act has less damaging consequences, or even a non-criminal act instead of a criminal one. Malign displacement involves a shift to a more serious offence or to offences that have worse consequences. Barr and Pease (*ibid.*: 199) also prefer the term 'crime deflection' to crime displacement, because it focuses on the *positive* achievement of moving a crime from a particular target.

URBAN DESIGN APPROACHES TO SAFER CITY CENTRES

This chapter is primarily concerned with *feelings* of safety in the public spaces of city centres and with urban design approaches that may make people feel safer when using city centres. It should therefore be recognised that actions to reduce the incidence of crime may be different from those to tackle fear of victimisation. Equally, however, one aspect of reducing fear of victimisation may involve bringing fear of victimisation more in line with the actual risk of victimisation, and reducing the incidence of offences may also have an effect on the fear of victimisation. Nevertheless, while approaches to safer city centres should reduce the actual incidence of offences, they should also aim to change people's perceptions of their environments. By doing so, they should encourage a greater sense of confidence and a corresponding diminution of the fear of victimisation, such that people are not forced to impoverish their lives through avoidance, precautionary actions and fear of crime.

If fear of victimisation and concerns about a lack of personal safety are regarded as a threat to the use of public space in city centres, those with responsibility for city centres need to consider the actions they can take to alleviate fear and create a greater sense of safety. The scope for action by those specifically concerned with the city centre, or with private properties within city centres, is usually necessarily focused on opportunity-reduction methods. There are four major urban design approaches to creating safer city centres (see Table 11.2), which we refer to as the fortress, panoptic, regulatory and animated approaches. These are not mutually exclusive and, in any particular location, the strategy adopted may include elements of more than one basic approach. As is apparent in the discussion below, the approaches overlap. Elements of all four approaches may be present in any particular location, and one or more of the approaches will usually provide the overall flavour and character of the strategy employed.

Table 11.2 Key features of urban design approaches to safer city centres

Safer city approach	Common features	Opportunity-reduction measures
Fortress	■ walls ■ barriers ■ gates ■ physical segregation ■ privatisation of territory ■ exclusion	■ target hardening ■ access controls ■ exit/entry screening ■ formal surveillance
Panoptic	■ control of public space ■ privatisation of space ■ explicit police presence ■ presence of security guards ■ CCTV systems ■ covert surveillance systems ■ erosion of civil liberties ■ exclusion ■ the 'police state'	■ exit/entry screening ■ formal surveillance ■ employee surveillance
Regulatory	■ management of public space ■ explicit rules and regulations ■ temporal regulations ■ spatial regulations ■ CCTV ■ the 'policed' state ■ ambassadors/city centre reps.	■ controlling facilitators ■ formal surveillance ■ employee surveillance ■ rule setting
Animated	■ people presence ■ people generators ■ activities ■ welcoming ambience ■ accessibility ■ inclusion ■ 24-hour/evening economy strategies	■ employee surveillance ■ natural surveillance ■ deflecting offenders

The fortress approach

The fortress approach entails the physical segregation and defence of territory with, for example, express access controls determining who can and who cannot enter. These define those who belong and the 'other'. The fortress approach is therefore expressly concerned with the control, and perhaps also the privatisation, of space. Nevertheless, by isolating and defending particular territories (and thereby certain social groups), it is inherently socially divisive. An example of this approach is the Renaissance Center in central Detroit, which consists of a hotel, office and shopping complex with integral parking. Both symbolically and in practice, it acts as a

Figure 11.1   Shopfront security grilles in Nottingham city centre

fortress with explicit access controls determining who can and who cannot enter. Similar trends are discernible at smaller scales. Although it is unlikely that the city centre will turn into a single fortress, it is quite possible that it will become a series of small fortresses. Many city centre retailers, for example, erect external solid security shutters out of hours to prevent burglary, ram-raids and criminal damage, with each retail unit effectively becoming a mini-fortress. While the retailer's concern is understandable, a high spatial concentration of shutters can generate fear. The 1994 circular *Planning Out Crime* (DoE 1994: 6), for example, warned that they give an area 'a "dead" appearance and contribute towards the creation of a hostile atmosphere' (see Figure 11.1).

The trend towards the privatisation of the public realm, whereby the control and management of what is ostensibly public space is transferred to private agencies, or where 'public' space is provided and controlled by private agencies, is also a manifestation of the fortress approach (e.g. see Punter 1990; Reeve 1996; Sternberg 1997). In this respect, it raises important issues of equity. As Carr *et al.* (1992: 361) observe: 'Increasing private control, because it tends to put space in the hands of those who view the physical environment as a means for creating profits, will tend to focus their attention on people with money to spend, and ignore or exclude the poor.' There are two important and related issues here. The first is a tendency towards the reduction of 'citizens' to 'consumers'. Although in public space citizens have rights, in most forms of private space their presence may be valued only in terms of their ability to consume. If they are not sufficiently affluent to be desirable as consumers or if their presence *might* deter other consumers, then their presence may be unwanted and they may be excluded. The second and related issue concerns the abrogation of some people's rights as citizens. One particular physical incivility that deters many people from using city centres is the presence of 'people who cause anxiety' and the attendant possibility – but not inevitability – of social incivilities. As a deterrent, the physical incivility is often sufficient. However, many people would prefer not to be confronted with the visual reality of certain social problems and would like them to 'disappear from sight' – in other words, to be excluded or, alternatively, to be denied entry.

As is apparent from the above discussion, there are a number of drawbacks to the fortress approach (Tiesdell and Oc 1998). First, it leads to constraints on the use, access and enjoyment of the environment. Second, its value must be measured against the actual crime risk: for example, in areas of low risk it may unnecessarily exacerbate fears of crime. Third, it protects only those individuals or groups that are 'inside' and, unless all targets in the immediate vicinity are equally protected, may simply lead to deflection on to other targets. Finally, it tends to protect property *directly* and people *indirectly*.

The panoptic approach

The panoptic approach is most vividly revealed through the combination of an explicit 'policing' presence and closed-circuit television (CCTV) systems. The idea originates in Foucault's discussion of Bentham's Panopticon – an all-seeing architectural form designed to keep prisoners under constant surveillance (see also Chapter 13). It is the extension of this concept to public spaces, city centres and even whole cities that ushers in the spectre of repressive social control in order to maintain public order. The agency with an explicit law enforcement function and responsibility for the maintenance of public order is the 'public' police. In quasi-public space, however, there is an increasing prevalence of private security guards – the 'private' police – whose main functions are to deter potential offenders and to intervene when and if required. There are, however, distinct differences in the public perception and the reality of the public and private police. The latter do not have the same priorities as the public police, their main concern being to protect the interests of the firms hiring them. As Fyfe (1995: 768) observes: 'private policing is about "policing for profit", acting in the private interests of clients in order to prevent losses.' As their only function is to enforce security, an over-concentration of security guards can itself generate fear. Formal policing may also be enhanced by electronic hardware such as CCTV. Such cameras, as is also recognised elsewhere in this book (see Chapters 1 and 10), are an increasingly common feature of the landscape of English city centres.

As is noted below, the regulatory approach concerns the regulation and management of city centres. As such, the regulatory and panoptic approaches are closely related and, indeed, can be thought of as different points along the same continuum. Whether the city centre landscape is regulatory or panoptic may also be open to individual interpretation: one individual's regulated city centre, for example, might be another's panoptic (Figure 11.2). In general, the crucial distinction is between a

Figure 11.2   CCTV control room, Coventry

socially authoritarian 'police state' and a 'policed' state that protects the freedoms of its citizens. Concern regarding the panoptic city is primarily about the former and, hence, the panoptic approach is malign and expressly about 'control', while the regulatory approach is more benign and primarily about 'management'. A very obvious control presence can be oppressive and raise fears about 'Big Brother', particularly if there are doubts about its legitimacy or if it does not enjoy public trust, confidence or respect (see Lyon 1994; Davies 1996). As noted in Chapter 10, the threat lies not in the presence of police or cameras *per se* but in the use and abuse of the information gathered. As Graham *et al.* (1996: 17) noted, while CCTV systems have not yet given their controllers 'the power of all-seeing, Orwellian "Big Brother", they may support the emergence of a large number of "Little Brothers"' (see also Davies 1996).

The regulatory approach

The regulatory approach concerns the regulation and management of city centres and public space. Local authorities have traditionally been responsible for managing and maintaining the public space of city centres. While they still have an important role, given their declining powers and financial abilities, a solution to the problem of the burden of upkeep of public space has been to promote a broadening of the sense of proprietorship and 'ownership'. In the UK, the most common vehicle for this has been the creation of town centre management (TCM) partnerships including various city centre stakeholders. Although TCM partnerships are usually more accountable to the various city centre interest groups, such as property owners, they are often less accountable in the 'democratic' manner of local authorities and may be swayed towards a more commercial agenda. This raises both the potential for problems of exclusion and the threat of the fortress city.

Local authorities and other managers of the public realm may seek to manage crime and the fear of crime by the opportunity-reduction technique of deflecting offenders, that is, by deliberate benign displacement. They may also introduce new rules and regulations akin to the opportunity-reduction technique of 'rule setting', where regulations establish an explicit standard of acceptable public behaviour or, since regulations are usually framed negatively, an explicit standard of *unacceptable* public behaviour. This thereby removes potential 'ambiguity' (Clarke 1992). The best examples of such regulations are the bylaws banning the public consumption of alcohol that have been introduced in a number of English city centres (see Ramsey 1990; Oc and Tiesdell 1998). In 1988, for example, Coventry became the first British city to introduce a bylaw prohibiting the public consumption of alcohol in the area within the ring-road. Examining the bylaw's impact, two surveys were carried out by Ramsey (1990), one shortly before the bylaw took effect and another a year later. Both surveys recorded significant popular support for the bylaw. Police statistics for recorded crimes in the city centre showed that although the key categories of assaults, robberies/thefts from the person and criminal damage were unaffected by the bylaw, they did reveal a reduction in incivilities. Ramsey's research also noted some displacement with small groups of drinkers congregating in a park outside the ring-road. Discounting these other social costs, there has been the positive achievement in making the city centre perceptibly safer.

The introduction of bylaws in the interests of the wider public good inevitably raises issues of personal freedoms. In principle, the acceptability of laws and regulations is dependent on interpretations of whether rule makers are seen as benign or malign (Muncie 1996: 18–19). In the former, the social order is deemed to be consensual, 'society' creates 'rules' and 'crime' is the infraction of that society's legal, moral or conduct norms. In the latter, the social order is considered to be conflict-based, whereby the 'state' is able to criminalise, and hence 'crime' refers to social and political processes that criminalise certain behaviours. In general, liberal societies establish a framework in which individuals can pursue their own purposes without undue interference from either the state or from other citizens. The freedom to pursue those purposes carries the implicit moral responsibility not to engage in anti-social behaviour. Hence, freedom may not be maximised simply by removing all law, since that leaves the individual unprotected from the invasion of his freedom by his neighbours. Laws and other regulations therefore aim to provide a system of constraints that, while limiting freedom in some respects, maximises freedom overall (Scruton 1982: 180). In the context of this chapter, the key issue is the extent to which people might be prepared to trade certain rights of citizenship for greater feelings of safety. As limiting the freedoms of some may increase the freedoms of others, the debate concerns the extent to which restrictions or regulations are considered worthwhile. Many may favour greater regulation of the public realm in the interests of the greater good of public safety and order, but there is a danger of progression from bylaws prohibiting certain behaviours enacted in the wider public interest to bylaws enacted to prohibit public behaviour objectionable to certain (dominant) groups for narrower reasons. As a result, a regulatory approach may become a panoptic approach.

Public order is maintained explicitly by the police and implicitly both by those who work in the public realm and by the general public. Although surveys show that many people want a more visible police presence (e.g. Guessoum-Benderbouz 1997), the image and ambience of an expressly 'law and order' city centre, while reassuring to some, may be undesirable and adversely oppressive to others. Equally, an overabundance of police may be fear-generating rather than reassuring. In the regulatory approach, however, it is the implicit management of public space that is important; Jacobs (1961: 41), for example, argued: 'No amount of police can enforce civilisation where the normal, causal enforcement of it has broken down.' As the presence of other people in public spaces is reassuring to many people, some cities employ people as 'city centre ambassadors'. While such ambassadors have an implicit 'control' function, it is neither overt nor their main function; in this respect, the regulatory approach parallels the animated approach. Similarly, those working in public places – bus conductors, car park attendants, receptionists, caretakers and shop owners, managers, or assistants – have a general responsibility for the security of the property and for supervising public behaviour in the places where they work. Furthermore, in terms of the public realm's ambience, their presence is preferable to that of security guards, since security and safety are not their primary function. Significantly, however, many of these staff have been replaced by technology in recent years or have simply never been replaced (Worpole and Greenhalgh 1996). As well as removing a vital managerial and control function, this creates a further de-peopling of the public realm.

One of the key areas for increased city centre management in the UK occurs with regard to alcohol and, increasingly, drugs. Alcohol is a major factor in offences and incivilities in many city centres. Research for the Nottingham crime audit, for example, estimated that alcohol was a factor in 88 percent of incidents of criminal damage and 78 percent of all assaults (KPM 1991). This issue often requires a judicious mixture of both greater regulation, usually spatial, and more relaxed regulation, usually temporal. As noted previously, some cities have introduced bylaws prohibiting the consumption of alcohol in public places in their city centres. This measure also deals with the physical incivility of winos and others drinking in public places and the possibility of social incivilities. Experience in cities pursuing 'twenty-four hour city' strategies, however, indicates that changes in licensing hours can be beneficial; the staggering of opening hours, for example, diffuses the impact of people being discharged from pubs (Lovatt 1994). Nevertheless, the latter also suggests a possible tension between the regulatory and animated approaches. Intended to stimulate the animation of the city centre, this apparent relaxation of regulation is usually accompanied by the introduction of other regulatory measures.

Another regulatory strategy is the use of what Barr and Pease (1992: 207) call 'fuse' areas. Given that it is often unlikely that an illegal activity will be eliminated entirely, it may be advantageous to concentrate certain activities in particular areas rather than having them spread evenly. Thus, for example, unruly pubs may be usefully concentrated in such areas for those who might prefer or desire such environments, while allowing others to go elsewhere. Barr and Pease (*ibid.*) argue that there are cases, admitted or otherwise, in which policing problems, such as prostitution, are controlled or reduced by having known fuse areas (see Chapter 14). This form of management strategy permits oversight and regulation and, if public outcry becomes too great, creates conditions for a focused operation to appease it.

The animated approach

A common claim is that 'peopled' places are safer places; a human presence in public spaces is reassuring for many people, and the presence and activity of people will often attract other people. Jane Jacobs (1961: 45), for example, famously wrote about the need for 'eyes on the street'. Similarly, *Planning Out Crime* (DoE 1994: s.14) states: 'One of the main reasons people give for shunning town centres at night is fear about their security and safety; one of the main reasons for that fear is the fact that there are very few people about.' Although there are few empirical studies to support this, research among women in Toronto by Wekerle and Whitzman (1995) found that deserted places were the most feared. By creating an appropriate ambience and activity, the 'peopling' of public places results in some concerns about personal safety being 'crowded out'. The deterrent effect of concentrations of people is to increase the risks to an offender by increasing the possibility of their being seen, apprehended or prevented from carrying out their intended crime. The approach works primarily through the opportunity-reduction technique of surveillance. Surveillance can be formal by, for example, the police or private security guards, who have an explicit security function. There is also a more general surveillance by those working in public places. Given that offenders will often be aware of the small risk of being seen by the first two groups, much of the deterrent effect of surveillance is provided by the general public going about their normal activities. In terms of urban planning, one way of creating natural animation and surveillance is through design. This may be accomplished, for example, by ensuring that occupied buildings overlook car parks or by planning controls that encourage or require a mix of uses within a particular area and mixed-use developments, thereby achieving the natural animation of the public realm. In this respect, there is a certain overlap with the panoptic approach.

It is also important to emphasise the need for either *sufficient* surveillance or *sufficient* density of people, or both, to ensure safety. Initially, at least, it may be difficult to provide sufficient density and concentration of activity to enable the whole city centre to be adequately animated. Concentrating activity into particular areas or 'safe corridors' can provide the necessary density of people in certain places. However, it is not suggested that certain areas of cities should be designated and signposted as 'safe corridors'. Instead, the intention is that the creation of safe, or safer, corridors should explicitly inform planning policy. Such activity corridors need to be good-quality environments, with good lighting and natural surveillance, close to car parking and public transport. Over time the area of these corridors can be expanded, provided that sufficient pedestrian density is maintained (Trench *et al.* 1992; Oc and Trench 1993). As the boundary inevitably defines an 'inside' and an 'outside', in creating safer corridors, there is, or can be, some overlap with the fortress approach. Furthermore, as crowds provide greater opportunities for some forms of crime, such as pickpocketing and various forms of social incivility, the composition of the 'crowd' is also important.

One of the keys to 'better-peopled' places is the spatial concentration of different uses, especially residential uses (Oc and Trench 1993). Many cities have also focused on the need to develop or enhance their 'evening economy' (Montgomery 1994; Bianchini 1994). The twenty-four hour city concept, for example, is a useful mechanism to consider city centres in a more positive and expansive manner (O'Connor 1993; Heath 1997). As postwar functional zoning policies have increasingly been abandoned, land-use planning controls are no longer a major obstacle to creating mixed use, and many cities positively encourage mixed use developments. *PPG1: General Policy and Principles* (DoE 1997), for example, prioritises mixed use as a key dimension of future urban development. Increasing the residential population of urban areas is also part of the UK government's aim to accommodate a significant proportion of the projected housing need, as demonstrated, for example, by the recent Urban Task Force (1999) report *Towards an Urban Renaissance*.

Land-use planning can also have a positive role in stimulating the public's animation of the public realm by encouraging the most interactive uses to claim the appropriate street frontages. MacCormac (1983), for example, discusses the 'osmotic' properties of streets: the manner in which activities within buildings are able to percolate through and infuse the street with life and activity. He notes that certain uses, such as large office blocks and multi-storey car parks, have little relation to people in the street, whereas there are others with which they are intimately involved, such as street markets, street cafes, shops and bars. The sense of human presence and vitality within public spaces is dependent on these relationships. Montgomery (1995: 104) argues that the animation of city centres can also be stimulated through planned programmes of 'cultural animation' to encourage people to visit, use and linger in urban places. Programmes usually involve a varied diet of events and activities, such as lunchtime concerts, art exhibitions, street theatre, live music and festivals, across a range of times and venues. Furthermore, from a more regulatory perspective, there is also increasing use of specialists for the staging of major events, so that they are 'positive celebrations' rather than 'dysfunctional events' characterised by high levels of disorder and violence (Homel *et al.* 1997: 266).

The overarching issue, however, is that for urban spaces to become peopled and animated, the public realm has to offer what people want and desire, and to do so in an attractive and safe environment. There are at least two areas where positive action is necessary: on the supply side to increase the available range of activities; and on the demand side to broaden the range of age, gender, social and ethnic groups using the city centre. A problem in many city centres is that there are not enough evening and night-time activities to attract a wider range of people (COMEDIA 1991). Two mutually reinforcing processes have happened concurrently. Avoidance of the city centre by many social groups means that the facilities used by those groups have ceased to be viable in that location and have either closed or moved elsewhere. Simultaneously, the growing domination of a city centre by a particular age group often means that the available range of facilities increasingly focuses on that group's needs. To remedy this situation, the challenge is to find out what the missing social groups want and need from the public realm and then encourage or facilitate its provision. Inevitably, there is a 'chicken-and-egg' situation as to which comes first: more people or more for them to do.

The desire for a safer city is not limited to the affluent and more mobile citizens, who in any case have the choice to reject the city centre. Large numbers of other citizens, without the luxury of choice, also want a safer city centre. Thus, as well as supplying what the missing social groups want from the public realm, policies should also help them to gain access to it. This involves increasing access by addressing the factors that deter such groups from using the city centre. Carr *et al.* (1992: 138) distinguish three types of access – physical, visual and symbolic. Physical access, or whether the space is physically available to the public, is addressed through safer transport and parking (see Trench 1996). Visual and symbolic access often interact and are further complicated by issues of 'emotional' and 'psychological' access. Visual access or visibility is important in order for people to feel free to enter a space. Symbolic access involves the presence of cues, in the form of people or design elements, suggesting who is and who is not welcome in the space. Equally, environments, individuals and groups perceived as threatening or as inviting may affect entry into a public space. Yet ultimately the major hurdle to increased animation of city centres may remain the perception of a lack of safety and the fear of victimisation. Such concerns can distort people's images of city centres and, despite tangible changes in those centres, popular perceptions of a lack of safety may be particularly enduring and resistant to change. Improvements in the public realm are therefore dependent on how they alter individual perceptions of the safety of the city centre.

LEEDS

As noted previously, the four urban design approaches to safer city centres identified above are not mutually exclusive. One or more may be present, explicitly or more often implicitly, in the urban design strategy of any given city centre and will usually provide an insight into the strategy's underlying ethos. To illustrate this point, the remainder of this chapter considers a case study of the city centre of Leeds, an area that clearly shows the interaction of the regulatory and animated approaches.

Located in the north of England, Leeds is a city of 700,000 people. It has always had a strong and diverse economy and is also a major regional shopping, cultural, leisure and educational centre. The city centre is also home to two national museums, a national opera company and three regional theatres as well as a wide and increasing range of night-clubs, cafe bars and restaurants. In functional terms, the city centre divides along Park Row. To the east is the retail core and to the west the main office area. The retail core consists of a grid of streets within which are embedded a number of Victorian shopping arcades and more modern shopping centres. The city centre lacks a main square or thoroughfare, and there are few large concentrations of pubs, bars and cafes. As such activities tend to be priced out of the main retail area, they tend to be located around the fringe of the retail core. Following the upheavals of comprehensive development and motorway building during the 1960s and 1970s, the city centre saw relatively little change during the 1980s, but a surge of development and improvements took place from the late 1980s onwards. Significant growth occurred in the city's financial services and it developed its status as a legal centre. The city is now arguably the prime financial and law centre in England outside London.

The animation of Leeds city centre is epitomised by its emphasis on the twenty-four hour concept. The twenty-four hour city initiative was launched in 1993 through the inspiration and vision of Jon Trickett, then council leader, and Leeds is now one of the leading proponents of the concept in the UK. Intended to develop the city centre economy around the clock, the initiative included a series of coordinated policies, including relaxing licensing restrictions, encouraging more residential accommodation within the city centre, providing programmes of events and entertainment, and promoting safety (Hamshaw 1998: 2). The term 'twenty-four hour city' is misleading, however, given that the accent has really been upon 'stretching' the city centre's 'working life' and encouraging more people to make use of it in the evenings and on Sundays (Jones *et al.* 1999: 164).

Concurrent with the start of the twenty-four hour city initiative, the Leeds city centre management (CCM) team was established. While its work focused initially on economic vitality and the quality of the physical environment, more recently it has become concerned with vitality and ambience – a crucial dimension of which is the sense of personal safety. Rather than being the main catalyst or instigator of change, the CCM team has responded to trends and has adopted what can be considered to be a growth-management approach.

The twenty-four hour city initiative was a deliberate attempt to 'Europeanise' the city. The city's planning strategy, set out in the 1993 *Unitary Development Plan* (UDP), was explicitly based on a notion and image of the 'European' city. The importance of the city centre was emphasised: 'an integral element in the promotion of Leeds as a major European city is the life and vibrancy of the city centre, and its usage for a wide range of activities throughout the 24 hour day' (Leeds City Council 1993: s.12.6.3). With regard to safety, the UDP (*ibid.*: s.13.1.8) set out seven objectives, including the following: 'to improve safe and secure access for all to and within the city centre.' Safety was discussed further in a section devoted to 'Access for All', which, *inter alia*, emphasised the contribution of environmental factors and the physical public realm to perceptions of safety. It noted: 'Spaces need to be designed to be *safe and accessible for all* . . . to provide good lighting with the absence of dark areas, to provide a variety of alternative routes, *and to be self-policing by achieving a continuous presence of other users*' (*ibid.*: s.13.4.20, emphases added). The final statement indicates an animated approach to community safety.

The quality of the physical environment is a pre- or co-requisite for animation, and physical revitalisation is frequently necessary to provide a more welcoming ambience and to attract people. Although the 1970s pedestrianisation of the central retail core had been a successful pioneering scheme, by the late 1980s its condition had deteriorated and from 1990 onwards a major new programme was carried out to upgrade the pedestrian environment with an emphasis on style and safety. The city council also developed a policy to resist solid external shop window shutters, with grant assistance provided to improve shop frontages and in particular removing grilles. An early 1990s survey, for example, had found a high spatial concentration of shutters, which, by creating a hostile atmosphere out of shopping hours, can generate fear.

The most significant physical change has been the revitalisation of the previously rundown Waterfront in the southern part of the city centre. During the late 1980s and 1990s, this became a mixed-use area including offices, bars and restaurants, housing, and museums and galleries. Development started in the mid-1980s, continued under the auspices of the Leeds Development Corporation (LDC), established in 1988, and has continued since the latter's disbandment in March 1995. Another revitalised area is that around the Corn Exchange, between the core retail area and the Waterfront. Now known as the Exchange Quarter, this area has now seen cafe bars and housing replace a secondary retail area with rundown buildings.

The first phase of a CCTV system was installed in 1996 to assist regulation and management, with now nearly fifty cameras covering the city centre and Waterfront area. The next phase of the CCTV initiative is the installation of further cameras in the Markets area. Funded by the city council, the agency operating the CCTV system is called Leedswatch. Developed in consultation with the police, the management system includes a code of practice for operators. There are also strict guidelines on the use of both the cameras and any tapes produced. There has been no attempt to conceal the cameras – the city council wants people to know that the cameras are there. People are reassured by knowing of their existence but otherwise do not regard the cameras as intrusive. According to Jones *et al.* (1999: 165), the installation of CCTV cameras has led to a 25 percent decline in reported crime within the city centre since 1996. There has also been a rolling programme of street lighting improvements, which has brought brighter lighting to all the city centre's main and side streets. The new lighting has also improved the efficiency of the CCTV system. There has also been a programme of architectural lighting to enhance buildings.

As noted previously, the spatial concentration of different uses is one of the keys to producing better-animated places. Within the city centre, the UDP set out an approach based on quarters. While it identified and defined a number of functional quarters within the city centre, it also encouraged a range of uses in all areas. The planning approach therefore explicitly favours mixed use, both in terms of mixed-use developments and a mix of uses within an area. The main objective was to achieve a greater mix of uses throughout the city centre, to avoid the creation of large single-use areas, which may be dead at certain times, and to contribute to a livelier and more vibrant city centre (Hamshaw 1998: 3–4). One of the most visible signs of the changing use of the city centre has been the dramatic growth in the number and variety of cafes, bars and high-quality restaurants. While the period 1991–94 saw just six new developments or extensions to existing bars and pubs, between 1995 and 1998 there were fifty new developments. By March 1999, a further eight developments were under construction, and a further fifteen had planning approval (Matley 1999: 5). Many of the new developments re-use vacant space, such as secondary shopping areas and, in particular, old banking halls and large historic buildings. The new bars also both exploit and create new market niches, providing a more diverse and sophisticated range of venues, including tapas bars, comedy clubs and jazz clubs. The impression is that bars and cafes are 'more civilised', less crowded and offer better service. The city also has a policy of encouraging pavement cafe licences as a key element of its twenty-four hour city initiative. The Leeds night-club scene has also expanded significantly. In 1991, there were thirteen clubs and by 1999, thirty-eight (*ibid.*).

Increasing the city centre's residential population was also a core aspect of the drive to become a European city. The UDP (Leeds City Council 1993: s.13.6.12), for example, stated that: 'In accordance with Leeds' European City aspirations, one of the main ways in which life and variety can be increased in the city centre, and extended throughout the day, is by the introduction of more housing into the centre, and retention of existing housing.' The revival of housing in the city centre began with the Regent Court scheme in Briggate in 1983 and subsequently the Victoria Quays complex on the Waterfront, completed in 1987. Further residential development occurred on the Waterfront at the end of the 1980s and was given an additional boost in 1988 by the establishment of the LDC in the southern part of the city centre. Although residential development had started prior to the LDC being established, its financial resources and remit to promote the physical regeneration of its designated area gave the process a boost. During the late 1980s and early 1990s, the Waterfront saw a mix of conversions and new building for rent and for sale. Although the CCM team and the city council have been instrumental in promoting the concept of city centre living, city centre living elsewhere has been slower to take off. However, there have been a series of joint ventures between the city council, housing associations and universities. Many of the residential developments have taken the form of living-over-the-shop schemes, with cafe bars or other facilities on the ground floor. In the early 1990s, census figures showed that the city centre had a population of 900 people. By the end of 1997, it was estimated to be about 4200 and during 1998, there were commencements of 100 additional units and applications for a further 700.

While greater provision of facilities assists the animation of a city centre, this 'natural' animation can be supplemented and enhanced through planned programmes of 'cultural animation', which encourage more people to visit, use and linger in urban places (Montgomery 1995). In recent years, Leeds has implemented a major programme of cultural animation, which has included Leeds Christmas lights; Valentine Fair; an international film festival; the Leeds marathon; the Network Q Rally; cycle racing; 'Rhythms of the City'; the Breeze youth festival; 'Jazz on the Waterfront' and other Waterfront events; an annual festival of street entertainment;

and one-off events such as the Briggate street party. The city has encouraged Sunday openings and 'Sundays in the City'.

While there has been significant success in developing an evening economy and animating the city centre, Leeds has also witnessed issues that have confronted other cities, such as Nottingham (see Oc and Tiesdell 1998). A major problem is that the evening economy's relatively narrow focus leads to criticisms that it is primarily alcohol-based and youth-oriented. Many cities have therefore recognised the need to bring in a wider cross-section of people, including families. As Jones *et al.* (1999: 164) observed:

> While a popular, and perhaps prejudiced, view of the '24 hour city' is of late-night pub/club, drinking and dancing culture dominated by young people, local authorities have been keen to stress that stretching a city centre's working life means attempting to offer a range of activities to a range of age groups. This ideally should include, for example, family entertainment, restaurants, and not only better use of existing leisure facilities but also the creation of opportunities for new ones.

As well as supplying what the missing social groups want from the public realm, people may, for various reasons, be unable to access it. As noted above, the issue of access is multidimensional. It can often be important to improve the environmental image and ambience of city centres to make them more welcoming or, at least, less intimidating to a wider range of social groups. Safe travel is also important (Trench 1996). The UDP (Leeds City Council 1993: s.12.6.3), for example, highlighted safe transport as a prerequisite for city centre vibrancy and vitality. It noted: 'many people, especially women, feel particularly vulnerable at the point at which they leave or take private or public transport and, with rising levels of thefts from and of cars, car parks can often be intimating places.' During the 1990s, many of the city's bus and rail stations, car parks, bus stops and taxi ranks have been upgraded to provide a more comfortable, better-lit and safer environment.

The city centre has become animated during the evening and into the night, and it is estimated that about 60,000 people use the city centre at weekends. Despite the twenty-four hour city initiative, the city centre remains largely used by white collar office workers commuting into the city for the day from a large hinterland. During the mid-1980s, there had been a perception that the city centre was unsafe. At that time, it was also used during the evening by a predominantly 16–21 years age group. While there are still significant numbers of this age group, they have been complemented by a significant number of older people. In part this has been the result of a broadening of the range of facilities to supply greater choice. The increasing animation of Leeds city centre reiterates the necessity of managing or regulating rather than simply permitting a *laissez-faire* animation. Alcohol and drugs are one of the key areas for increased city centre management. Jones *et al.* (1999: 165), for example, describe how the move towards a twenty-four hour city poses a number of management challenges. They comment that 'if the extension of the evening economy is largely based on extending the opening hours of licensed premises, there will be understandable concern that this will serve only to promote an alcohol- and drug-centred youth culture.' Similarly, Homel *et al.* (1997: 264) have noted that although a vibrant entertainment industry brings considerable economic benefits, it can also create major problems if 'entertainment is equated with alcohol consumption' and licensed venues are concentrated in a small area. Clarke (1992: 24–5), for example, notes that 'psychological disinhibitors', such as alcohol and drugs, facilitate crime by undermining the 'usual social or moral inhibitors', or impair perception and cognition so that offenders are less aware of breaking the law and more generally engage in disorderly or uncivil conduct. Hence, some forms of animation often create spillover effects, such as boisterous crowds of young men, which are detrimental to the enjoyment of other groups.

This issue requires a judicious mixture of control and relaxation. Leeds has been in the forefront of the campaign to challenge and change the United Kingdom's licensing laws. To investigate licensing issues, a series of workshops were held with police, magistrates, the leisure industry, planning officers, etc., and, within the bounds of the existing licensing laws, a joint approach was agreed. Late night cafe and restaurant licences in the city now permit opening between 11pm and 5am. The police support the new licences and the special hours, and overall there has been a coordinated approach (Matley 1999: 4). The safety benefit of staggered opening hours is to diffuse the impact of people being discharged from pubs and clubs. In Leeds, when licensing hours became more relaxed, there was a perception that it would lead to problems by having slow drip of trouble in a wider range of places over a longer period instead of a shorter, concentrated period. Yet this has not been the experience. Using crime pattern analysis, the police are able to identify and anticipate crime and disorder hot spots and have resources available. According to the city council, it is the 'more relaxed and cosmopolitan regulations' that have established Leeds' nightlife as one of the most vibrant and popular in the UK (Leeds City Council 1998; see also 1999).

However, this issue does raise a possible tension between the regulatory and animated approaches. The idea was not just to open through the night for the sake of it, because certain management issues also had to be addressed. Forty-seven percent of disorder offences and 50 percent of arrests at flashpoints, for example, occurred between 11pm and midnight on Friday and Saturday nights (Matley 1999: 4). Hence, although intended to stimulate the city centre's animation, the apparent relaxation of regulation is usually accompanied by the introduction of other regulatory measures, such as doorman registration schemes and other measures to encourage bar and night-club owners to operate more responsibly. The development and promotion of responsible host policies encourage licensees, landlords and nightclub operators to take responsibility more generally for the conduct of their patrons both within and outside their premises. According to Homel *et al.* (1997: 266) this involves 'a focus on the way licensed venues are managed; the "re-education" of patrons concerning their role as consumers of "quality hospitality"; and attention to situational factors, including serving practices, that promote intoxication and violent confrontations.' In principle, the emphasis in most cities has been to shift the focus from 'alcohol' towards 'entertainment' by emphasising food, music or other attractions. In partnership with the police and night-clubs, for example, the CCM team has also developed a strict code of practice to reduce the use of drugs.

For night-clubs joining the scheme, CCTV must be provided, all door staff must receive training, and the club must reserve the right to search customers and confiscate any drugs found in their possession (Hamshaw 1998: 7–8). Clubs must also have a CCTV in their foyer and entrance areas and, if there is an incident, the police must be given a copy of the tape.

Although surveys often show that many people want a more visible police presence, the image and ambience of an expressly 'law and order' city centre, while reassuring to some, may be undesirable and adversely oppressive to others. Equally, an overabundance of police may be fear-generating rather than reassuring. As noted previously, the implicit management of public space is also important and, as the presence of other people in public spaces is reassuring, some cities employ people as 'city centre ambassadors' and, although these might have an implicit 'control' function, it is neither overt nor their main function. In Leeds, a city centre patrol service was introduced in September 1997. During office hours, two liaison officers assist or attend to a wide range of issues, including community safety issues such as aggressive begging, nuisance and other social incivilities, and the condition of street and spaces (i.e. physical incivilities). A wide area of the centre is patrolled, with emphasis given to the retail core. Although the patrols aim to provide a civilian support service to other agencies, they have no powers of arrest or enforcement themselves. In late 1997, the police introduced special afternoon patrols to work with the liaison officers.

As recognised earlier, many factors or issues that deter people from using city centres are not technically crimes. One particular physical incivility is the presence of 'people who cause anxiety' and the attendant possibility, but not inevitability, of social incivilities. As a deterrent, and because the possibility of social incivilities is ever-present, the physical incivility is often sufficient. However, many people would prefer not to be confronted with the visual reality of certain social problems. As such, there are two main areas of nuisance and incivility in Leeds city centre: active or aggressive begging and prostitution. The city centre has forty to fifty active beggars, and police action is focused on disrupting aggressive rather than passive begging. There is also a problem of evening prostitution around the Calls and Sovereign Street area. At one level, such prostitution is a physical incivility. At other levels, it is a social incivility and, in certain instances, can be a crime. Prostitutes can be actively soliciting, while punters can be actively seeking; both result in the harassment of other city centre users. Both the nuisance and prostitution policies raise issues of exclusion and, as a police inspector stressed, it cannot simply be an issue of moving them out of sight or away from the public realm.

CONCLUSION

This chapter has outlined four urban design approaches to creating safer city centres. The seductive, albeit short-term, appeal of the fortress and panoptic approaches is that they are positive actions; something is being seen to be done, but they are often private-minded behaviours. The consequence of such individualistic actions may be that while it becomes safer for some, it may become increasingly unsafe for others. Furthermore, city centres that are made safe but are no longer appealing or generate fear and intimidate potential users defeat their very *raison d'être*. Thus, while elements of the fortress and panoptic urban design strategies may have their place, there are other ways of making the landscape of the city centre feel safer, such as the regulatory and animated approaches, which offer more expansive and positive notions of public space and city centres generally. Leeds has been used to illustrate these approaches to creating safer city centres and, as the case study demonstrates, these approaches frequently reinforce and complement each other. The approaches also have their own challenges; the animated approach, for example, is dependent on a culture change that favours a more urban lifestyle than has previously been common in England. Achieving a major part of the government's target for 60 percent of new housing on brownfield sites in or around city centres would, however, provide a stronger economic base for well-animated city centres. While the animation of the public realm is linked to the general revitalisation of the city centre, there is the problem that to be perceived as safe the public realm must be animated; and to be animated, it must be perceived to be safe. A major hurdle to increased animation in city centres is, therefore, the continuing perception of a lack of safety and the fear of victimisation, which can put an exaggerated perspective upon people's images of city centres. Despite tangible changes in a city centre, popular perceptions of a lack of safety may be particularly enduring, and ultimately improvements in the public realm are crucially dependent on how they alter individual *perceptions* of safety. A survey of residents in Leeds, for example, found that those who use it tended to perceive it as safer than those who did not. In general, it was younger people who used and felt safer in the city centre and older people who tended not to use the city centre and felt less safe there. Users who were more than forty years old tend not to have a positive perception of the city centre, appear to have fixed attitudes and may have organised their lives without any need for the city centre.

Although this chapter has focused on urban design approaches, it should also be noted that the creation of safer city centres requires actions that address all dimensions of the criminal act. Hence, as well as opportunity-reduction measures, wider actions are need to encourage the acceptance of moral responsibility and self-control, and to promote better personal behaviour and conduct within the public space of the city centre. This is a general societal problem and beyond the scope of situational and management measures.

ACKNOWLEDGEMENTS

The authors would like to thank the members of the Leeds city centre management team and the West Yorkshire Constabulary who were interviewed for the purposes of the research reported in this chapter.

CHAPTER 12

A CONSTABLE IN THE LANDSCAPE
John Baxter and Paul Catley

The purpose of this chapter is to examine the place of the constable in the modern policing and social landscapes of Britain. Those landscapes contain a wide variety of communities, ranging from comfortable middle-class suburbs to inner city and declining zones enduring social and environmental deprivation. These different communities experience different crime rates and will require, or expect, different levels and styles of policing. The more difficult the problems to be solved, the more drastic the solution that may be required.

In policing terms, 'zero tolerance' has been promoted as just such a solution, and it is this form of policing that will be discussed in this chapter against the background of what might be called the traditional values and rhetoric of British policing. The conclusion reached is that it has formed yet another accretion to the policing repertoire, which has been growing since the formation of the 'new' police in 1829, and that while not providing a panacea, zero tolerance can make a valuable contribution in some situations. It is argued that the style of policing is not new, nor does it necessarily run counter to previously established models of policing. However, the approach brings with it certain risks that may outweigh its short-term benefits.

Raise the topic of the 'traditional British bobby' and a variety of responses will be heard depending on such factors as the age group of the respondents, their ethnic origin and whether they live in a rural or an urban environment. A traditional feature that can be isolated from the literature is visibility: whether the police constable is seen standing on a corner, walking his beat or directing traffic. The constable is also portrayed as being for the law-abiding citizen: a benign figure, friendly, accessible and if not immediately available only a few minutes away. As a part of this image, the role of the police is shown as a consensual arrangement under which the citizen tolerates the occasional interference of authority because it is a small price to pay for a peaceful community where life, limb and property will be protected.

It will be seen from the first part of this chapter that this image was established during the formative years of the police force in the nineteenth century and remained part of the portrayal of the police until the 1960s. The second part demonstrates that in our own times it is difficult to reconcile this uncomplicated image with the practice of modern policing and that there are now a diversity of styles and methods determined by the specific needs of the community to be policed. Policing, like any other service, is subject to changing fashions and catch phrases. A cursory glance over the last decade or so at a journal like *Police Review* will reveal a series of crazes, preoccupations and themes that have fallen from view. In the mid- to late 1990s, the fashionable phrase has been 'zero tolerance'. Put briefly, this is a method of policing that allocates resources to a particular problem area with the intention of clamping down on *all* infringements of the law. Offences such as riding cycles on pavements, litter offences, graffiti, begging, soliciting, drug use, drug dealing, burglary, mugging and criminal damage all fall to be enforced as part of the campaign. Zero tolerance theory requires blanket enforcement, including the maximum use of stop-and-search powers, so that with such a hard policing regime the prosecution of 'quality of life offences' improves the area generally and encourages further rehabilitation.

The theory and the effectiveness of this coercive style of policing will be explored later in this chapter, but in the interim it is necessary to identify its place within the policing context and the democratic expectations of exercising powers and discretion in accordance with standards of legality and reasonableness. In so doing it will be possible to determine whether or not the new methods of policing are changing the status and role of the constable on the street and whether the constable is expected to perform several, sometimes contradictory, roles. In terms of public perception the police will continue to promote a variation of the traditional image, which stresses force efficiency and hopes to keep the majority of the community on its side. That tradition and the changes that are taking place will now be considered.

THE POLICING TRADITION

It is often forgotten that the police force is a relatively recent British institution. The chaotic arrangements of the seventeenth and eighteenth centuries were unable to cope with the law and order problems presented by the huge conurbations spawned by the Industrial Revolution. In particular, the policing elements of feudalism, such as tything and later the system of parish constables and watchmen, which provided for self-policing, would soon prove to be outmoded for the demands imposed by the new urban communities. Yet this history of inefficient self-reliance would make the introduction of public policing a major challenge for the public authorities (Critchley 1973: 27). As with so many policing issues, it was London that led the way in replacing the old, ineffective system, and the Metropolitan Police Act of 1829 marks the birth of the modern force. The Act was seen as an essentially foreign import, and its promoter, Home Secretary Sir Robert Peel, had to ensure that the police would eventually earn the respect of the populace and pacify those critics who saw the new police as oppressive (*ibid.*: 27–8). When the Metropolitan Police Force was established, its function was to be crime prevention by means of patrols by uniformed constables. It was not until 1842 that a detective branch began to emerge (Porter 1987: 5). The English police force evolved gradually, based on principles of legality and the recognition of individual liberty. Critchley (1973: 28–9) has observed that they were to 'go out of their way to win the good will and cooperation of the public, advertising themselves as a service rather than a force, and that they were answerable to the ordinary courts of law for any breach of the law.' They were to have, quoting Peel's first instructions, 'perfect command of temper'. Public approval was acquired by keeping in touch with public sympathies through appointing those of similar background.

This shared experience and the fortunate combination of an absence of unpopular laws with a period of comparative social stability have been cited as some of the reasons for the success of Peel's creation (*ibid.*: 31–5). It should not be forgotten, however, that this success was determined by the public face of the police as presented by the constable on the beat and that the police institution was developing behind this façade. As the nineteenth century progressed, the detective branch would eventually become the Metropolitan Criminal Investigation Branch, founded in 1878. Growing economic difficulties and the Fenian bombings of the 1880s were among the factors that led to the formation of the Metropolitan Special Branch in 1881. These less public functions did not need to be so liberal in their approach since they could work behind the scenes. While Peel's constables were to be recruited from the working class, this new type of desk policeman could be drawn partly from the entirely different British experience of colonial government (see Porter 1987; Rose 1996).

Until the 1970s, the public face of the police service was still very much influenced by the benign images inherited from the liberalism of Peel's early model. That model was still being presented in schoolchildren's literature in 1938, over a century later. In *About Policeman* by Richardson, a booklet in a series entitled *Introductions to Citizenship*, the illustrations show city streets with a policeman on point duty being 'always very helpful to people with perambulators' and portrayed helping children to cross the road, 'being work which Automatic Traffic Signals cannot do.' However, it is with regard to the behaviour expected of the constable on the beat that Richardson (1938: 4, 33) echoed most strongly the standards that were sought by Peel:

> He is always ready to answer questions and help people. It is part of his work. He has been taught at the Police School always to be polite to people, and when speaking never to use slang.

> He has been taught never to lose his temper and to be careful to speak the truth.

> After an accident, or if he has caught someone doing something wrong, a policeman must always be able to say at the police station exactly what happened.

> It might be difficult for him to remember, so he at once writes a brief account in his notebook. He must write clearly and exactly. He would get into great trouble at the police station if he tore a leaf out of his notebook.

These ideals still represent in a condensed form the expectations that a society governed by the rule of law have of the British police. The constable is expected to be on the side of the law-abiding citizen, and to be tolerant, honest and reliable in dealing with him or her. It is a high expectation to have of any occupation, but particularly so in the everyday world of the constable dealing with those whose honesty must necessarily be questioned; where the duty of proving guilt, while paying due regard to the presumption of innocence, must appear at times to be both contradictory and obstructive. Yet it is because of the expectations we have of the police that the intervention of the police into the everyday life of law-abiding citizens is tolerated. It is at this point that the notion of policing by consent becomes important.

POLICING BY CONSENT

Part of the emerging transaction between the originally sceptical British public and the new police was that their presence and actions would be tolerated or receive consent, provided that they kept within the rules of both legality and decent behaviour. Tolerance was to be expected on both sides of the pact. It is this understanding that explains why police powers were augmented only sparingly (see RCPPP 1929: paras 15–16). However, the traditional liberal approach is not easily reconcilable with efficiency and may easily fall victim to contemporary needs.

This reluctance to enact new powers continued until the Police and Criminal Evidence Act 1984 (hereafter referred to as PACE). PACE marked a shift from the balance struck in the nineteenth century whereby the lack of powers enjoyed by the police were compensated for by the cooperation and consent that was forthcoming from the general population. The Royal Commission on Criminal Procedure, which reported in 1981, took the view that more police powers were needed and that the compensation to the public for granting them would have to be effective safeguards (RCCP 1981). PACE followed this new arrangement, so coercive powers plus safeguards became the model for the many new powers that the legislation implemented.

The gaps that the new provisions filled had not suddenly been discovered. For example, until the new legislation came into force the police had no power to search private premises for evidence of a suspected murder. More often than not, though, the police would acquire the consent of the property owner or could carry out a search knowing that a challenge in the courts would be unlikely. This was a consequence of the policing by consent tradition. The Royal Commission recognised that this was not always satisfactory and decided that specific laws should be provided to give the police powers of compulsion subject to legal safeguards (e.g., see RCCP 1981 on the need for a balance (1.11–1.13), stop-and-search (3.24–3.28) and powers of arrest (3.75–3.88)). The result was to reduce the area of policing by consent and to extend police powers of stop-and-search and to search premises, and their powers of arrest. The theory behind this readjustment is attractive in its tidiness if nothing else, but the weakness arises with those powers such as arrest and stop-and-search, which are exercised summarily by constables operating in the community. To a great extent, the safeguards are dependent for their effectiveness upon the police themselves rather than any independent check by means of judicial intervention – as applies with regard to warrants for the search of premises. Indeed, practical considerations make it difficult to find any other alternative once police powers are extended, but those difficulties make it even more important that abuses of the process or failings in the system be recognised and addressed.

It has become clear that there is a major difficulty with the exercise of stop-and-search powers in that the ethnic minorities bear a disproportionate share of their use (Brown 1997: 16–27). With the removal of the choice to cooperate with the police, the possibility exists of cooperation being withdrawn in other areas. The police have traditionally relied on the public providing information in order to assist investigations. The danger that such assistance will be lost in situations where the police no longer have the support of the general public, or of significant groupings within the public, cannot be ignored. One consequence could be police demands for yet further powers to make up for the lack of cooperation from an increasingly alienated community. This could then further alienate the community and lead to a dangerous downward spiral as policing by consent is abandoned, only to be replaced by ever-expanding police powers. A model of policing for the next century based around surveillance cameras, listening devices, telephone taps, scrutiny of e-mails, increasing powers of stop-and-search, where offenders are electronically tagged and those deemed 'dangerous' are incarcerated, is increasingly possible. In such a world, the police could be less dependent on the public for 'tip-offs' in their fight against crime and the public less inclined to help.

It will be seen later that PACE was the start of a process of adding to police powers. This has continued across a change of governments, with politicians placing great faith in police powers while seeking to maintain a belief that the community has a role to play in supporting the police through institutional arrangements. This has replaced the previous unstructured and unregulated hope that communities will cooperate and allow the police to do what the police think is acceptable. This preference for coercive powers as a solution also prepares the way for a slide into coercive policing styles such as zero tolerance when the police feel they must be seen to be doing something. The dilemma of balancing police efficiency and reputation against individual expectations was summed up many years ago by Ben Whitaker (1964: 20) when he said:

> On the one hand, today, we value efficiency and demand results; on the other we suspect strong authority and defend the rights of the individual more than ever before. We ask the police, as the agents of force necessary even in peacetime democratic society, to fulfil our responsibilities which we acknowledge to be essential, if unpleasant. But we also use their uniforms as the most readily identifiable targets for our resentments at these same tasks.

CHANGING METHODS AND IMAGES

In a chapter by Richardson entitled 'Police and Modern Inventions', a set of photographs records an 'operator' using 'wireless apparatus inside a police van' and a sports car leaving a police station with the caption: 'A police car starting out' (1938: 60, 63). Thirty years later it was the combined application of these two inventions that saw the police becoming distanced from the community as reactive policing came to be seen as a more efficient use of resources than the traditional beat system. Losing touch with the society it was supposed to serve had implications for the police, both in terms of the support received from the community and in the attitudes that the police adopted towards ordinary people. The policing style that was seen as an antidote to this potential fault line became known as 'community policing' (see Bennet and Kotch 1993).

One of the factors contributing to the need for stronger links with communities was the ethnic and cultural diversity brought by postwar immigration. This change continues to pose a major challenge for the police institution, as evidenced by the Stephen Lawrence case and subsequent inquiry (Macpherson 1999). The inner city riots of 1981 focused attention on issues surrounding the policing of ethnic minority communities. Lord Scarman's subsequent inquiry (1981) raised the profile of community policing, which is now well established as part of the policing repertoire. It appeals to the idea that policing by consent and gaining the cooperation of the community is important, while also supporting the idea that better information is needed in order to catch criminals. It seeks to establish a dialogue between the police and the community, but it also raises questions about police accountability, since the police will not wish it to involve any loss of operational control.

One of the consequences of the Scarman Report was that the attention given to community policing led to PACE making statutory requirements for consultation machinery. In theory, the style is intended to be non-confrontational and inclusive, putting the police back into the community. Constables are allocated particular beats, where they are expected to become known and act as a channel for consultation and communication as well as being an enforcement officer. However, the Home Office review of PACE after the first ten years (Brown 1997) concluded that there had been particular difficulties with regard to the consultation machinery established in terms of recruiting participants from ethnic minorities. In addition, meetings were often irregular and poorly attended, with the police tending to dominate agendas and being reluctant to share information. This contrasts starkly with a view of community policing that sees openness as an essential condition for creating trust between the police and the community (Stevens and Yach 1995). Nevertheless, although community initiatives vary from force to force, versions of it are now widespread.

During the same period, the term 'Neighbourhood Watch' came to describe schemes that involved residents in forwarding information to the police and receiving warnings and guidance in return. The now familiar signs on lampposts declaring that you are in a Neighbourhood Watch area are for many the only visible sign of such a scheme, and the success of such schemes has been questioned. Home Office research concludes that implementation has been variable. Schemes are most popular in middle-class areas with low crime rates and they are most problematic in inner city areas with high crime rates (Laycock and Tilley 1995: 7–8). However, Neighbourhood Watch schemes and community policing represent attempts to return to the notion of self-help, the main feature of enforcing the law in earlier times, which, by reason of scarcity of resources and the nature of modern society, may be due for a revival. Against that, such schemes face the spectre of community apathy and may be difficult to operate in areas where a community is difficult to identify or has become dysfunctional due to fear of crime or actual criminality.

ZERO TOLERANCE

In the mid-1990s, zero tolerance captured the headlines as the latest answer to dealing with such problems. Popularised more or less simultaneously on both sides of the Atlantic, the style of policing has been characterised as that adopted by Bill Bratton in New York and by Ray Mallon in Cleveland (UK). Both Bratton and Mallon have disseminated information about their policing styles with evangelical zeal (see particularly Bratton 1998; Dennis and Mallon 1998a; also Giuliani and Bratton 1994; Bratton 1995; Horowitz 1995; Gibbons 1997; Dennis and Mallon 1998b). In the context of British policing the choice of such a title for a policing method should cause some alarm. The traditional public expectation of a police constable has been that he or she should be fair and tolerant. John Alderson (1973: 39) observed that:

> The police, like laws, reflect the nature of the society they serve. Corrupt societies deserve, and get, corrupt police. Totalitarian societies acquire omnipotent police. Violent societies get violent police. Tolerant societies get tolerant police.

Given that 'few people, least of all those brought up in and professionally trained to respect British traditions, are in favour of intolerance, or would admit it if they were' (Dennis 1998: 1), the appeal of a policy labelled 'zero tolerance' may initially seem surprising. The explanation can perhaps be found in the rapid rise in crime figures throughout the 1980s, coupled with the increasing view that crime was too big a problem to be tackled. Explanations for crime were turning away from simple messages about individual responsibility. The problem was increasingly presented as a complex one: crime was the product of society. To combat the problem of crime, then, the problems of society had to be tackled. The message was equally bleak whether those problems were identified as the breakdown of moral values, the growth of single parenthood and increased welfare dependence or whether they were characterised as relative deprivation, racism, poor housing and inadequate education. Crime could not be brought under control unless the ills of society were first cured.

In this setting, the message of zero tolerance had an immediate appeal. It ignored the complexities of the situation and presented a simple message of hope. Not only crime, but also the minor incivilities that detracted from the quality of life, could be stopped. All the police had to do was get stuck in, confront those who behaved badly and order could be restored. The message had a naive simplicity akin to the 'Just Say No' campaign against drugs. Bill Bratton (1998: 42), widely credited with introducing zero tolerance policing to New York, has reservations about the term but acknowledges that:

> Phrases such as 'zero tolerance' send powerful messages. That is why they catch on so
> quickly. Zero tolerance conveys a forceful message about the importance of civility and
> order in complex societies and about the need for police to restore and maintain order.

The Labour and Conservative parties clearly heeded that message in the run-up to the 1997 British General Election. Home Secretary Michael Howard and his opposition counterpart Jack Straw each went to New York to see the new style of policing in action. Labour pledged in its manifesto that 'our zero tolerance approach will ensure that petty criminality among young offenders is seriously addressed.' This was followed by Straw's announcement at the Labour Party Conference in September 1997 that he wanted 'zero tolerance of crime and disorder in our neighbourhoods.'

Broken windows

The blueprint for the new style of policing that was commonly adopted in the 1990s and known as zero tolerance is commonly taken to be Wilson and Kelling's article 'Broken Windows' (1982; see also Bratton 1995; Gibbons 1996; Sullivan 1997; Neyroud 1998; Pollard 1998). Wilson and Kelling noted that foot patrols did not reduce crime. This conclusion, based on research from the Newark (NJ) patrol experiment, aroused little surprise. The switch to patrol cars, which had taken place on both sides of the Atlantic, had been presented as promoting efficiency. Patrol cars would be quicker to reach emergency calls, could pursue villains more effectively and were altogether a more modern and appropriate means of policing. The officer on the beat was easily avoided by the criminal, and the chances of the officer happening upon a robbery or burglary were remote. In terms of British television dramas, PC George Dixon of *Dixon of Dock Green* was a figure from a bygone age, replaced by his motorised successors in *Z Cars* and *The Sweeney*.

The findings on fear of crime were more surprising. Residents living in areas with foot patrols felt more secure than residents of other areas. Citizens in areas where there were foot patrols tended to believe that crime had been reduced and had a more favourable view of the police. In crime reduction terms, foot patrols were not working, but they were thought to be working. The change in attitudes worked both ways. The New Jersey research found that officers walking the beat had greater job satisfaction than their colleagues in cars and had a more favourable attitude towards the citizens of the areas that they policed. The lessons of this research can be seen in the shift towards community policing and the moves to reinstate the 'bobby on the beat'.

Other lessons of the 'broken windows' thesis attracted less initial interest. Some of these other lessons were taken to heart and formed the basis of the new-style policing in New York and Cleveland. These related to what have become known as quality of life offences. Kelling (1998: 4), observing the foot patrols in Newark, noted how officers would enforce 'order' on the streets:

> police were involved in developing neighbourhood 'regulations': that is they negotiated
> with the 'good' citizens of the community as well as the troublesome residents of a
> community, the standards for a neighbourhood.

Drunks and addicts could sit but were not allowed to lie down. Alcohol could be drunk in the side streets, but not on the main streets. Beggars could approach people walking past but not those waiting at bus stops. Teenagers who were too noisy would be told to be quiet. This police action enforced a code of behaviour on the streets; in some instances there were laws that could be enforced to support the police actions, in some cases there were not. However, Kelling noted that some sort of consensus existed between the police and the whole community as to what was and what was not acceptable behaviour.

There is nothing to suggest that traditional community policing could not arrive at this sort of consensus. Where Wilson and Kelling went further was in their assessment of the significance of order and the consequences of disorder. Their thesis takes its name from the consequences of a window being broken and it being left unrepaired. The broken window works as a metaphor for all sorts of minor behaviour that can lead to the descent of a neighbourhood into crime and disorder. As Wilson and Kelling (1982: 32) observed:

> A piece of property is abandoned, weeds grow up, a window is smashed. Adults stop scolding rowdy children; the children emboldened, become more rowdy. Families move out, unattached adults move in. Teenagers gather in front of the corner store. The merchant asks them to move; they refuse. Fights occur. Litter accumulates. People start drinking in front of the grocery; in time, an inebriate slumps to the sidewalk and is allowed to sleep it off. Pedestrians are approached by panhandlers.

From a policing point of view the changes described may seem low-level. No serious crime is occurring. The local community may be worried about how their neighbourhood is going, but is it really a policing issue? Kelling (1998: 5) discussed the format of police–community meetings:

> Police would want to talk about the number and locations of burglaries and robberies, and citizens would want to talk about issues such as scantily-clad prostitutes who were embarrassing husbands in front of families, youths who were drinking in parks and denying them to families, aggressive beggars who were intimidating and threatening citizens, abandoned vehicles, graffiti and other such minor disorderly conditions and behaviour. For many citizens, this gap between citizens and police meant that either police just didn't get or they just didn't care about neighbourhood problems.

Community policing initiatives meant that the two groups were meeting, but this did not necessarily mean that they were coming to agreement as to what should be done. According to Kelling's research in the USA, the police would have three major reservations. First, whilst sympathetic to local concerns, they would see 'serious crime' as the main problem and the one towards which most effort must be addressed. Second, emergency calls had to be answered quickly, and this meant police had to be in cars. Third, they would not want to get involved in policing problems such as drugs and prostitution because of the dangers of police corruption.

This third objection was perhaps less relevant to the British situation than in the USA. The police there had become so worried about their officers being corrupted by organised crime, especially that involving drugs and prostitution, that forces such as the New York Police Department had adopted policies whereby their officers virtually ceased to police such activities (Bratton 1998: 33). In Britain, however, the first two objections were highly relevant. Dealing with petty crime and minor disorder was not seen as a core policing activity. In the 1990s, as the police became subject to the same efficiency drives that had been applied to other public services, the measures of efficiency initially selected were those that were easy to measure, such as response times to 999 calls (the equivalent of US 911 calls). These 'efficiency drives' kept police at headquarters or in their cars, even though communities wanted them back on the beat.

The 'broken windows' thesis did not stop at simply identifying how a community could enter a downward spiral in which petty crime and minor disorder became widespread. The thesis proposed that such areas would become 'vulnerable to criminal invasion'. The bleak picture portrayed by Wilson and Kelling (1982: 32) was:

> That drugs will change hands, prostitutes will solicit, and cars will be stripped. That the drunks will be robbed by boys who do it as a lark, and the prostitutes' customers will be robbed by men who do it purposefully and perhaps violently. That muggings will occur.

In this way the concerns of the community meet the concerns of the police. The community is worried about the minor incivilities. The police are worried about serious crime. If the minor incivilities are not dealt with the result is likely to be serious crime. One broken window left will not just lead to more broken windows; it could lead to the collapse of the neighbourhood.

This message was taken to heart by Bratton when appointed Commissioner of the New York Police Department (NYPD) in 1994. Bratton (quoted by Gibbons 1996: 19) stated:

> Something had to be done about quality of life offences, they were causing fear. We had aggressive begging, noise, graffiti, public drinking and urination. They're not serious themselves, but they raise the fear levels of those who witness them and contribute to more serious crimes.

The linchpin of the strategies adopted was entitled 'Reclaiming the Public Spaces of New York' (NYPD 1995: 8). The strategy acknowledged the failures of the previous twenty years in the shape of the low priority given to problems such as noisy car stereos, street prostitution, under-age drinking and squeegee pests. All were identified as factors that could blight a neighbourhood, and these problems were identified as a major factor in making New York 'probably number one in the perception and fear of crime' (*ibid.*). Given New York's enormous crime problems, with over 2000 murders in 1990, it was perhaps surprising that Bratton, in conjunction with the new mayor Rudolph Giuliani, decided to emphasise such minor crimes as being a major focus of attention. The rationale was that street disorder 'affects people's perception and fear of crime. Controlling disorder can also change the behaviour of criminals, who are less likely to commit crimes in a well-policed, orderly environment' (*ibid.*)

The results were astounding. In Bratton's twenty-eight months as Commissioner of the NYPD violent crime was reduced by 38 percent and the murder rate by

51 percent (Bratton 1998: 40). Crime in New York fell from just over 600,000 offences in 1993 to well under 400,000 offences in 1996 (FBI 1994: 97; 1998: 104). There has been a subsequent debate as to the reasons for the decline (Horowitz 1995; Bowling 1996; Gresty 1996; Bratton 1998; Pollard 1998), but the figures appear to provide ammunition for the proponents of the 'broken windows' theory. The police became more involved in maintaining order on the streets and dealing with minor incivilities. Not only did total crime fall generally, but more serious violent crime fell dramatically. It seemed that Wilson and Kelling's thesis was proven. Community concerns about street crime and police prioritisation of serious crime could be combined in a strategy that would combat both problems. Dealing with street crime could reduce serious crime.

At the same time that Bill Bratton was trying out the 'broken windows' philosophy on the streets of New York, the Cleveland Constabulary was adopting a similar approach. Superintendent Ray Mallon has become the face of zero tolerance policing in Britain. In April 1994 he took on the role of leading the fight against crime in Hartlepool. His aim was that 'the police would "return peace to the streets" by controlling minor situations in the interest of "decent" and "respectable" citizens' (Dennis and Mallon 1998a: 66). Mallon (quoted in Gibbons 1996: 20) added:

> I told officers to get intimate with this sort of behaviour [anti-social behaviour] and yobs. Where appropriate, they should arrest people, but also just confront them. When they see 10 yobs on a street corner and get out of the patrol car and confront them and let them know we're there. It's a simple strategy that has led to crime figures tumbling down.

His comments support the 'broken windows' thesis in that low-level disorder is being addressed, with the types of behaviour that worry some citizens, such as youths congregating on street corners, being addressed by the police. However, this raises civil liberties concerns. The fact that youths congregate on street corners is not in itself criminal. The recommended police response to their presence is confrontational and could easily amount to harassment. The policing style envisaged is also interesting in that it appears to be halfway between the patrol officer on foot patrol and the officer in a patrol car. While Mallon envisages that his officer will be in a patrol car, he expects him to be very ready to leave his car to deal with quality of life concerns. This presents a proactive vision of policing from a patrol car. The officer is not simply driving around waiting to be called to an emergency. On the other hand, the officer in the patrol car does not seem able to fulfil all the beneficial roles envisaged in Wilson and Kelling's model or in the community policing vision of the beat officer. An officer in a car will not be as approachable as an officer on foot. His links with the community that he patrols will not be as close and his relationship with the community might not be as positive.

The comment on crime rates is also worthy of consideration. When Ray Mallon was appointed to lead the fight against crime in Hartlepool, crime had risen by 38 percent in four months; by the time he left two and a half years later, it had been virtually halved (Chesshyre 1997: 22; Cradon 1997: 13). He was transferred to Middlesbrough, where he promised to cut crime by 20 percent in 18 months: the target was achieved in six weeks (Romeanes 1998: 39). Zero tolerance policing was extended throughout the Cleveland Constabulary and in 1997 Cleveland, with an 18 percent drop, recorded the greatest fall in crime rate of any of the forty-three police forces in England and Wales. The crime reduction figures are spectacular, although zero tolerance is only part of the explanation (*ibid.*).

As in New York, the Cleveland style of policing places considerable emphasis on so-called quality of life offences. In both cases, the police were responding to public concerns. In Cleveland, 75 percent of calls to the police involved 'quality of life complaints, including nuisance neighbours, graffiti, litter and youngsters riding motorbikes on pavements' (Dean 1997: 345). This focus on quality of life offences has other benefits not touched upon in 'broken windows'. By focusing on all petty crime, more serious crime could also be addressed. In New York, where the carrying of weapons in public was commonplace, offenders might think twice if they knew that any minor offence could lead to them being searched. Part of the decline in the homicide rate in New York can begin to be understood in this way. Young men who know they are liable to be searched by the police if they commit any minor misdemeanour may choose to leave their guns at home. If they then get into an argument on the streets and do not have a gun to hand, the altercation is therefore much less likely to end in death. The impact of such strategies is likely to be much less dramatic in Britain, where shootings are rare and the carrying of guns on the streets much less common. However, at a lower level it could affect some potential offenders: drug dealers, car thieves or burglars wary of the risk of being stopped and searched may decide to leave the tools of their trades at home, and as a result some crimes may be avoided.

A further benefit of the zero tolerance style of policing cited by Ray Mallon is that it enables the police to intervene in a juvenile's criminal career at a very early point. 'Criminals, like house burglars, have a career path which starts with anti-social behaviour so we decided to concentrate our resources on tackling that' (quoted in Gibbons 1996: 20). On the assumption that an individual can more easily be deterred from a life of crime if caught early, the policing of minor incivilities give police the opportunity to divert these young offenders away from more serious criminal careers.

Mayor Giuliani's re-election in New York with a significantly increased majority and Ray Mallon's popularity in Cleveland, evidenced by his winning Radio Cleveland's Personality of the Year competition, suggests that the style of policing has proved popular. Zero tolerance as a new school of thought was able to strike a chord with the general public because it was seen to be doing something about problems that seemed insurmountable.

Zero tolerance does not stop at law enforcement being stepped up. Its distinguishing feature is that all infringements of the law should be dealt with in some way, and in the eyes of many it translates as the maximum use being made of stop-and-search powers. In an editorial in *Police* (anon. 1997a), it was stated that zero tolerance meant 'making full use of existing stop and search powers.' This certainly seems to have been the case in Cleveland, with the numbers recorded as having been stopped and searched going up from under 2500 in 1993 to over 48,000 in

1997 (Home Office 1994b; 1999b). Mallon (quoted in Chessyre 1997: 24) is reported to have said:

> We've got to get more assertive and stringent. If there's someone on the street at 3.00 a.m., stop 'em, search 'em. If you see a car, stop it, check it. Don't accept it's all right. Be confrontational. If you're not happy, lock 'em up.

This approach seems a long way from that characterised by PC Dixon. It also seems dangerously close to ignoring the safeguards enshrined in PACE, which noted that an officer should only carry out a search where there are 'reasonable grounds for suspecting that he will find stolen or prohibited articles' (PACE s.1 (3)). Arrests should generally take place only if the officer has reasonable grounds for suspecting that an arrestable offence has been or is being committed and has reasonable grounds for suspecting that the person he is arresting is guilty of that offence (PACE s.24). The codes of practice, published pursuant to PACE, state that reasonable grounds for suspicion cannot be based on stereotyped images of those thought most likely to commit the offence and require that the officer must have a genuine suspicion and reasonable grounds for forming such a suspicion.

DISCUSSION AND CONCLUSION

The policing style is also reminiscent of Operation Swamp '81 – the police operation criticised by the Scarman Inquiry (1981) as one of the factors that led to the Brixton riots. In Brixton, 943 searches had been carried out in the week preceding the riots. The operation involved a large number of random, arbitrary and discriminatory searches (Christian 1983: 21). In his conclusions as to the causes of the Brixton riots, Lord Scarman concluded that they 'were not premeditated. They began as a spontaneous reaction to what was seen as police harassment' (Scarman 1981: para 8.11). Concern about a return to such inner city unrest is a common feature of British police responses to zero tolerance policing. An editorial in a police magazine commented: 'Before we get too carried away, we would do well to remember that we have been here before. It was something very akin to 'zero tolerance' policing which led, if Lord Scarman is to be believed, to the rioting in Brixton and elsewhere in the early Eighties' (anon. 1997a: 11). Metropolitan Deputy Commissioner Brian Hayes (1998: 20), assessing zero tolerance policing, expressed the view that:

> If policing styles are seen to be unnecessarily insensitive, if resentment is created by the apparently arbitrary use of PACE searches, issues of slips to produce driving documents or the constant 'moving on' of groups of ethnic minority youths, then any short-term gain may well jeopardise longer term order and tranquillity.

A major concern for the police is that their strategies are going to be perceived to be unfair. If the policy targets certain groups within society either deliberately or inadvertently then those groups may well object. If the objections are simply coming from burglars and muggers, society may not be worried or may see the complaints as evidence of the success of the policy. However, if objections are more widespread and are coming from law-abiding groups within society then the police are heading towards serious problems. Historically stop-and-search powers have been used disproportionately against the black community (Brown 1997: 16–27), and there is a danger that, if an increase in stops follows this pattern, it will exacerbate racial tensions. This is of particular concern following the Lawrence Inquiry, which focused attention on institutionalised racism not just in the Metropolitan Police but in police forces throughout Britain. Paul Whitehouse, Chief Constable of Sussex Police, has written of the danger that if the police are 'perceived by a minority community to have become agents of neighbourhood bigotry, then the danger to community relations and public order are obvious' (Whitehouse 1997: para 4.1b). Similarly Brian Hayes (1998: 20) noted that: 'The firm and sometimes aggressive styles of policing associated with zero tolerance may gradually lose support for the police if certain groups feel they are being victimised.'

In addition, in concentrating on street crimes, zero tolerance is focusing on crimes predominantly associated with the underclass in society. People who are without work and possibly without homes are more likely to be on the streets as they lack alternative places to go. Clamping down on low-level disorder is likely to target socially and economically disadvantaged groups, who lack the community representatives who can voice their anger and frustration at such treatment. They provide an easy target for police and politicians alike. In May 1994, John Major, in a speech in Bristol, stated that 'Beggars are offensive and could drive shoppers away from cities. The law should be used vigorously to deal with them' (quoted in Burke 1998: 89). His attack was taken up by the Labour Party, with Jack Straw criticising 'street beggars, winos, addicts and squeegee merchants' (quoted in Crowther 1997: 3) and Tony Blair justifying his support for the targeting of aggressive beggars by the police (anon. 1997b). Burke's comment (1998: 88) that 'targeting and criminalising the deprived and disadvantaged is simply unfair' has gained little political support.

Indeed, politicians of both major parties have sought to increase the police's armoury of powers to deal with street crime. This in itself might not seem surprising, but what is significant is the form that some of the recent changes have taken. Zero tolerance can be seen as a means of imposing central influence on the way individual constables exercise their discretion. In other words, if they are to respond with 'zero tolerance', this new model of police powers would seem to be consistent with the centralisation of discretion. This is done by relieving the constable on the street of the need to apply the criterion of reasonable suspicion before making certain stops and searches, as is required for the stop-and-search powers and powers of arrest found in PACE. The new model places this stage of the process with a senior officer within the command structure. The decision is made at a desk rather than on the street. This procedure was first adopted for section 4 of PACE, which provides for the imposition of road checks in connection with the investigation of 'serious arrestable offences' as defined by the Act. For example, if there are reasonable grounds to suspect that a person has committed such an offence and is likely to be within the area, an authorisation may be issued allowing constables to stop vehicles at random within the area specified. There is no need to suspect that any particular vehicle is carrying such a suspect. To some, this may be seen to be an inoffensive provision, since any search and arrest will be a matter for the constable's discretion at the scene of the road check. In deciding whether to search, the officer should be assessing whether he has reasonable grounds to suspect that either a search or an arrest is justified in accordance with his powers under section 1 or section 24 of PACE.

Ten years later, this method was applied to actual searches rather than merely to stopping vehicles by section 60 of the Criminal Justice and Public Order Act 1994. Under this section, a police inspector is now permitted to authorise his constables to stop and search individuals for offensive weapons if he anticipates that incidents involving serious violence may occur imminently, or if he reasonably believes that people will be carrying dangerous or offensive weapons, in the specified locality. At street level, the constable may carry out the search as he or she thinks fit with no need to suspect the individual. Following the London Docklands bombing of 1996 (see Chapter 7), the new model was again adopted in connection with outbreaks of terrorism. Under the terms of the Prevention of Terrorism (Additional Powers) Act, anyone within the area specified could be stopped and searched for articles that can be used in connection with terrorism. Again, no specific suspicion of the person searched is required. Similarly, following the murder of headmaster Philip Lawrence, the Knives Act 1997 extended section 60 to apply more specifically to the carrying of knives. As a result, the opportunities for the police to interfere with individual liberty have increased without any accompanying safeguards requiring reasonable suspicion of the person to be searched being imposed.

The logic of this type of legislation in a zero tolerance context is obvious. Each of these new powers may be justified within the context of its own debate. A future Home Secretary or private Member of Parliament may see these new powers as an encouragement to catch the mood of the times and secure further extensions to existing search powers. By such *ad hoc*, incremental steps, each arguably justifiable, major encroachments could be made into individual liberty. Creating more opportunities for senior officers to give constables blanket authorisations with the encouragement of a zero tolerance campaign in a targeted problem area will be a tempting progression for any Home Secretary and many other back-benchers who recognise the popularity of 'law and order' crusades. However, the shift towards central authorisation changes the nature of the constable from an official who must exercise discretion according to reasonable criteria into an operative carrying out orders. Furthermore, when combined with a policy of zero tolerance, the operative is positively encouraged by the rhetoric to be 'intolerant'. Any power to be exercised by a constable at street level that does not have any requirement of reasonable suspicion may lead to some groups concluding that they are being unfairly singled out, with the consequent damage to community relations. If people do not know why they are being stopped or think they are being stopped for no good reason, then the exercise of the power may well prove counter-productive and may harm the basis of policing by consent on which so much policing still rests.

Such changes in the law are steps towards the realm of arbitrary stops. While the majority might accept such changes as justifiable, those against whom they are applied may well respond very differently if the power is seen to be unreasonable, unfair or even illegal. It should not be forgotten, for example, that stop-and-search powers tend to be used particularly against young black men. If they are to be used increasingly without reasonable grounds, whether legally or illegally, whether under the old or new models, the impact on race relations is likely to be very serious. Combine this mixture with zero tolerance and the concerns regarding racism that we have already noted, then this much publicised style of policing starts to look like a high-risk strategy that could easily turn out to be a step too far.

New additions to the police armoury are found in the Crime and Disorder Act 1998. Section 14 provides local authorities with the power to establish local curfew schemes, and section 16 provides the police with powers to remove suspected truants from the streets and take them back to school. The consultation paper *Tackling Youth Crime* (Home Office 1997: para. 113), published prior to the Act, states that 'unsupervised children gathered in public places can cause real alarm and misery to local communities and can encourage one another into anti-social and criminal habits.' The problem epitomises the sort of quality of life infractions that zero tolerance sets out to combat. The 'alarm and misery' of the populace is recognised as a legitimate concern. 'Anti-social' and 'criminal' behaviour are linked, the distinction that one is quite possibly lawful and the other unlawful obfuscated. Both types of behaviour are seen as being appropriate targets of police intervention. The power provided in the Act, enabling local authorities to impose local short-term curfews for children aged under 10, may be seen as a fairly limited step in this direction. However, the tenor of the consultation paper suggests that it may be only a first step in controlling the activities of children on the streets.

The Act also introduces anti-social behaviour orders. These are based on the premise that persistent petty offences such as verbal abuse, depositing litter and causing excessive noise can seriously affect quality of life and that it is the law's role to provide protection. While aimed more at offences committed by neighbours rather than street offences, the rationale for the policy is clearly in accord with the 'broken windows' thesis and fits in with the general approach of zero tolerance. The orders may be sought by either the police or the local authority and can relate to behaviour already covered by the criminal law, such as criminal damage. It can also apply to transgressions that may well not constitute criminal behaviour, such as noise nuisance or conduct or language not amounting to an offence under either the common law or any other statutory provision. Just as the rationale for the curfew orders muddies the distinction between criminal and non-criminal behaviour, the same is true of anti-social behaviour orders. Ostensibly the provision is a civil matter, but, unusually for a civil action, the police can institute it. Further muddying is caused by the fact that a breach of such an order is treated as a criminal offence.

Recent legislation is clearly extending police powers to enable them to deal with a wide range of behaviour, whether on the streets or between neighbours. This ties in closely with the ethos of zero tolerance and fits the model of policing in which the police's role is to maintain order. In this respect, the difference lies not in the principle but in the degree of enforcement and the addition of the hard edge of policy in substitution for the flexibility of the constable's discretion. Even minor threats to the quality of life of law-abiding citizens are not to be tolerated. Yet it has never been part of policing theory to tolerate crime as such. Police discretion allows enforcement through the courts to be kept in check, with consequent saving in time and money. Warnings and cautions may all be acceptable methods of applying the law with regard to certain types of offence in certain types of community. 'Turning a blind eye' when on the beat no doubt happens in the interest of good community relations or administrative convenience. Students of the English legal system soon realise that without such discretion the administration of the criminal justice system would become unworkable, and the police themselves would become very unpopular.

The rhetoric of zero tolerance suggests elimination of this discretion. If this means that a constable is expected to act differently when dealing with different communities, then it must be recognised that equality before the law is being sacrificed in order to achieve a wider public objective. In one area a person might be arrested for a minor offence, whereas in another area the police might turn a blind eye to such behaviour. Furthermore, if powers of arrest or stop-and-search are exercised with no real intention of processing the case through the courts, the application of zero tolerance short-circuits the traditional process by imposing discipline extra-judicially by a method that may well be seen by some significant minorities as harassment, and in turn, by their lawyers, as an abuse of the process. In this way, the constable becomes an operative in a programme of community catharsis, where other policies have failed to pull the area out of the quagmire of urban decay.

It may be that not only police powers but policing itself will be crafted into a new model in some police forces so that, in the areas to be changed, we will have to prepare for centrally imposed street policing. The police will round up children on the streets during school hours and take them back to school. Young children on the streets late at night will be similarly rounded up and returned to their homes. Fears about knives on the streets will lead to anyone being searched, while bad neighbour disputes become a policing task through the quest to eliminate anti-social behaviour.

The fact that the police have been given these new powers and tasks does not necessarily mean that they will be quick to use them. Many police officers are wary of the new model of policing that is being offered. Charles Pollard (1998: 55) points out the limitations of such policing:

> One can go so far in applying 'Zero Tolerance' but the time will come when it is not enough, and it is seen not to be enough. Then 'Zero Tolerance' will become positively counter-productive. It may then be too late. Firstly the police will have lost touch with the community. Confidence will have drained away. Tensions will have risen. It will then need only a spark to ignite serious disorder, as happened in Los Angeles following the Rodney King case. We know about these things in England too. They happened in our own cities in the 1980s and we have learned hard lessons of our own.

It was claimed at the beginning of this chapter that zero tolerance was the latest policing method to hit the headlines. After the initial excitement and controversy the dust seems to be settling, leaving a clearer view of an area of law enforcement that carries a simple message. This is that where anti-social behaviour and criminal offences are ignored, there is a natural development for things to get worse. In such situations, zero tolerance offers a possible solution. Where the proponents of zero tolerance may have allowed their proselytising and the press to get the better of them is to allow the 'let's get tough' school of thought to overshadow a wide range of other initiatives. Such initiatives include partnerships between the police, business and the local authorities, and problem-oriented policing, with its analysis of underlying problems with a view to removing the causes of crime. Perhaps the time has come for greater public attention to be given to such initiatives.

CHAPTER 13

DESIGNING CONTROL AND CONTROLLING FOR DESIGN: TOWARDS A PRISON PLAN CLASSIFICATION FOR ENGLAND AND WALES?

Simon Marshall

The prison represents society's ultimate institutional landscape of defence. As David Canter (1987) remarks: 'it is hard to think of any other building that is in itself a solution to a social problem.' They promote 'geographies of exclusion' (Sibley 1995) by bounding and excluding deviance behind walls, defending society against the perceived risk from serious crime. By doing so, they exemplify what Sibley describes as society's attempts at the 'socio-spatial control of deviance' – most notably demonstrated in Jeremy Bentham's (1791) plan for a Panopticon prison, famously explored in Michel Foucault's *Discipline and Punish* (1977; see also this volume, Chapter 11). The prison ensures that criminals are prevented from gaining access to individuals and places within society, a tool for the 'purification' of places. As such, the prison may be viewed as a spatial tool used institutionally to regulate liberty. Yet despite the belief that prison design can directly influence the behaviour and control of prisoners, the lack of literature on the built environment of prisons is one of the most cited gaps in the history of penal research (Home Office 1984b; 1985; Mott 1985; Canter 1987; Ditchfield 1990; Marshall 1994; 1995; 1997a).

The chapter has three aims. First, by drawing on a multidisciplinary study carried out for the Home Office looking at order and control in medium-security prisons in England and Wales (Marshall 1995; 1997b), it identifies the applied problems of examining control, in particular, the problem of controlling *for* prison design. Second, by drawing on previous attempts to classify prison form and the three main schools of urban built form classification (Conzenian, Canniggian and Marchian), it examines the theoretical problems in creating a prison plan classification. Finally, it briefly considers some of the lessons *from*, and problems *of*, applying the plan classification and looks at the potential implications of the classification for future prison management policy.

CONTROL IN PRISONS

The major political and historical event that marks the path of prison history is the riot.

Berkman 1979: 34

Modern society has become increasing preoccupied with risk management (Beck 1986). The main perceived risk to the public posed by incarcerated prisoners is the possibility that they might escape. The most important function of prison for the public is therefore physical security or custody. The Prison Service, in contrast, is committed to both the custody and care of prisoners, having to ensure the safety of everyone in prison. It must therefore attempt to rationalise both external security and internal order. Routine order in prison is achieved through an unspoken consensus between the staff (being in a minority) and the inmates (being the majority) that the rules by which their daily lives are governed are imposed and enforced fairly in pursuit of the legitimate task of management. As Ditchfield (1990: 1) remarked: 'Control is ensured by virtue of management's *authority* rather than its *power*.' The maintenance of order therefore requires a clear management distinction between the tasks of control (the imposition of order) and security (Morgan 1994: Sparks *et al.* 1996). Strong physical security will not ensure control; indeed, evidence suggests that it may in fact incite disorder (Woolf 1991).

The level of control imposed in prison may be viewed as a risk assessment between providing 'humane containment', on the one hand, and any potential threat to security which that might entail, on the other. One way in which this risk is managed is the security categorisation of prisoners into Categories A, B, C and D, according to the likelihood of their escape and subsequent threat to the public in the event of such an escape (Mountbatten 1966). 'Category A' prisoners pose the greatest threat should they escape and 'Category D' the least. In addition, the same categorisation is also applied to the establishments designed to hold the different categories of inmate. Mountbatten's classification was intended as an indicator of security and *not* control. He suggested that once the security categorisation had been applied, control of prisoners should be a matter for the management 'regime'. Ironically, however, even Mountbatten admitted that a security classification was an imprecise method of allocating prisoners, given that security was the prime consideration for only a small minority of prisoners.

While prisons, given the nature of their population, are unquestionably more susceptible than most institutions to problems of order, it should be emphasised that such problems are not endemic to normal prison life; indeed, despite media portrayals, history shows that riots remain relatively rare occurrences (Marshall 1997a). Nevertheless, on 6 October 1993, a riot broke out at Wymott Prison, near Preston in Lancashire. The incident lasted almost twenty-three hours and resulted in millions of pounds worth of structural damage, making it the second most serious incident in English prison history, after the well-publicised rioting at Strangeways Prison three years earlier. The incident was merely the latest, and most serious, example of an increasing tendency in modern penal history for disorder to occur not in the austere maximum security prisons or the overcrowded local prisons but in the relatively

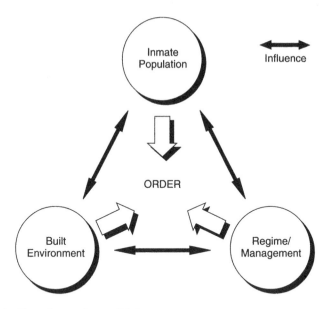

Figure 13.1 The prison order equilibrium

relaxed regime of the medium-security or 'Category C' prisons. In the report of the official inquiry into the incident that followed, the then Chief Inspector of Prisons, Judge Stephen Tumin, called for research to be carried out to look at improving long-term control in the Category C Estate (HMCIP 1993).

Accepting his recommendations, the Prison Service commissioned two studies on control, based upon a tripartite conceptualisation of order (see Figure 13.1). The first study, *Control in Category C Prisons* (Cookson *et al.* 1994), examined the role of the inmate population in maintaining order. The second, *Control in Category C Prisons: Built-Environment and Regime Study* (Marshall 1995), examined the relationship between the physical design of prisons (built environment) and their management (regime) and considered their impact on the creation and maintenance of order. The Marshall study marks a departure in approach from previous research, as ironically penal research has traditionally ignored the prison itself. This failure of empirical studies to control systematically for the influence of the built environment of prisons must cast considerable doubt on the validity of the findings of previous studies on control. However, as Ditchfield (1990) notes, this is largely because of the permanency of design in comparison with the more ephemeral features of the prison, such as the population or regime. It is almost impossible to hold other factors constant while the design of establishments is varied. Nevertheless, unless research is able to control for the marked physical variation between establishments – in terms of age, size, location, and accommodation type – it will be impossible to identify the influence of the built environment in the maintenance of control.

CLASSIFYING PRISONS?

> Good design can have beneficial effects, but it is not always clear why or in what way.
> The first difficulty lies in defining how to measure a successful design.
>
> Fairweather 1995: 19

It is important to establish some form of classification system if we are to make sense of the relationship between the built environment and control problems for prisons. This point was powerfully made by Di Genera (1975: 1):

> It is obvious that the reality of an institution is the result of the interplay between physical structures and the treatment programmes organised within them ... The research focused on the consideration of the architectural structure as an independent variable which, to a considerable extent influences the type of activity that can be organised within it. Although a bad structure may be used in the best way and a good structure may be misused, it seems that a study confined to the physical container of possible treatment programmes is worthwhile itself for the advancement of correctional systems.

Yet considering the obvious benefits in terms of providing a structure for design evaluation, there have been remarkably few attempts at prison design classification. The few examples that do exist fall broadly into one of two categories: academic design classifications and operational classifications.

In general, *academic* prison classifications have tended to outline detailed plan-type categorisations but have made little attempt to apply them to the real world beyond a single perfect exemplar of each type. For example, in 1970 the Law Enforcement Assistance Administration of the US Department of Justice commissioned a series of research projects aimed at avoiding the mistakes of the past in developing a new prison-building programme. Two resulting studies were William Nagel's *The New Red Barn* (1973) and an accompanying monograph by Norman Johnson entitled *The Human Cage: A Brief History of Prison Architecture* (1973). While Johnson's monograph still remains the authoritative account of the origins of the main forms of penal design and includes many plans of design 'types', it is primarily descriptive and makes little attempt to conceptualise these designs or infer properties telling us what is unique to each design. Nagel, on the other hand, drawing upon Johnson, does summarise the basic plan types and provided examples and histories, but fails to define the different types in terms that would allow them to be applied further. Similarly, Gill (1972), in a discussion of the philosophy of prison architecture, attempts to categorise what he considers to be the ten US 'plan types', with dates of origin and an example of each. Again, there is very little discussion of the types; in particular, there is little analysis of the defining characteristics of each type, making application and continuation of the typology extremely problematic.

In 1975, the United Nations gathered together six specialists from Italy, Sweden, France, the USA and the UK to examine the international state of prison design in the mid-1970s (Di Genera 1975). In addition to a number of essays on aspects of design, the report collated plans from fourteen countries – Canada, Denmark,

Finland, France, Germany, Israel, Italy, Japan, Mexico, the Netherlands, Poland, Sweden, the UK and the USA. Although some attempt was made to summarise the plan types examined, once more no attempt was made to define the characteristics of each type. The same criticisms may be levied at later UK classifications by Canter *et al.* (1980) and Fairweather (1989). However, the concentration of these studies on the genealogy of prison designs has largely resulted in a failure to consider their application. More importantly, they fail to provide what Kropf (1995: 11) terms a 'labelling convention', an agreed system for identifying distinct forms in a consistent and systematic way. Instead, categories have been identified primarily by means of studying 'pure' origins supported by 'perfect' examples. As such, they are likely to remain largely academic in so far as they do not lend themselves to practical application.

By contrast, *operational* design classifications may broadly be defined as those carried out by penal systems for operational needs. They tend to be the antithesis of academic studies in the sense that while they are typically applied to a broad number of examples, they have a tendency to employ poorly defined categories that all too often are not mutually exclusive. Unlike academic classifications, operational classifications are seldom published, so there are few well-known examples. Two English classifications, however, do deserve particular mention: the 'design suitability scale' and the Directorate of Works energy classification.

The 'design suitability scale' was developed by the Prison Service for its own use. It is not a plan classification *per se*, although it is an attempt to classify the prison estate in England and Wales. Under the classification, establishments are awarded a ranking (1–5) based upon the 'suitability' of their design for their purpose, where 1 represents a well-designed establishment and 4 a poorly designed establishment. The classification was employed to control statistically the influence of design in an unpublished Home Office study of rates of assault in Category C prisons (Ditchfield 1995). The study concluded that design 'had no significant correlations with any other variables and did not appear as a significant predictor. Nor was it part of any significant factor' (*ibid.*: 2). In view of the complexity of the categorisation system, this is hardly surprising. The classification groups too many factors together, including location, design, layout and function, and it is too rigid, assuming too much about particular designs. For example, the classification does not take account of the fact that establishments commonly have more than one type of cell block design. The precise nature of the gradation between the categories is also unclear.

Similarly, the Prison Service Directorate of Works (DoW) developed a design classification for the purpose of comparing energy requirements. While the classification is closer to a true 'plan' classification than the design suitability scale, once again the categories are confusing and not mutually exclusive, mixing dates of origin and previous land use, for example 'postwar-1980' and 'rebuilt camp'. Time-based categories are always problematic, particularly in making later additions.

These typologies are clearly based upon the specific characteristics of establishments at the time they are created and not defined categories or classes. They do not, therefore, lend themselves easily to either continuation or adaptation in the future. However, unlike the academic classifications, which rarely aim to be more than taxonomic, operational classifications are based upon a need to apply a typology to the real world, even if they rarely aim to provide a template for further use. Whatever their failings, all of these existing classifications, whether academic or operational, at least provide a detailed genealogy of existing prison forms and a common set of terms for those forms. What is evidently lacking is a hybrid typology that is simultaneously taxonomic, identifying and defining the characteristics of generic plan types, and applicable to all existing and subsequently added establishments. Since penology has been unable to provide a suitable methodology for such a typology, attention might well focus instead on the wider study of built form classification for a workable solution.

THE CLASSIFICATION OF URBAN FORM

Efforts to categorise and comprehend urban form have occurred simultaneously in a number of disciplines: geography, planning, urban design, architecture and urban history. Primarily, although not exclusively, the study of urban form is embodied by the geographical subdiscipline of urban morphology, defined as 'the systematic study of the origin, growth, form, plan, structure, functions, and development of a town of the urban habitat' (*Longman Dictionary of Geography* 1985). Broadly speaking, despite a proliferation of terminology, we may identify three largely independent strains or schools of urban classification. The first is based upon the works of M.R.G. Conzen (1958; 1960; 1962; 1966; 1968; 1975; 1988), which is strongly associated with the Urban Morphology Research Group at the University of Birmingham (see Slater 1981; 1990; Whitehand 1981; 1987a; 1987b; M.P. Conzen 1990; Whitehand and Larkham 1992; Kropf 1994). The second strain is the Italian architectural school of building typology, developed through the works of Saverio Muratori (1959; 1963), Aldo Rossi (1970; 1982) and, in particular, Gianfranco Canniggia (1979; 1984), which is associated with the work of the Joint Centre for Urban Design at Oxford Brookes University (see Samuels 1985; also Kropf 1994). The final strain is based on Lionel March's (1972) work on urban form and is particularly associated with the Martin Centre at Cambridge University and the Centre for Configurational Studies, Faculty of Technology, the Open University. Taken together, these might be described as the geographical (Conzenian), architectural (Canniggian) and mathematical (Marchian) schools of urban morphology. Despite these strands having been developed in complete isolation from one another, all share a number of common themes. In particular, each provides an essentially structuralist conceptualisation of the built environment based upon hierarchical principles. Consideration of each of these schools may therefore provide a suitable aetiology for constructing a prison plan classification.

Conzenian town plan analysis

Conzen argued that the townscape represents intellectual heritage, 'the cumulative aspirations and labour of successive generations' (1958: 53). Therefore, by analysing the town plan it may be possible to go some way towards identifying what it is that creates the 'spirit of a place', its *genius loci*. Central to Conzen's analysis was the concept of the morphological frame, 'an antecedent plan feature, topographical outline or set of outlines exerting a morphological influence on subsequent more or less conformable plan developments on the same site' (1969: 127). Conzen proposed a tripartite division of the townscape into three distinct form categories: town plan (itself subdivided into the four-element complexes of street system, plot pattern, buildings and blocks), building form and land use. Through this inclusion of the element of use, Conzen's typology adopts an evolutionary or morphogenetic approach, appreciating the interdependence of the town plan and the economic, physical and social background within which it is developed.

There are obvious parallels to be drawn between many of Conzen's ideas and the issue of control in prisons. First, Conzen's attempt to analyse the intangible nature of the *genius loci* is in many ways very similar to penologists' desire to identify the equally intangible nature of a 'prison climate' (Moos 1968; King 1972; 1985; Montgomery 1974; Tosh 1977; Thornton 1985). Second, just as Conzen (1966: 61) believed that the town plan is a 'static link' to the past, the physical incarnation of our intellectual heritage, so too academics have long argued that prison designs reflect the ideologies of those who created them. Indeed, prisons have been referred to as 'frozen penology' (e.g. Johnson 1973; Fairweather 1975). Conzen may therefore offer something that has widely eluded penologists: a systematic method of controlling prison design statistically, thereby illuminating the nature of control. Third, Conzen's work was primarily concerned with medieval town centres, where the frame is extremely pronounced in the form of the town wall (albeit subsequently removed or its course used as a ring road). While urban morphologists have criticised this theoretical construct as limiting (Bandini 1985; Samuels 1985), there are fewer problems in transferring Conzen's ideas to the prison, its walled or fenced perimeter also creating a morphological 'frame' for subsequent development. Finally, Whitehand (1981) suggests that Conzen's greatest contribution was in introducing a systematic methodology and a much-needed conceptualisation of built form into the previously methodologically eclectic indigenous strain of British urban geography, a situation not unlike that at present in penology.

Canniggian building typology

While Conzenian analysis offers much in the way of a basic framework for a prison plan classification, its application is dependent upon reconceptualising the prison as a microcosm of the town, which, while possible, offers a problematic basis for such a methodology. In particular, Conzenian analysis offers little for studies aimed at a level of resolution greater than the individual building. Building typology or the architectural approach to urban morphology, while receiving little attention in the English-speaking literature, primarily due to the late availability of translations (Samuels 1985), has been considerably more successful in its application to the modern townscape. The work of Canniggia (1979; 1984), in particular, offers a workable complement to that of Conzen (Kropf 1993).

Canniggia, like Conzen, seeks to understand the built environment through examination of temporal relations or the process of formation (derivation) and spatial relations at any given time (copresence) in association with the notion of use or purpose (Canniggia 1979). As for Conzen, Canniggia's copresence is analysed through a tautological hierarchical structure (i.e. defined in terms of the elements within it). This hierarchy has four levels: elements, structures of elements, systems of structures and organisms of systems (*ibid.*: 73). Unlike Conzen, however, Canniggia applies this structure not only to the town as a whole but also to individual buildings. Thus building materials (elements) form walls, roofs and floors (structures of elements), which form rooms, stairs and corridors (systems of structures), which ultimately combine to form buildings (organisms of systems). The number of characteristics used to delineate a type (i.e. how detailed the classification) is defined by the level of specificity employed and the minimum scale defined (i.e. either an individual building or building materials). Unlike in Conzen's form complexes, Canniggia treats use as a function of every element and not as a separate issue, which alongside the importance of derivation again makes building and urban typology morphogenetic in nature.

Again, like Conzen, Canniggia has much to offer a typology of prisons. The greater level of specificity offered by Canniggia's typology offers more scope in identifying the differences between otherwise seemingly identical establishments, the maximum level of specificity being the 'unique' prison. In addition, Canniggia's notion of an elementary cell as the common building block of subsequent development is analogous to the basic unit of accommodation around which prisons are built, for example cells and dormitories. Similarly, his concepts of 'base type', the original notion of a type (e.g. Georgian), and 'leading type', the current notion of a 'type' used as the basis for new building (e.g. postmodern), helps in the analysis of modern prisons. Additions to original prison designs or base types are common, each wave employing a different leading type. The built environment of prisons may, therefore, be stratified into the aspirations of previous architectural periods.

Marchian built forms

> Buildings are complex artifacts. Most are unique. Generalisations about buildings are not easy to make. It may help to look instead at built forms which are not buildings.
>
> (March 1972: 56)

Lionel March first put forward his mathematical built-form and land-use typologies in *Urban Space and Structures*, co-edited with Leslie Martin (Martin and March 1972). March's concept of built form is a purely structuralist form of urban analysis based upon the geometry of buildings and plans. Buildings are also treated as being divorced from their origins and site. In this sense, built form is essentially morphographic or descriptive in nature. Like Conzen and Canniggia, March draws attention to factors other than just form, but these are not situational or location-specific, instead consisting of factors such as cost, heat loss and effective use of land.

The advantage of March's typology is the degree to which geometric conceptualisation lends itself to computer-based analysis, in particular using geographical information systems (GIS), and therefore to large-scale application. Brown *et al.* (1991), for example, have attempted to analyse the residential stock of the entire United Kingdom, while Steadman (1994) has created a database of all the non-residential building stock in the UK for use by the Department of Environment in formulating energy policies. Built-form analysis is again based upon representing buildings hierarchically to differing degrees of complexity to enable general statements to be made about buildings with similar shapes without loss of precision and reliance upon anecdotal evidence. In addition, built-form classifications are taxonomic in nature, defining categories in such a manner as to deal with both older adapted buildings and new construction. Each building is recorded in geometric terms by its envelopes, (e.g. walls and roof types), each floor being represented as a succession of floor polygons to which non-geometric attributes such as usage and materials are added.

Steadman (*ibid.*) further distinguishes between principal forms, or the main usable spaces of buildings, and parasitic forms, or subservient spaces such as corridors, extensions, canopies, plant rooms and porches. While, as Steadman notes, the distinction between 'served' and 'servient' spaces is often far from clear, what such a differentiation does offer is the possibility of treating specific aspects of buildings separately. Steadman further distinguishes between open-plan space (e.g. warehouses and offices), halls (e.g. large meeting spaces) and cellular space (e.g. the basic repeated unit such as hotel rooms, offices or classrooms). Finally, Steadman draws attention to the problem of composites or hybrid forms, suggesting that these must be treated as separate from the principal forms.

While in comparison with both the architectural and geographical schools of typology the concept of built forms might appear shallow, it does offer many useful concepts for a prison typology. First, it has been applied to non-residential and more importantly institutional land uses. Second, the notion of principal forms allows attention to be concentrated on the most significant areas of the prison (administrative offices and the officers' mess, for example, do not have a day-to-day bearing upon control). Third, the idea of a repeated basic unit is especially applicable to the structure of prison houseblocks, being based around either cells, cubicles (non-secure rooms) or dormitories. Finally, the idea of hybrid forms is extremely useful for old prisons that have had more recent large-scale additions in a different style or leading type.

While the preceding discussion of the three main strands of morphology has shown that no single branch – geographical, architectural or mathematical – is entirely suited as a structure for a prison typology, together they provide a body of approximately analogous concepts and terminology that does offer a basis for a classification.

THE CATEGORY C PRISON PLAN CLASSIFICATION

The Category C prison plan classification is an attempt to identify basic 'generic' prison plan types in order to initiate development of a tool for meaningful analysis of prison design. In doing so, the aim is to enable comparative study of the Category C estate at a very broad level, while still realistically reflecting the uniqueness of each individual building or prison design. It is therefore hoped that the plan classification will enable a meaningful comparison of control capabilities by 'determining underlying systems that might be respected or extended in new projects' (Samuels 1985: 8). As Samuels (*ibid.*: 9) emphasised: 'the test of a building's appropriateness is not simply its capacity to fulfil a function but also its ability to contribute to the qualities of its context.'

The Category C classification is based upon three precepts. First, establishments are specifically classified according to *visible geometric information* available on prison block and cross-sectional plans. It is, strictly speaking, descriptive or morphographic, in much the same way as March's built forms, and it does not explicitly reflect dates of construction, location, original land use or current function. However, such factors are often implicit in plan forms. For example, former military camps, which constitute 30 percent of the Category C estate, have common layouts, and specific prison designs are more popular at particular points in history. Second, the classification concentrates solely on *residential units*, these being the principal forms of the prison and the basic building units (e.g. cellular space or elementary cells) of prisons. Furthermore, the majority of inmates' time is spent in residential units defining inmates' activity space (the area in which they may demonstrate a degree of locational decision making). Unsurprisingly, they are therefore the principal loci of disorder (Atlas 1984) and thus of primary interest from the point of view of control. In addition, the absence of explicit consideration of parasitic forms (e.g. boiler houses and kitchens) is intended to reduce the complexity of what is necessarily already a complex typology. Finally, each establishment is classified according to the 'essence' or principal characteristics of its *original built form*, what Johnson (1973: 42) refers to as a prison's 'paternity'.

On this basis, therefore, where the plans of two establishments appear to be marginally different but look to conform to the same basic design principle, they are considered as being variations upon the same generic type. This avoids the problem, inherent in typologies, of endlessly creating additional categories where there are only meagre differences in order to try to represent subjects exactly. This is particularly problematic where small numbers of objects are being classified (as in the case of the Category C estate, with only thirty-five establishments), since many groupings may render findings almost meaningless as each class will contain only one or two items. However, while every effort has been made to avoid the creation of a 'composite' or 'other' category, such a category is largely inevitable given both the range of Category C prison origins (e.g. from purpose-built prisons to country houses, a Norman castle and a former harbour citadel) and the number of additions to prisons over the years. The classification consists of two hierarchical tiers (form outline and internal configuration), within which four levels of resolution are addressed: houseblock composition, block layout, building type and cross-section. Each of these will now be examined in more detail.

Form outline

Houseblock composition   Establishments are classified according to the overall composition of the houseblock plan into three groups. These are linked plans, where individual blocks or wings are physically connected, for example, by corridors; non-linked plans, in which individual blocks or wings are free-standing and not physically connected in any way; and composite plans, which contain both linked units and free-standing blocks, typically as a result of blocks being added to an existing linked plan.

Block layout   Establishments are further classified according to their general layout, or relative position. They may be radial, telegraph pole, nucleated, courtyard or irregularly linked designs for linked plans, and camp, campus or free designs for non-linked plans. These design types owe their origins to historical designs and previous classifications (e.g. Fairweather 1975; 1989; Johnson 1973). Many are terms commonly recognised and used by penologists, such as radial or telegraph pole. Table 13.1 summarises the morphological origins of plan types (for a general history of the prison, see Morris and Rothman 1996). Table 13.2 shows some examples of the composition and layout classifications as applied to the Category C estate.

Building type   Establishments are classified according to the geometric form of the buildings that make up the block layout. Each layout type may consist of any combination of building types. Again, many of these building forms are widely recognised designs, such as the rectangular block (New Jail, Milan, 1624), the T-shaped block (Milan House of Correction, *c.* 1756), H blocks (*c.* 1780) and cloverleaf blocks (Louisiana State Penitentiary, Angola, *c.* 1955). Due to the complexity that this creates for graphical representation, both this and the final tier in the hierarchy (cross-section), are shown separately from composition and layout (Table 13.3).

Internal arrangement

Finally, establishments are classified according to landing type or the internal cross-sectional composition of each building type. Houseblocks are either floored (i.e. the building is divided into floors with no line of vision between each landing) or galleried (i.e. the building is open to the roof with landings above ground level only partially floored or galleried). Examples of different cross-sectional plans in the Category C estate are also shown in Table 13.3.

A demonstration of how this classification works is given in Table 13.4, which compares two pairs of establishments with similar composition and layout. While both Everthorpe and Wymott, and Erlestoke and Stafford, are similar in composition and layout, on closer inspection it can be seen that they differ in terms of the type and internal layouts of the cell blocks. The classification therefore enables generalised comparisons while recognising that, in specific terms, establishments

Table 13.1 Morphogenesis of prison plan types in England and Wales

Generic plan type	Plan background	Origin(s)	Early UK examples	Design advantages	Design disadvantages
Radial (Victorian) Cellblocks 'radiate' from a 'centre' facilitating centralised supervision 	The radial plan was the first internally distinctive prison that differed from other institutions. The form 'evolved' from the segregated cellular design of the Maison de Force, Ghent (1772) and the circular and crescent plans of Jeremy Bentham's Panopticon (1790), The Edinburgh House of Correction, and the Virginia Penitentiary, Richmond (1797–1800). English Architect William Blackburn's 'cruciform' and fan array prisons of the late C18th and early C19th are clearly another influence. The design was however popularised after it became associated with the Quaker's 'separate system' in the USA following William Penn's penal code of 1682. The 'Separate System' demanded strict isolation, hard labour, and moral introspection which was reflected in English emigré John Haviland's design for the Eastern State Pen. (1829) consisting of wings of cells radiating out from a central hub from which a single officer could supervise the entire prison. Haviland later refined the design at Trenton (1833). Visitors travelled from all over Europe to see Haviland's design, which was further refined by Sir Joshua Jebb at Pentonville 'the model prison' (1842), becoming the most copied prison in the world, dominating C19th global prison design, ironically with the exception of America, which favoured the Auburn design.	*Belgium (1772)* Ghent prison, (1772) (Architect Jean Jacques Philippe Vilain) *UK (1784–1818/1842)* Suffolk County Gaol, Ipswich (1784–1790), Suffolk House of Correction, Bury St Edmunds (1803–1805) Abingdon Gaol (1804–1812) Kent Penitentiary, Maidstone (1816–1818) Pentonville, London (1842) *USA (1821–1836)* 'Cherry Hill' Eastern State Penitentiary Philadelphia, Pennsylvania (1821–1836) (Architect John Haviland) New Jersey State Prison, Trenton (1833–1836) (Architect John Haviland)	Usk, Gwent (1838) Armley, Leeds (1840) Pentonville, North London (1842) Reading (1844) Preston (1844) Wakefield (1847) Strangeways, Manchester (1868)	Designed for separation, the radial system provides the greatest possible supervision over a large number of inmates by a minimum number of staff. As the design originally envisaged inmates would spend the majority of their day in a cell, where they would also be required to work, cells tend to be large by modern standards.	The radial design with its associated regime was eventually abandoned as it was found to drive inmates mad, was costly to build and costly to maintain. The original design does not incorporate much space for association or other time out of cell. Difficult to adapt for the requirements of a modern penal regime. Large cells are generous by modern standards and are therefore lend themselves more to the 'doubling' up of inmates and so are more prone to overcrowding. Though spurs may be gated, the essentially open-plan design means fire and disorder can spread rapidly and wings tend to be noisy. Design does not lend itself to differential regimes.

Telegraph pole (telephone pole, satellite, fish spine, double comb) Cell blocks are arranged at right angles to a central spine on the ends of crossbar corridors (like satellites) (medium / high security)	The telegraph pole, so called because of the resemblance of its plan form, marked a break from the C19th 'systems' being an architectural solution to increased inmate mobility within the perimeter (vocational training, support services, education etc.). The design was first used in hospitals employing arcaded walkways and corridors between surgery and wards. The first prison built to this plan was Du Cane's Wormwood Scrubs, for many years the largest prison in Europe. Fresnes, France (1898) was a more pronounced example with enclosed corridors and is generally accepted as the origin of this plan type. The design was further refined in the USA in the early C20th with both accommodation and service blocks being linked to the spine. It was, however, the work of Alfred Hopkins at Lewisburg (1932) which popularised the telegraph pole, becoming almost the only design for medium and high-security prisons in the USA for nearly 40 years. The design was further refined at Terre Haute 'the final word' (1940) introducing corridor zones and at Angola (1955) fence zoning the grounds. The UK variation is based upon the designs of Angola (1955) and Marion (1963) with 'H' or 'cloverleaf' blocks bisected by corridors connecting to a spine containing all the centralised services. The design has been used as far as Japan (1923) and Brazil (1937–42)	*UK (1844)* Wormwood Scrubs, London (1844) (Architect Sir Edmund Du Cane) *France (1898)* Seine Department Prison, Fresnes les Ringis, near Paris (1898) (Architect Francisque Henri Poussin) *USA (1914–1963)* Minnesota State Penitentiary, Stillwater (1913–14) Kilby State Prison, Montgomery, Alabama (1922) Lewisburg Federal Penitentiary (1932) (Architect Alfred Hopkins) El Reno Federal Reformatory (1934) Terre Haute, Indiana Federal Penitentiary (1940) Louisiana State Penitentiary, Angola (1955) Marion US Penitentiary, Illinois (1963) Everthorpe (1958) Brockhill (1965)	Allows for spatially differentiated regimes to be run in each houseblock. Plan allows access to central facilities by different groups to be controlled. Allows spatial segregation of land-use functions without the need for inmates to circulate outside by controlled circulation along corridors. In the event of a riot or other disturbances, inmate groups may easily be isolated in houseblocks.	The design is repetitive and can prove disorienting. The design tends to produce long corridors (In the Texas Maximum Security Prison (1957) staff were forced to use bicycles to get from one end of the spine to the other) and makes central supervision difficult, spreading staff thinly.

Table 13.2 Category C prison plan classification (composition and layout)

Composition	Linked plans					Non-linked plans			Composite plans
Layout	radial (Victorian)	telegraph pole (satellite)	nucleated (dog leg)	courtyard (hollow square)	irregularly linked	camp (army)	campus (new generation)	free	composite
	Cell blocks 'radiate' from a centre allowing centralised supervision	Cell blocks are arranged at right-angles to a central block on the ends of corridors (like satellites)	Cell blocks are arranged on the corners of a central block (like the legs of a dog)	Cell blocks are arranged at right-angles to form a self-enclosing courtyard pattern	Cell blocks are linked in an irregular manner	Cell blocks are arranged in a uniform formalised pattern	Cell blocks are arranged in a casual pattern, typically around a physical feature	Cell blocks are arranged at random within the establishment	Cell blocks are a mixture of linked and non-linked plans
Category C examples	Usk (1844)	Wymott (1979)	Coldingley (1969)	Shepton Mallett (1625)	Lancaster (1196)	Send (1962)	Lindholme (1985)	Latchmere (1948)	The Verne (1947) (irregular / free)

Table 13.3 Category C prison plan classification (building and landing type)

Establishment	Building and gallery type											
	T		I		L		✻		H		other	
	floor	gallery	floor	gallery	floor	gallery	floor	gallery	floor	gallery	floor	gallery
Erlestoke	Y				Y							Y
Everthorpe				Y								
Stafford		Y		Y								
Wymott				Y			Y					

Table 13.4 Plan classifications for four Category C establishments

Hierarchical tier	Example 1 Everthorpe	Example 2 Wymott	Example 3 Erlestoke	Example 4 Stafford
Building composition	linked plan	linked plan	non-linked plan	non-linked plan
Layout	telegraph pole	telegraph pole	free	free
Building types	T blocks	✻ blocks	T, L and irregular blocks	T and I blocks
Cross-section	galleried	floored	galleried and floored	galleried

are in fact different and that these small differences may actually make a significant difference to control.

APPLYING THE PLAN CLASSIFICATION

> There may be useful learning from establishments of the same design which have different approaches to, and success in, maintaining low levels of security acceptable to the Service.
>
> (HM Prison Service 1990: 56)

History suggests that traditionally, linked-plan prisons, with their secure movement between blocks, have tended to be used for high-security establishments, while non-linked-plan prisons, with their freedom of movement around the site, have been favoured for low-security establishments. The Category C estate's need for semi-secure establishments, therefore, creates a dilemma by being neither totally secure nor open. As a result, the estate is an eclectic mix of plan types, with the composite plan being the best example of a compromise solution. As such, it might be expected to share many of the same problems and benefits as the linked and non-linked plans. The major advantages and disadvantages of linked and non-linked plan types

	Advantages	Disadvantages
Linked plan	**SECURE MOVEMENT** Easy to control the movement of inmates between buildings owing to secure corridors and walkways	**NOT NORMALISING** Overly deterministic and restrictive movement around the establishment
Non-linked plan	**NORMALISING** Greater potential freedom of movement around the establishment	**NON-SECURE MOVEMENT** Difficult to control the circulation of inmates around the establishment due to the freedom of movement

Figure 13.2   The control implications of plan linkage type

are summarised in Figure 13.2. Unsurprisingly, these appear to be diametrically opposed. As Figure 13.2 shows, the main advantages relate to regime management, in particular in relation to the 'normalising' effect of the environment – diminishing as opposed to stressing the differences between prisons and non-custodial environments – while the major disadvantages relate to containment of disorder.

Many plan types are broadly time-specific: the radial dating from the 1800s, the nucleated and camp from the 1960s, and the campus from the 1980s. This might suggest that their design was linked to a particular penal ethos that has subsequently been discredited and as a result the designs discontinued. In contrast, since both the telegraph pole and free layouts appear to be popular recurring designs, this might suggest that they are based upon sound functional principles. In fact, the manner in which the Category C estate developed suggests that forms are purely the result of site origin (Marshall 1997a). Only 27 percent of the Category C estate was purpose-built for use as prisons. The majority of the estate consists of buildings converted from other functions. The irregular plan can therefore be seen to have evolved out of necessity and not out of any particular penal theory. The same is true of the camp plan, with 30 percent of the estate using former military sites. Equally, the fact that the majority of composite plans are the result of free-layout additions to telegraph-pole plans might be taken as a reflection upon the suitability of that design. Therefore, the extent to which any prison plan type is the result of informed policy remains questionable.

The Category C estate is a fairly even mixture of houseblock designs and galleried and floored landings, many establishments combining several blocks of different designs. As Table 13.4 shows, the difference between two establishments' designs may be as specific as plan cross-section, as in the case of the difference between Everthorpe and Wymott. However, in terms of control this distinction has a very important influence (see Figure 13.3). Galleried blocks aid good daily control through ease of supervision but are less effective in the event of a serious disorder since there is little internal segregation. Floored designs, in contrast, do not assist daily control, requiring staff-intensive supervision, but aid control in the event of a serious disorder because they have greater internal subdivision.

	Advantages	Disadvantages
Galleried	**NON-STAFF-INTENSIVE** Good visibility of almost the entire wing from a single vantage point on the ground floor.	**DISTURBANCE DIFFUSION** Difficult to contain disorder in the event of a disturbance owing to the open-plan design of the wing.
Floored	**DISTURBANCE CONTAINMENT** Easy to contain and isolate disorder to a minimum area due to compartmental nature of the wing.	**STAFF-INTENSIVE** Poor overall visibility, only being able to view a small part of any wing at a time. Creates low, dark, narrow corridors.

Figure 13.3 The control implications of cell block cross-section

The maintenance of order would therefore appear to rest on design risk assessment, a trade-off between designs that facilitate good daily control and designs that facilitate management of serious disorder. In practice, if supervision is effective then this need not be the case as disturbances are seen and reacted to quickly. The floored design suffers from serious flaws. By not being conducive to daily supervision, it is more likely that a minor disturbance could form unobserved and become serious. In contrast, the galleried design makes it easier for staff to observe entire wings and therefore deal with minor incidents before they develop into anything more serious. These fundamental differences may help to explain why the riot at Wymott, a floored design, in September 1993 resulted in more damage than the incident at Everthorpe, a galleried design, in January 1995, despite both establishments being telegraph-pole-plan designs. In practice, however, there is no single cause of rioting, and clearly other factors were also involved.

TOWARDS A PLAN CLASSIFICATION FOR ENGLAND AND WALES?

> The classifications and typologies which now proliferate in professional discourse represent themselves as morally neutral, rational means of fitting offenders to appropriate regimes rather than simple expressions of moral worth or community judgement.
>
> (Garland 1990: 187)

Before considering the potential implications of the use of the classification, it is important to address its limitations. There is growing criticism in criminology of what has been termed the 'demoralisation' of penal discourse; the discussion of 'clinical' models of imprisonment seemingly divorced from the realities of imprisonment (see Christie 1982). Increasingly, Foucault (1977: 9) argued, 'punishment leaves the domain of everyday perception and enters that of abstract conscience.' Such criticism is easily levied at a typology such as the prison plan classification. However, as the criminologist O'Malley (1987: 84) argues in defence of a structuralist approach, 'we do not retreat into theory, it is the means whereby we move beyond the realm of outward appearances.' The purpose of the classification is to allow more meaningful discussion and analysis of prison design than is possible by looking at unique individual establishments.

Information on the built environment is both exhaustive and dynamic, but type must be discrete and limited to be practical. As Steadman (1994: 11) argues: 'classification as an activity cannot be carried on in the abstract . . . taxonomic categories are always coloured by the purposes for which they are made.' The prison plan classification was developed in the context of examining order. As such it is only a first step towards a true classification of all prisons in England and Wales or even, ultimately, internationally. For example, the classification does not specifically include higher-category prisons, although in practice these tend to draw on the same basic designs with enhanced physical security. Equally, at present the classification covers only plan forms used in England and Wales and does not include examples of non-British types such as metropolitan correction centres (tower-block urban prisons such as those found in New York and Chicago), or panopticons, as found in the Netherlands and the United States.

However, the classification does make a valuable first step towards offering a meaningful way of comparing prisons as opposed to treating all prisons as homogeneous shells. The structural framework of the classification is such that it should allow for easy expansion to suit particular needs and thereby lend itself to consideration of any number of topics. By increasing the level of resolution, analysis could be continued to the maximum specificity of the 'unique' prison. Consideration could be given to different arrangements of spurs on landings, the composition of the individual units of accommodation on each spur (cells, cubicles and dormitories) and differences in building materials (as Canniggia does). Similarly, by widening the scale of analysis, the existing classification could be broadened to consider an establishment's perimeter (its morphological frame), or indeed its surroundings and locational setting.

CONTROLLING FOR DESIGN AND DESIGNING FOR CONTROL

The aim of producing a typology of Category C prison plans was to develop a research tool to enable the prison built environment to be considered systematically and its effects ultimately controlled. As such, it has identified important implications of prison design for risk management and staffing. It was also an attempt to break away from the simplistic deterministic approach of past design studies, such as that by Canter *et al.* (1980). However, the importance of the built environment of prisons should be neither underestimated nor overstressed. Bentham, with his panopticon design, was one of the earliest advocates of the power of penal architecture; 'the Gordian knot of the poor laws not cut but untied – all by a simple idea in architecture' (Bentham 1791: 1). All too often since, penal research has fallen back upon environmental determinism when confronted with unexplainable differences.

To quote the much-cited poem by Richard Lovelace (1618–1657) *To Anthea from Prison*, 'stone walls do not a prison make, nor iron bars a cage.' Good design can facilitate order, but it cannot create it; just as poor design may hinder control, but it does not preclude it.

The classification has a further potential role to play by assisting in the design of prisons that facilitate order and aid control. As Canter (1987: 227) concluded: 'the major need now is for post-occupancy evaluation of existing, especially recently built, establishments. It is indefensible that new public money is poured after old without a systematic evaluation of the successes and failures of previous designs.' Never was George Bernard Shaw's adage 'we learn from history that we learn nothing from history' more true than in the case of prison design. Analysis of Category C plan types has shown that designs have been repeated for decades, perpetuating known design faults. As we enter the new millennium, the building programme in England and Wales shows little sign of slowing. Perhaps by encouraging the systematic analysis of existing designs, we may reduce the future risk of serious disorder in prison.

CHAPTER 14

POLICING THE PUBLIC REALM: COMMUNITY ACTION AND THE EXCLUSION OF STREET PROSTITUTION

Phil Hubbard

PROLOGUE

On a warm June day in 1994, a large crowd of residents gather around midday on the corner of Cheddar Road in Balsall Heath, an inner city district to the south of Birmingham city centre. Given the predominantly male, and largely South Asian, composition of this crowd, passers-by may have mistaken it as a group making their way to the nearby Willows Lane Mosque, although a closer inspection of the home-made placards and signs they hold would suggest otherwise. Hastily daubed with slogans – 'Kerb-crawlers, we have your number', 'This is a green light zone not a red-light zone' – these signs proclaim the purpose of this self-ordained community picket, to displace street prostitutes and kerb crawlers from the streets of Balsall Heath. Some cars slow down to see what the fuss is about, before their embarrassed drivers speed away, often to cheers from the growing crowd of residents. Occasionally, women emerge from local houses, and, recognised by the pickets as prostitutes, are subjected to verbal abuse – 'Go home slag', 'We're putting you sluts out of business' – although the presence of community elders, and the reminder that this is a peaceful protest, seem to temper the exuberance of the younger protesters, who begin to pick up stones to hurl at prostitutes and their clients. Later that afternoon, the pickets disperse, their initial excitement in the confrontation wearing off, but a dedicated core remains, huddled around braziers, maintaining their vigil long into the night.

INTRODUCTION

This description of community protest, in the form of direct action by residents, is one that has become increasingly familiar in recent years as *individuals* adopt strategies designed to create new certainties in an era of fluidity, flux and uncertainty. Such 'defensive' measures have been particularly noted in the field of ecological and environmental protest. Highly publicised and mediated protests against road building, hazardous industrial works and nuclear dumping sites, for example, often involve highly symbolic appropriations of space (Routledge 1997b). By adopting tactics well beyond those conventionally associated with community groups (such as drawing up petitions, writing to politicians and organising public meetings), these activists employ forms of action that may bind individuals together in 'neo-tribes' whose values oppose dominant modes of thought and action (Maffesoli 1996). In some instances, this results in unlikely alliances as different class, ethnic and lifestyle interests come together, however fleetingly, to oppose an apparent threat to a particular community. Recent protests against motorway construction in Britain, for example, have seen green activists, new age travellers and retired army colonels standing alongside one another against the bulldozers, attempting to prevent the desecration of the idyllic, and idealised, English countryside (Cresswell 1997).

Quite why there should have been an explosion of popular activism and grass-roots politics has been the subject of much popular conjecture as well as academic debate. On one level, this do-it-yourself activism has been portrayed as a manifestation of people's dissatisfaction with the efficacy of conventional party politics and state-sponsored planning systems in resolving socio-spatial conflicts (McKay 1996). In a broader sense, however, it has been postulated that this dissatisfaction is merely a facet of the postmodern condition where social life lacks security, certainty and order (Giddens 1988). With the order formerly guaranteed by the modern state breaking down in the face of economic and political turbulence, many commentators have therefore suggested that contemporary society is characterised by contingency, ambivalence and ambient risk. Perhaps the most forceful and widely cited version of this argument is contained in Ulrich Beck's (1992) description of a *Risk Society*. Beck argues that, with the global media making the public aware of an ever-increasing number of risks, each individual is constantly faced with choices, which force them to face the likelihood of being affected by a threat. In such circumstances, he maintains that risk has become the dominant *leitmotif* that structures forms of sociality, running palpably through all areas of social life. In caricature, he suggests that this has resulted from the breakdown of the stable modes of social regulation associated with industrial capitalist processes and their replacement by the more diffuse and amorphous flexible production systems associated with post-industrial accumulation processes. Forms of (national) state regulation, financial management and welfarism appear increasingly unable to provide certainty and order in the face of global fluctuation and instability, while transnational organisations like the United Nations, the World Bank and the International Labour Organisation seem distant, obscure and out of touch (see also Doel and Clarke 1997). This perceived 'breakdown' and rupture of the organisations, laws and procedures that managed risks for people in industrial society are held to have prompted individuals to confront *reflexively* local problems and manufacture new certainties in an era characterised by global fluidity, flux and uncertainty.

Beck's thesis of a risk society not only appears apposite in explaining the dramatic increase in the number of community protests occurring in many Western nations but also in accounting for the ever-widening range of targets that are resisted, opposed or attacked in such protests. It is no longer the case that community groups seek to manage risks simply by resisting those hazardous developments that generate fear of environmental pollution, since they also do so by isolating, excluding and eradicating those social pollutants that threaten their ordered, but increasingly unstable lives. For example, many geographers have also begun to document pervasive opposition to human service facilities such as centres and hostels for the mentally ill, the disabled and people living with AIDS (Takahashi and Dear 1997; Wilton 1998). In these instances, people's reasons for opposing such facilities are much less apparent than when they seek to oppose noxious facilities like nuclear power plants or factories. Certainly, there may be concern that such facilities will generate traffic, noise and parking problems (Wilton 1998), but the overwhelming evidence is that community members oppose the siting of these facilities because of less tangible, but equally real, anxieties and fears about populations regarded as *different*.

Developing Beck's thesis, it might therefore be argued that individual urges to minimise risk now not only encompass desires to prevent those intrusive developments that threaten the perceived ordering of space but are also manifest in desires to exclude threatening 'others' from residential and public spaces. An extreme example is the way that residents may come together to oust known or suspected paedophiles from their community, but the desire to exclude marginal groups is much wider than this. Exclusion has, as Sibley (1995) noted, become a pernicious trend in the urban West as powerful social groups seek to purify 'their' residential space. Indeed, this 'exclusionary urge' has been most vividly demonstrated in the way that city space, often regarded as democratic and open, has become increasingly regulated. As a result, groups and individuals whose lifestyles are viewed as incompatible with so-called 'normal' ways of behaving have had their access to urban space limited. As one of the many marginal social groups whose presence in the city has been under considerable scrutiny in recent years – a list that also includes the homeless, rowdy teenagers, beggars and the mentally ill – prostitutes find themselves subject to a range of informal social control mechanisms designed to limit their participation in a variety of urban spaces. The collection edited by Fyfe (1998), for example, offers a useful overview of the way that the streets of Western cities have become implicated in the dynamics of social exclusion, with even acts like smoking, eating and drinking prompting debates about appropriate behaviour in public space.

Although intuitively attractive, Beck's thesis has its critics, and his depiction of the late modern era as fundamentally distinct from other stages of modernity obfuscates many important continuities in the way people deal with risk. In this respect, while Beck presents a mesmerising account of how risk is produced and managed in contemporary societies, it overdramatises the rupture between contemporary experiences of risk and those that existed in other times and other spaces. We may well be in a period characterised by a heightened sensation and perception of risk. Yet this should not blind us to the fact that risk and fear have always constituted fundamental human emotions, with the action of the state often regarded as insufficient in assuaging individual and popular fears. There is, however, a sense in which Beck's ideas are useful for conceptualising the community protest against prostitution witnessed in Balsall Heath, and that concerns the idea that the regulatory frameworks and institutions associated with modernity have unravelled *en masse.* Traditionally, prostitution has always prompted social concern. Although it essentially involves a private contract negotiated between two consenting individuals, the state and the law have rejected this contractual interpretation of sex work in favour of legislation designed to express moral condemnation of prostitution as well as to protect 'respectable' populations from the sight and sounds of those engaged in prostitution. Yet, as has been described elsewhere (Symanski 1981; Lowman 1992), the interpretation and enactment of these vice laws has been inherently geographical, with the legal codes, norms and understandings surrounding sex work varying from locale to locale. This means that the state's urge to maintain the distinction between moral sexuality and the immoral prostitute has been translated into a series of *spatially specific* vice laws, which in turn have informed the policing of prostitution in Western towns and cities. Thus, the *de facto* criminalisation of prostitution has been implicated in the construction of a characteristically nefarious geography of red-light districts and zones of erotic entertainment. Seen to disturb the social order of specific communities, the state and the law have thus tended to distance prostitution away from the wealthier, whiter and more politically articulate residential neighbourhoods, reinforcing the distinctions between 'good' and 'bad' citizens in the process (Hubbard 1999).

Recently, however, these modes of regulation have been inexorably unravelling in the face of critiques of vice laws and policing strategies. Specifically, vice legislation has repeatedly been exposed as being of dubious value in offering a meaningful, long-term solution to the 'problem' of prostitution, publicly criticised by many senior police officers as entailing a huge outlay of resources for little demonstrable return (Sharpe 1998). Moreover, vice laws have also been opposed on broadly humanitarian and libertarian grounds by a growing number of campaigners – including prostitutes' unions, legal experts and feminist groups – who instead support the liberalisation of prostitution laws. Recognising that many women and men will either want to or be forced to work in the sex industry, advocates suggest that liberalisation offers a more pragmatic way of dealing with the problems associated with sex work. They also suggest that it would offer more protection to prostitutes, clients and those living in the vicinity of sites of sex work. In effect, some have gone so far as to suggest that the modern forms of vice regulation established in Western nations have actually exacerbated the problems associated with sex work by attempting to contain prostitution in specific areas. Here there appear to be important parallels with Beck's idea that the frameworks and institutions established by the modern state are themselves implicated in the chronic production and reproduction of risks in all areas of social life (see also Hinchcliffe 1997). By failing to predict, map and manage risk, these outmoded institutions add to a pervasive atmosphere of risk by uniformly failing to create certainty and predictability; risk thus takes many forms and can occur in many spaces. Doel and Clarke (1997: 20) have sought to outline the consequences of this in terms of the *ambient* fear that saturates all social spaces, with scares about food safety and environmental pollution rendering even routine and banal events like eating, drinking, walking and breathing potential hazardous:

> In short, we are all hostages to a stroboscopic pulsation of haecceities, to a statistical indifference that inaugurates a socius of total anonymity and total responsibility . . . Each occurrence has the potential to change everything. Hence the need to vigilantly monitor even the banal minutiae of everyday life, forcing the real to become hyperreal: Street Watch, Neighbourhood Watch, Business Watch, Vehicle Watch, Body Watch, Pet Watch, Bay Watch and the whole paraphernalia of remote-sensing and passive surveillance.

According to Doel and Clarke, in the risk society insecurity is everywhere, no longer confined to fear of those exceptional risks (of murder, genocide, war, famines) that may have preoccupied peoples in the past. Fear of death has been transformed into fear of life.

In this chapter, I want to offer an interpretation of the events in Balsall Heath and other British cities where there have been protests against prostitution in the 1990s, placing these in the broader context of debates surrounding the relationship between civil society and the state as it is played out in the public realm. To do this, I want first to describe the nature of these protests in more detail, identifying the form and content of the different geopolitical strategies that have been used in the attempt to displace prostitutes and kerb crawlers from residential and public spaces. In doing so, I will particularly seek to highlight how such protests engage with contemporary ideas about crime prevention and community safety, blurring the distinctions of police and policed in the process. Second, I want to argue that the attempt to purify residential spaces by removing prostitution represents part of a more general process in which threatening 'others' are having their access to public space challenged. In this respect, I want to highlight the *gendered* and *sexualised* nature of these protests. In doing so, I wish to stress that public space, far from being democratic and open, is, and has always been, characterised by exclusions made on the basis of gender and sexual difference, with the figure of the prostitute central in this process. Throughout, I thus draw on a range of contemporary geographical literature that has sought to relate ideas of citizenship, family life and community space to contemporary debates surrounding identity and social exclusion. A central theme in this literature is that the regulation, ordering and imagining of urban space is a crucial means through which social identities and distinctions are created and maintained, but that this space is produced through complex contestation and social negotiation between *different* publics. These ideas lead me to finish by commenting on the fractured nature of community protests, acknowledging that attempts to exclude prostitutes from specific areas are as much about 'who belongs' as 'who doesn't belong'.

(DIS)PLACING PROSTITUTION: COMMUNITY PROTESTS AGAINST SEX WORK

Although prostitution has always provoked considerable public anxiety and moral condemnation in the urban West (Symanski 1981), recent events in British cities have reawakened debates about the appropriateness of sex work in different locations, particularly the streets of residential areas. In many of these cities, sex work has been long established in often notorious red-light districts, with the police often seemingly content to use existing British vice laws such as the 1959 Street Offences Act and the 1985 Sexual Offences Act to contain prostitution within areas where they are able to monitor and control the situation (Sharpe 1998). Ostensibly, this is in line with the spirit of laws that have the twin aims of preventing 'the serious nuisance to the public caused when prostitutes ply their trade in the street' while penalising the 'pimps, brothel keepers and others who seek to encourage, control and exploit the prostitution of others' (cited in Edwards 1997: 928). In practice, this creates a paradoxical situation. Although prostitution may not be illegal, it is impossible for female prostitutes to work outside unofficially recognised areas of sex work without breaking a number of laws. As Duncan (1996) asserts, in some cities prostitutes are accordingly 'placed' out of sight, located off-street in brothels, massage parlours or private flats where their sexuality may be performed apparently unfettered by the state and the law. Yet in other cities, sex work can be encountered fully in public view, on the streets of marginal areas where it is subject to the disciplining and regulating gaze of the state and the law. In such circumstances, the sex worker's occupation of public space may therefore be highly circumscribed, subject to monitoring and regulation by the police and agents of social control.

The concentration of sex work in often economically marginal inner city locations, has often been resisted by local people. Until recently, protests had mainly been restricted to petitioning and lobbying of local councillors or members of parliament, prompting occasional high-profile vice 'crackdowns' by police. Overwhelmingly, such short-lived campaigns of punitive policing have made little difference to the location of red-light districts. A recent survey of vice squads suggesting that sex work has tended to remain concentrated in areas of no more than 1 square mile in most British cities (Benson and Matthews 1995). The events in Balsall Heath, notorious as the centre of Birmingham's sex industry since the nineteenth century, thus signalled the adoption of a new approach by local residents, who felt that the police were failing to protect them from the 'public nuisance' associated with street prostitution. By adopting a more direct form of protest, in which community pickets attempted to disrupt the work of street prostitutes and hand the registration numbers of kerb crawlers' cars to the police, the residents of Balsall Heath were asserting their own right to regulate what they saw as the streets of 'their' community, effectively taking the law into their own hands. These approaches had been tried in North America, such as Vancouver's notorious 'Shame the Johns' campaign (Larsen 1992), but the protest in Balsall Heath was remarkable in terms of its scale and intensity. Upwards of 150 pickets were strategically positioned throughout the area to cause maximum disruption to both street prostitutes and 'window workers' – those women who worked from rented residences and advertised their services by sitting in the bay windows of these terraced houses.

The impact of this picket was demonstrated by an immediate decline in the number of kerb crawlers in the area, with an 80 percent reduction in traffic in the days following the establishment of the picket. This decline in trade appeared to prompt many prostitutes to leave Balsall Heath to work 'beats' in other Midlands towns, probably expecting to return once the pickets had ceased. However, with a 'hard core' of thirty or so prostitutes continuing to work in Balsall Heath, the pickets persevered in their attempt to remove street prostitution throughout the winter of 1994–95. The local police were initially suspicious of the motives of the protesters, whom they had often referred to as vigilantes. Nevertheless, by the summer of 1995 they had coopted the protest by recommending that the pickets be recognised as part of the Home Office's experimental Street Watch campaign, pioneered in a number of rural settlements. This process entailed the registration of fifty pickets with police and the issuing of official incident sheets to the participants (Casey 1995). This formalised the picketing and patrol process, which although scaled down, still involved the regular monitoring of streets associated with street prostitution. In addition, a number of unofficial protesters still maintained a presence on the streets, although these were discouraged by the official Street Watch campaign. Overall then, the net result of these protests, and the extensive publicity that they generated, was substantial. By the beginning of 1996, 'window working' was no longer prac- tised in the area, and there were only a few women working the streets, mainly on the fringes of Balsall Heath at times such as early mornings when pickets did not regularly patrol. Although it is difficult to estimate prostitute numbers, figures from Birmingham's SAFE-HIV project outreach sessions suggest there has been a two- thirds reduction since the numbers of street sex workers peaked in 1989. As few as one new incoming prostitute is now encountered in each four-week period (see Hubbard 1997). With a sizeable proportion of those who worked Balsall Heath now known to be working off-street for escort agencies or working the streets in other cities, there is little sign of prostitution returning at anything like its previous level.

Given the publicity given to events in Balsall Heath in the national press, it was perhaps not surprising that it spawned imitations in red-light districts elsewhere. For example, in Streatham (London) the Woodfields Residents Association took part in street demonstrations over nine nights in July 1995 to express residents' frustration with police inaction against prostitution. Although the protest had little direct impact on prostitute numbers, it was recognised as instrumental in influen- cing the police's decision to establish a more permanent vice squad in Streatham. Likewise, the community protests in the Hengistbourne district of Bournemouth, coupled with the installation of closed-circuit television (CCTV) cameras, largely succeeded in displacing sex workers from an established area of street prostitution (Campbell *et al.* 1996). Further afield, the residents of two adjoining neighbour- hoods joined forces to protest against the designation of an official zone of street prostitution (or *tippelzone*) in Heerlen, which had been established by the Dutch police and social services following the model established in Amsterdam. Picketing the *tippelzone* during its evening opening hours, these protesters disrupted the work of prostitutes, and after 100 days of picketing succeeded in persuading the Christian Democrat-dominated city council that the zone should be closed down (Visser 1997).

In this case, the numbers of pickets and prostitutes involved was low, and the protest generated little publicity. However, the same could certainly not be said for Manningham in Bradford, where the screening of the television drama *Band of*

Gold resurrected many complaints about police inaction against prostitutes, despite 1361 arrests for soliciting in 1990 alone (the highest figure for any vice squad in the country). This prompted a street picket similar to the one in Balsall Heath, organised here by the elders of the Jamiyat Tablighul-Islam Mosque on Southfield Square. Against a backdrop of exacerbated tensions between pickets (who were mainly Asian), prostitutes and the police, these protests initially escalated into substantial unrest over the weekend of 9–11 June 1995 – subsequently described as the worst outbreak of rioting in Bradford's history. Nonetheless, following negotiations between Toller Lane police and local residents, 'rules of engagement' were identified. Small, well-organised groups of pickets taking registration numbers were now permitted by the police, who sent follow-up warning letters on the basis of this information (Gibbons 1995). Unlike previous attempts at removing street prostitution from Manningham through punitive policing, this community protest, maintained continuously over a number of months, did succeed in moving street prostitution away from residential areas. Yet the fact that women prostitutes are now soliciting in industrial areas to the south of the Thornton Road in the neighbouring Listerhills district again demonstrates that such actions tend to displace rather than remove the supply or demand for sexual services. The more recent move of some prostitutes to working from the local university campus, where improved street lighting designed for student safety offers a more secure working environment, supports this displacement thesis (*Yorkshire Post*, 1 October 1996). The fact that some former sex workers from Bradford are also reported to be working in Doncaster, Norwich, Keighley, Leeds and Nottingham, among other towns and cities, further suggests that prostitution is not necessarily an opportunistic crime that can be 'designed out'. Prostitutes often adapt to changing local circumstances by simply moving to other areas, albeit frequently under the duress of pimps.

CRIME, ANXIETY AND THE 'CLEANSING' OF SPACE

Although, as noted above, these community protests against prostitution were initially dismissed by the police and local authorities as examples of vigilantism, they have subsequently been referred to as successful community safety initiatives and cited as exemplars of a partnership-based approach to crime control. As Hughes (1997) has stressed, the rhetoric and language of partnership has featured strongly in recent crime-control debates following the 1991 Morgan Report, which called for the 'active citizen' to play a key role in surveillance and crime prevention (Morgan Committee 1991: see also this volume, Chapter 10). While the recommendations of this report were acted on selectively, the Home Office's (re)launch of three complementary community safety programmes in 1995 (Neighbourhood Watch, Neighbourhood Constables and Street Watch) demonstrated a clear commitment to involving the public in crime control. The public were encouraged to take responsibility in policy areas previously controlled by the state (Fyfe and Bannister 1996). Yet while the principles of Neighbourhood Watch and Neighbourhood Constables were already well established, Street Watch represented a more contentious and ill-defined attempt to involve the public in community safety. Although described by Home Secretary Michael Howard as 'walking with purpose', it was criticised by the Police Federation, which felt it would promote vigilantism (Casey 1995). Nonetheless, the principles of Street Watch dovetailed neatly with concurrent neo-liberal political thinking about the role of the 'active citizen'. From that perspective, the cooption and redesignation of the residents' protest in Balsall Heath as a Street Watch campaign lent legitimacy to this initiative, showing that the ostensible 'citizenisation of crime control' could successfully promote community safety.

The popularity of such partnership programmes is understandable when viewed against a climate of rising public anxiety about crime and victimisation, itself fuelled by sensationalist media exposés of the 'mean streets' of Britain. These programmes potentially promote social cohesion while ostensibly enhancing police accountability at the local level. By extending active citizenship and giving residents a sense of propriety, such schemes thus represent an integral part of a wide range of crime-prevention initiatives designed to lessen crime and the fear of crime in British cities, fitting alongside other strategies such as CCTV, street lighting and offender tagging. Indeed, the fact that these community pickets have had more impact on the location of prostitutes in their respective cities than any police intervention over the last thirty years suggests that multi-agency approaches involving cooperation between police, local authorities and the public may be successfully pursued as a long-term solution to the regulation of prostitution, particularly when allied with traffic-management schemes (Matthews 1993). In this respect, it is perhaps also worth noting that encouraging residents to participate in voluntary surveillance schemes represents a cheap alternative to the more traditional methods of intensive policing that have characterised postwar efforts to suppress and contain prostitution (Hubbard 1997).

However, for a variety of reasons, it is difficult to see how Street Watch initiatives could be replicated in other areas with respect to a wide variety of crimes and incivilities. Indeed, rather than couching discussion about community pickets in terms of debates surrounding crime and incivility, it is potentially more useful to locate such defensive actions in terms of the aforementioned geographical debates on risk, exclusion and NIMBY-ism (the 'not-in-my-back-yard' syndrome). Traditionally, NIMBY-ism has been explained as a reaction to perceived negative externalities, as exemplified by the negative impact that the environmental and aesthetic pollution brought by new development might have on local property prices. Such economic impacts are, to some extent, quantifiable. What is less measurable is the extent to which the NIMBY syndrome can be taken as evidence for a more complex mixture of popular anxieties about populations regarded as 'other'. It is only more recently that geographers have also begun to explore how such protests reflect the desires of more powerful (and typically white, middle-class, heterosexual) social groups to exclude those whose lifestyles do not accord with what they regard as desirable or acceptable. For example, drawing on psychoanalytical theories about the importance of preserving self-identity – literally, the boundaries of the self – Sibley (1995) has argued that the urge to exclude threatening 'others' from one's proximity is connected to deeply engrained, often subconscious desires to maintain cleanliness and purity. Following Douglas' (1966: 41) argument that 'dirt is matter out of place', Sibley explored the ways in which this abject fear of the self being defiled or polluted is projected, or 'mapped' on to specific individuals or groups who are depicted as deviant or dangerous. He thus argues that spatial exclusion has been the dominant process used to create social boundaries in Western society, and the key means by which hegemonic groups have marginalised and controlled those who do not match their ideas of what is an acceptable way of living or behaving.

Rather than being provoked through fears for personal safety or fears of crime victimisation, it could therefore be argued that the protests in Balsall Heath, Manningham, Streatham and other red-light districts demonstrate the apprehension and anxiety that many people experience about commercial sex work, from which they subsequently seek to distance themselves. Combining contemporary work in cultural geography, as well as Beck's (1992) ideas on the production of certainties in late modern society, it might therefore be argued that street prostitutes are being made scapegoats – identified as a polluting influence, detrimental to local amenity and quality of life. This disordering, in turn, may prompt a form of 'cleansing' to maintain the purity of residential space. This desire was reflected in discourses surrounding these protests, with diverse metaphors of displacement converging to produce a seemingly common-sense view that commercial sex work is unnatural and deviant, linking it to wider problems of criminality, deprivation and environmental pollution. Hence, although the nuisance and noise created by prostitutes and kerb crawlers has often been identified by residents as a major reason to protest – and there are numerous examples of non-prostitute women being pestered by kerb crawlers – it is difficult to accept that the level of this nuisance alone would be sufficient to motivate the widespread community protests witnessed. Rather, as a police-commissioned survey of Manningham's residents' attitudes to prostitution (Walker *et al.* 1994: 22) concluded:

> Most of the discontent about prostitution arises from indirect rather than direct contact. Indeed, both prostitutes and kerb-crawlers seem to be able to discriminate with a fair level of accuracy between 'players' and 'non-players' ... it follows that what really animates residents is the offence of knowledge of the trade rather than any direct encounters or harm. It makes them feel uncomfortable that indications of sexual activity are visible in what to them are inappropriate times and locations.

The suggestion here is that the majority of soliciting and kerb crawling is carried out in a relatively unobtrusive manner, with few negative consequences for local residents, who, although they may be aware of the transactions going on in their neighbourhood, are rarely confronted by prostitution *per se.* Rather, it is the labelling of their neighbourhood as one in which street prostitution occurs that tends to create anxiety among residents. As current geographical studies of incivility suggest, while prostitution itself may not have a substantial impact on neighbourhood quality, like graffiti or litter, it may signify to people that an area is vulnerable to crime, setting in motion a dynamic of neighbourhood decline (Herbert 1993).

It is this fear of street prostitution instigating other forms of criminality and immorality that appeared as crucial in motivating local residents to seek to displace prostitutes and kerb crawlers from their streets. The organiser of Balsall Heath Street Watch, Raja Amin, argued that one of the main aims of the campaign was to 'stamp out the other crimes associated with prostitution' (personal communication), claiming that prostitutes constitute an integral part of a criminal subculture. This was also a pervasive theme in press reporting of the campaigns, with sometimes salacious media stories making a connection between the presence of prostitutes on the streets and the occurrence of other crimes, particularly drug-related ones. Invoking the widespread myth of the 'junkie whore' (Symanski 1981), it was argued that the actions of protesters were 'hitting hard at the strong and volatile link between vice and drugs as dealers began to lose their customers, the prostitutes' (*Birmingham Evening Mail*, 27 July 1994). Similar stories of drug dealers from Moss Side, Manchester, forced to give up trafficking to Manningham, were also given prominent coverage in Bradford's local press, while other stories played up the undercurrents of violence inherent in street prostitution, suggesting that the connection between commercial sex, violence and crime is inevitable (*Yorkshire Post*, 13 May 1995).

In this sense, the local and national media appeared to play a significant role in supporting these campaigns, actively perpetuating a series of myths about red-light districts that reproduced stereotyped ideas of the prostitute as 'polluting'. These ideas were echoed in the attitudes expressed by residents, pickets and even some prostitutes that selling sex is immoral, deviant and dangerous. Such ideas may be partly related to the 'moral panic' that has surrounded the transmission of HIV-1 (Campbell *et al.* 1996). Based on the assumption that sex workers have a high number of sexual contacts, prostitutes have often been portrayed as a pivotal group in the transmission of HIV-1, an assertion supported with reference to the high incidence of HIV infection among prostitutes in parts of sub-Saharan Africa. British studies have suggested that, because of awareness of safer sexual practices as well as the comparatively low epidemiological risk of female-to-male transmission, such assertions are erroneous. For example, one reputable study of 228 female prostitutes in London revealed that only two were infected with HIV-1, while 98 percent of all prostitutes claimed to use condoms always with commercial clients (Ward *et al.* 1993). Nevertheless, the story of a child blowing up a discarded condom left by a prostitute was cited by protesters in both Birmingham and Bradford as a direct spur to their campaigns. Descriptions of prostitutes as 'dirty street scum' or 'diseased human scavengers' were not uncommon (*Birmingham Evening Mail*, 27 July 1995).

It therefore appears that the desire to exclude prostitutes from both Manningham and Balsall Heath and 'cleanse' the neighbourhood in the process relied upon the stereotypical social representation of sex workers as dirty, deviant, dangerous and diseased, and the drawing of a boundary between acceptable (ordered) and unacceptable (disorderly) behaviours. Cresswell (1997: 334) argued that:

> The notion that everything has its place, and that things (e.g. people, actions) can be in place or out of place is deeply engrained in the way we think and act ... when individuals or groups ignore this socially produced common sense, they are said to be out of place and defined as deviant. Frequently, this labelling of out-of-place is metaphorical, based on analogies which themselves refer to common sense expectations.

As this suggests, metaphors of displacement used to describe certain groups or behaviours as 'out of place' include consistent references to waste matter and dirt (*cf.* Douglas 1966), distinguishing such groups from 'normal' purified populations. As such, these metaphors of displacement cannot be dismissed as simple poetic flourishes, but, as these examples of community activism demonstrate, are often indicative of more deeply rooted exclusionary desires.

SEX, SPACE AND THE PUBLIC SPHERE

While these protests thus pose important questions about the nature of risk and anxiety in the late modern era, they also highlight a number of related debates surrounding the changing nature of public space. The idea that public space is becoming increasingly structured in favour of a narrowly defined social order is one that has preoccupied geographers and urbanists in recent years. Starting from the normative position that city life entails a 'being together of strangers' (Young 1990: 237), it has been argued that public space should be democratic and open, offering a site where difference can be acknowledged and celebrated. Zukin (1995: 260), for example, argues that public space is the primary site of 'public culture', where people may mingle and socialise in such a way that a shared citizenship is created – a sense of urban identity transcending class, gender and ethnic boundaries. This idealisation of public space suggests that the spaces of the street, the park and the city square represent truly democratic and open spaces where all may gather, free from exclusionary violence (Mitchell 1996). As such, drawing on the arguments of Habermas (1991), the terms 'public space' and 'public sphere' have tended to be used interchangeably, with open spaces seen as sites of critical public discourse, of reasoned argument, and, on occasion, of protest and resistance.

It is for such reasons that commentators like Davis (1990) have identified the emergence of carceral spaces, like the 'gated' enclaves and privatised malls of Los Angeles, as symptomatic of the death of public space in both a metaphorical and a material sense. The seeming replacement of open and democratic spaces by single-minded, sanitised spaces in which certain groups and individuals are seen to be 'out of place' has been described as representing a major threat to notions of citizenship defined in terms of equal access and participation in society (see Davis 1990; Sennett 1990; Mitchell 1995). This issue even reached the highest court in America, where protests over the privatisation of public malls provoked Supreme Court Justice Marshall (quoted in Kowinski 1982: 35) to observe that:

> As governments rely on private enterprise, public property decreases in favour of privately-owned property. It becomes harder and harder for citizens to find means to communicate with other citizens. Soon, only the wealthy may find effective communication.

In such carefully planned environments, strategies of surveillance ensure that the atmosphere of conspicuous consumption and style is not broken by those who are apparently not very interested in spending money. Accordingly, many commentaries on the nature of contemporary public space are thus switching their focus to the way that some groups are being excluded from 'the street'. Smith (1996), for example, has highlighted this in his recent excavation of the 'revanchism' evident in contemporary cities, where sometimes brutal and avenging attacks on minorities have become cloaked in the language of civic morality, neighbourhood security and 'family values'. According to him, manifestations of revanchism have become apparent throughout Western cities with the introduction of curfews designed to curb the alleged excesses of youth, public order acts banning congregation in open spaces and the espousal of 'zero tolerance' campaigns targeting street begging. Against this rising tide of concern about the deviancy of the streets, Erickson (1996), with presumably no sense of irony, has suggested that public space be zoned. Thus begging and prostitution would be permitted only in the 5 percent of public spaces designated as 'red zones', with other public spaces either having some limits on permissible behaviour (orange zones) or being reserved for 'unusually sensitive peoples' (green zones).

Effectively summarizing much of this literature, Sibley (1995) concluded that various individuals, distinguished by their behaviour, dress, smell or shape, are being expelled from these privatised spaces by groups who feel that their appearance and conduct do not conform to the well-ordered ambience of 'family' consumption. Following these arguments, it is tempting to conclude that Street Watch campaigns against prostitutes represent an extension of this surveillance and cleansing process into residential public spaces, where 'others' may disturb the ordered nature of family space. As Aitken (1998) has outlined, two main myths serve to govern family life in Western societies: namely, that the ideal type of family is one headed by heterosexual partners and that bounded suburbia provides the ideal space for this family. As such, the performance of the nuclear family seems to rest on the reproduction of exclusive, private space rather than public space. The exclusion of sexual 'others' from this space therefore not only includes gay and lesbian couples, who may experience considerable harassment in such spaces (Valentine 1993), but also heterosexual 'others', including lone parents, spinsters and prostitutes (as well as paedophiles, pornographers and bisexuals). As such, the spaces that are erroneously described as 'community' spaces are nothing of the kind. Rather they are spaces that increasingly exclude those people whose sexual identities do not tally with reproductive gender roles (whereby mothering and fathering are construed as the only representational modes for sexual activity as well as the ultimate goal of the sexual relation).

In this regard, a related example of how the presence of heterosexual immorality may provoke exclusionary actions can be found in New York city. Here Mayor Giuliani has sought to eliminate commercial sex establishments and businesses from Manhattan, especially Times Square, where the Disney Corporation's efforts to reinvent the area as a site of 'family' entertainment was seen as incompatible with its existing status as a site of 'adult' entertainment. At Giuliani's request, New York City Council approved amendments to its Zoning Resolution in October 1995. These were designed to 'guide the future use of the City's land by encouraging the development of desirable residential, commercial and manufacturing areas with appropriate groupings of compatible and related uses and thus to promote and to protect public health, safety and general welfare' (quoted in *New York Times*, 23 February 1998). In effect, this resolution will mean that legal sex businesses will be forced to close. The new law defines an 'adult establishment' as any business that has a 'substantial portion of its stock-in-trade' in 'materials which are characterized by an emphasis upon the depiction or description of specified sexual activities or specified anatomical areas.' In this instance, such establishments are characterised as 'objectionable non-conforming uses which are detrimental to the character of the districts in which (they) are located.' Specifically dictating that adult establishments should 'be located at least 500 feet from a church, a school [or] a Residence District', this law stands as a remarkable attempt by legislators to reaffirm socio-spatial order by seeking to maintain distance between 'obscene' (non-normative) and moral expressions of heterosexuality.

Current attempts to eliminate specific sexual identities and practices from the street thus expose a major contradiction in current conceptualisations of public space. Street surveillance and cleansing are often justified with reference to revers-ing the decline of the public realm, in other words a struggle to make public space safe and secure for all. Yet this in itself relies on a form of exclusionary politics that rests on an impoverished notion of citizenship. Staeheli and Thompson (1997) stressed that this is a notion of citizenship where the boundaries of community take on a moral form that bears little relation to legal or procedural conceptions. As such, it is a notion that denies access to the public realm to those who are seen as failing to act as responsible 'family' citizens (i.e. those labelled as *counterpublics*). In the light of such ideas, glib claims about the need to maintain the openness and democracy of the streets by removing those who threaten 'the public' seem to dissolve. Instead of a unified public, it is readily apparent that there are many publics, some with more power than others (see also Bell 1995).

Therefore, while idealised public spaces (like the Greek agora) have frequently been represented as truly open public spaces where 'the thickness and intensity of human feelings' can be celebrated (Berman 1986: 494), this line of argument em-phasises the fact that public spaces have *always* been unevenly accessible and struggled over in the realms of cultural politics. The observation that public space has always been permeated by exclusions made on the basis of gender, race, class, sexuality and so on is one that has been remarked on in the past (Mitchell 1996). Feminist researchers in particular have drawn attention to the importance of the private/public dichotomy in shaping women's lives, arguing that this dichotomy stubbornly remains a potent means by which gender and sexual difference are constructed, controlled, excluded and oppressed (Duncan 1996). Significantly, the figure of the prostitute has been recognised as paradigmatic in this process. Swanson (1995), for example, argued that prostitution designates a whole host of sexual, social and economic relations that problematise the reorganisation of social life in the city, simultaneously constituting an exemplar of both the dangers and fascinations of urban life. In Victorian England, for example, the boundaries between domest-icated femininity and unfettered, immoral sexuality were constructed and main-tained through the discursive identification of prostitutes as dirty and dangerous, a threat to bourgeois society. Inevitably, a variety of urban *flâneurs*, epidemiologists, philologists and moral reformers contributed to this process, their obsessive scrutiny of the habits and habitats of 'fallen women' contributing to a variety of pseudo-scientific discourses in which dominant (male) fears and fantasies were projected on to the figure of the *street* prostitute. As Walkowitz (1992: 233) has suggested:

> The prostitute was the quintessential figure of the urban scene . . . for men as well as women, the prostitute was the central spectacle in a set of urban encounters and fantasies. Repudiated and desired, degraded and threatening, the prostitute attracted the attention of a range of urban male explorers from the 1840s to the 1880s.

In particular, it was the way that these women blurred the distinction between private and public, effectively eroticising the streets through their adoption of specific forms of dress and behaviour, that provoked these respectable fears and anxieties.

While it is now widely accepted that men and women should be able to express their sexuality outside the traditional constraints of wedlock (Duncan 1996), it appears that contemporary society is still based on the heterosexualised exchange of material, physical and emotional values in a domestic, family setting (Giddens 1992). In this sense, the female prostitute who offers sexual services outside the stable and ordered institutions of the family and home remains a threat to heterosexual dominance, literally upsetting the mentality that reproduces the public/private space dichotomy. The idea that conspicuous sexuality, as exemplified by the street prostitute, disturbs social and spatial ordering is one that appeared to be an important spur to the community actions in Balsall Heath (Hubbard 1999), where street prostitution was considered an assault against 'public decency' in a largely South Asian residential area. Specifically, the sight of women adopting a predatory sexual role in public spaces is one that appeared to provoke considerable male anxiety, although many protesters claimed that it was the 'innocent' gaze of women and children that they were trying to protect through their protests. This antipathy to particular public displays of sexuality is crucial to the structuring of public space. As I have argued elsewhere (*ibid.*), women's ability to perform particular sexual roles in particular spaces and at particular times informs broader notions of their ability to do *anything*. In effect, gender, and sexual and bodily identities interact to shape participation in the public realm, and while public space may be enabling for those whose lifestyles accord with dominant heterosexual values, for others it remains largely constraining (Bell 1995). For example, women are often perceived as an intruding influence on the public spaces of city life when they use them as a space of social presence. This, combined with ideas of respectability that dictate that women should be escorted at night in public spaces (Duncan 1996), means that it is not surprising there is rarely any possibility of the female *flâneur*, only the prostitute (Swanson 1995). By placing restrictions on women prostitutes, therefore, 'community' protests and Street Watch campaigns might not open up the public realm to greater diversity or freedom of sexual expression, Rather they are in danger of creating a situation in which *all* women's attempts to use public space for love, friendship, fun and companionship are curtailed by aggressive protests that equate urban risk and disorder with fears about sexual 'others'.

CONCLUSION

With people's desire for 'ontological security' – a sense of well-being and identity – increasingly manifest in defensive protests designed to maintain social and spatial ordering (Beck 1992), it is perhaps unsurprising that there have recently been a number of high-profile protests against prostitutes in many British red-light districts. Prompted by fears of difference and diversity and fuelled by an unabated barrage of images of street violence, drugs and sexual depravity, these protests have nonetheless succeeded in highlighting the extent to which British prostitution laws are largely inoperable given the time and resources available for policing prostitution. They also suggest that the time is right to explore alternative approaches such as decriminalisation, legalisation or, as has been suggested in Birmingham and Sheffield, the designation of a toleration zone (Hubbard 1997). Moreover, given that these community pickets have had more impact on the location of prostitutes than any police intervention (as noted above), it might also be suggested that multi-agency approaches involving partnerships between the police, local authorities and the public should be pursued as a long-term solution to regulating prostitution in public spaces (Matthews 1986). Yet some important caveats need to be added to this conclusion. Not least among them is the fact that even when presented in terms of 'community safety' discourses, street picketing constitutes an inherently aggressive – and distinctly *masculine* – claim to public space that perpetuates a series of gendered and sexualised distinctions in terms of access to the public realm. Fischer and Poland's (1998: 192) more general critique of 'community' policing initiatives is worth considering in this regard:

> While the rhetoric of inclusivity and consensual community characterises much of the professional discourse on community, the normative and functional focus of 'community' groups in combination with the mundane resources required to organise and act . . . have resulted, in practice, in narrowly defined and selectively homogeneous types of 'community' groups. Typically, participation is highest among the middle class, property-owning, employed and mainstream population, whereas the more marginal elements . . . are effectively excluded.

This reminds us that claims to community are bound into complex modalities of power, with the consequent fragmentation of the public realm being accompanied by fear, suspicion, tension and conflict between different social groups. As such, claims that informal or formal surveillance (whether CCTV, Neighbourhood Watch or street picketing) can enhance security and minimise fear of crime victimisation need to be balanced against the observation that surveillance involves an extension of social control into the civil and public realm, which, unless suitably regulated, may encourage an increased intolerance of difference and diversity.

EPILOGUE

It is September 1995, a cool late summer's evening on the streets of Balsall Heath. A crowd is gathering on Calthorpe Park, opposite Cheddar Road, a street once synonymous with sex work. Fifty strong, clutching torches and lanterns, this is Balsall Heath Women's Group, intent on reclaiming the streets and the night. As the male protesters from the Street Watch maintain their distance, watching suspiciously, the women explain to a group of journalists why they are there. 'To be honest,' one argues, 'we'd rather have the prostitutes back . . . it's not safe for any woman on the streets with these pickets around.' The press take a few photographs, wish the women good luck and then disappear, leaving them to undertake their torchlit vigil around the streets of 'their' community, singing songs to keep the spirits up. A couple of hours later, they return home, their act of resistance – an attempt to destabilise the taken-for-granted nature of 'community' space – over. The next night though, the Street Watch protesters are still out, ever-watchful for the few prostitute women who continue to play a cat-and-mouse game with police and pickets, while the majority of the area's female residents remain, as ever, in the private spaces of their home, 'protected' from the street and the night.

BIBLIOGRAPHY

Adamson, I. (1981) *The Cruithin*, Belfast: Donard.

Affron, M. and Antiff, M. (eds) (1997) *Fascist Visions: Art and Ideology in France and Italy*, Princeton, NJ: Princeton University Press.

Aitken, S.C. (1992) The personal contexts of neighbourhood change, *Journal of Architectural and Planning Research*, **9**, 339–60.

Aitken, S.C. (1998) *Family Fantasies and Community Space*, New Brunswick, NJ: Rutgers University Press.

Aitken, S.C. (1999) Play, rights and borders: gender bound parents and the social construction of children, in Holloway, S. and Valentine, G. (eds) *Children's Geographies: Living, Playing, Learning and Transforming Everyday Worlds*, London: Routledge.

Aitken, S.C. and Prosser, R. (1990) Residents' spatial knowledge of neighbourhood continuity and form, *Geographical Analysis*, **22**, 301–25.

Alderson, J.C. (1973) The principles and practice of the British police, in Alderson, J.C. and Stead, P.J. (eds) *The Police We Deserve*, London: Wolfe Publishing.

Aleiss, A. (1994) A race divided: the Indian westerns of John Ford, *American Indian Cultural and Research Journal*, **18**, 167–86.

al Khalil, S. (1991) *The Monument: Art, Vulgarity and Responsibility in Iraq*, London: Deutsch.

Allen, M. (1994) Native American control of tribal natural resource development in the context of the federal trust and tribal self-determination, in Wells, R. (ed.) *Native American Resurgence and Renewal: A Reader and Bibliography*, London: Scarecrow Press.

Allen, T. (1999) Perceiving contemporary wars, in Allen, T. and Seaton, J. (eds) *The Media of Conflict: War Reporting and Representations of Ethnic Violence*, London: Zed Books.

Anderson, B. (1981) *Imagined Communities*, London: Verso.

Anderson, J. (1996) The shifting stage of politics: new medieval and postmodern territorialities? *Environment and Planning D: Society and Space*, **14**, 133–53.

Anderson, K. (1987) The idea of Chinatown: The power of place and institutional practice in the making of a racial category, *Annals of the Association of American Geographers*, **77**, 580–98.

anon. (1986) *Fodor's Israel 1987*, New York: Fodor.

anon. (1997a) Editorial: could 'Zero Tolerance' end in tears? *Police*, March.

anon. (1997b) Interview with Tony Blair, *The Big Issue*, 8 January.

Appleton, J. (1990) *The Symbolism of Habitat*, Washington, DC: University of Washington Press.

Appleton, J. (1996) *The Experience of Landscape*, Chichester: John Wiley.

Ardrey, R. (1966) *The Territorial Imperative*, New York: Delta.

Ashworth, G. (1987) Urban form and the defence functions of cities, in Bateman, M. and Riley, R.W. (eds) *The Geography of Defence*, Beckenham, Kent: Croom Helm, 17–51.

Atlas, R. (1984) Violence in prison: environmental influences, *Environment and Behaviour*, **16**, 275–306.

Bachrach, P. and Baratz, M. (1970) *Power and Poverty*, New York: Oxford University Press.

Baignet, M. and Leigh, R. (1991) *The Dead Sea Scrolls Deception*, London: Corgi Books.

Bairner, A. and Shirlow, P. (1998) The territorial politics of soccer in Northern Ireland, *Space and Polity*, **2**, 163–77.

Baker, A.R.H. (1992) On ideology and landscape, in Baker, A.R.H. and Biger, G. (eds) *Ideology and Landscape in Historical Perspective: Essays on the Meanings of Some Places in the Past*, Cambridge: Cambridge University Press.

Baker, A.R.H. and Harley, J.B. (eds) (1973); *Man Made the Land*, Newton Abbot, Devon: David & Charles.

Baker, D. (1983) *Race, Ethnicity and Power: A Comparative Study*, London: Routledge.

Baker, K. (ed.) (1993) *The Faber Book of Conservatism*, London: Faber and Faber.

Baker, M.H., Nienstedt, B.C., Everett, R.S. and Cleary, R. (1983) Impact of a crime wave: perceptions, fear and confidence in the police, *Law and Society Review*, **17**, 319–35.

Bandini, M. (1985) *Urban Morphology: The British Contribution*, unpublished report for the French Ministry of Urbanism, Housing and Transport commissioned by the Institute of Urbanism, University of Paris.

Bannister, J., Fyfe, N. and Kearns, A. (1998) Closed circuit television and the city, in Norris, C., Morgan, J. and Armstrong, G. (eds) *Surveillance, Closed Circuit Television and Social Control*, Aldershot: Ashgate Press.

Barr, R. and Pease, K. (1992) A place for every crime and every crime in its place: an alternative perspective on crime displacement, in Evans, D.J., Fyfe, N.J. and Herbert, D.T. (eds) *Crime, Policing and Place: Essays in Environmental Criminology*, London: Routledge.

Bayley, D.H. (1994) *Police for the Future*, New York: Oxford University Press.

Beck, U. (1986, 1992) *Risk Society: Towards a New Modernity*, London: Sage.

Beck, U. (1995) *Ecological Politics in an Age of Risk*, Cambridge: Polity Press.

Beck, U. (1998) *Democracy without Enemies*, Cambridge: Polity Press.

Beck, U., Giddens, A. and Lash, S. (1994) *Reflexive Modernisation: Politics, Tradition and Aesthetics in the Modern Social Order*, Cambridge: Polity Press.

Bell, D. (1995) Pleasure and danger: the paradoxical spaces of sexual citizenship, *Political Geography*, **14**, 139–53.

Bell, G. (1990) *The Protestants of Ulster*, London: Pluto Press.

Bender, B. (1998) *Stonehenge: Making Space*, Oxford: Berg.

Bennet, T.H. and Kotch, B. (1993) *Community Policing in Canada and Britain*, Home Office Research and Statistics Bulletin 34, London: HMSO.

Bennett, D.G. (1992) The impact of retirement migration on Carteret and Brunswick counties, North Carolina, *North Carolina Geographer*, **1**, 25–38.

Bennett, D.G. (1993) Retirement migration and economic development in high amenity non-metropolitan areas, *Journal of Applied Gerontology*, **12**, 466–81.

Benson, C. and Matthews, R. (1995) *The National Vice Squad Survey*, Centre for Criminology, Middlesex University.

Bentham, J. (1791) Panopticon: or the inspection house, in Bowring, J. (ed.) (1843) *The Works of Jeremy Bentham*, Vol. 4, Edinburgh: William Tait.

Berkman, R. (1979) *Opening the Gates: The Rise of the Prisoners' Movement*, Cambridge, Mass.: Lexington.

Berman, M. (1986) Take it to the streets: conflict and community in public space, *Dissent*, Fall, 470–94.

Bertell, R. (1985) *No Immediate Danger*, London: Women's Press.

Bianchini, F. (1994) Night cultures, night economies, *Town and Country Planning*, **63**, 308–10.

BIA (Bureau of Indian Affairs) (1999a) Answers to frequently asked questions, website: http://www.doi.gov/bia/aitoday/q_and_a.html, accessed 21 September.

BIA (Bureau of Indian Affairs (1999b) Statistical-Abstract-on-the-Web, website: http://www.doi.gov/nrl/StatAbst/StatHome.html, accessed 21 September.

Blakely, E.J. and Snyder, M.G. (1995a) *Fortress America: Gated and Walled Communities in the United States*, Cambridge, Mass.: Lincoln Institute of Land Policy.

Blakely, E.J., and Snyder, M.G. (1995b) Fortress communities: the walling and gating of American suburbs, *Landlines*, **7**(5), 1,3.

Blakely, E.J. and Snyder, M.G. (1997) *Fortress America: Gated Communities in the United States*, Washington, DC: Brookings Institution Press.

Blowers, A. (1984) *Something in the Air: Corporate Power and the Environment*, London: Harper & Row.

Blowers, A. and Leroy, P. (1994) Power, politics and environmental inequality: a theoretical and empirical analysis of the process of 'peripheralisation', *Environmental Politics*, **3**, 197–228.

Blowers, A., Lowry, D. and Solomon, B. (1991) *The International Politics of Nuclear Waste*, London: Macmillan.

Blowers, A. and Lowry, D. (1996) *Radioactive Waste in Germany: The Dimensions of Conflict*, Global Environmental Change, Working Paper Series 7, Open University.

Blowers, A. and Lowry, D. (1997) Nuclear conflict in Germany: the wider context, *Environmental Politics*, **6**, 148–55.

Boehmer-Christiansen, S. (1994) Politics and environmental management, *Environmental Management and Planning*, **37**, 69–85.

Boehmer-Christiansen, S. and Skea, J. (1991) *Acid Politics: Energy and Environmental Policies in Britain and Germany*, London: Belhaven.

Bohland, J.A. and Rowles, G.D. (1988) The significance of elderly migration to changes in elderly population concentration in the United States: 1960–1980, *Journal of Gerontology*, **43**, 145–52.

Bourassa, S.C. (1991) *Aesthetics of Landscape*, London: Belhaven.

Bourdieu, P. (1984) *Distinction: A Social Critique of the Judgement of Taste*, London: Routledge.

Bowling, B. (1996) Zero tolerance: cracking down on crime in New York City, *Criminal Justice Matters*, 25, Autumn, 17–31.

Box, S., Hale, C. and Andrews, G. (1988) Explaining fear of crime, *British Journal of Criminology*, **28**, 340–56.

Brake, M. and Hale, C. (1992) *Public Order and Private Lives*, London: Routledge.

Bratton, W.J. (1995) The New York City Police Department's civil enforcement of quality-of-life crimes, *Journal of Law and Policy*, **3**.

Bratton, W.J. (1998) Crime is down in New York City: blame the police, in Dennis, N. (ed.) *Zero-Tolerance: Policing a Free Society*, second edition, London: Institute of Economic Affairs.

Breed, G.F., Jr (1999) http://www.gatedcommunity.com

Brenner, N. (1999) Globalisation as reterritorialisation: the re-scaling of urban governance in the European Union, *Urban Studies*, **36**, 431–51.

Bright, J. (1991) Crime prevention: the British experience, in Stenson, K. and Cowell, D. (eds) *The Politics of Crime Control*, London: Sage.

Brogden, M. and Shearing, C. (1993) *Policing for a New South Africa*, London: Routledge.

Brown, B. (1995) *CCTV in Town Centres: Three Case Studies*, Police Research Group Crime Detection and Prevention Series, Paper 68, London: Home Office Police Department.

Brown, D. (1997) *PACE Ten Years On: A View of the Research*, London: HMSO.

Brown, F.E., Bruhns, H.R., Rickaby, P.A. and Steadman, J.P. (1991) *A Database and Classification System for the United Kingdom's Non-Domestic Building Stocks*, report to the Department of the Environment and Building Research Establishment, Milton Keynes: Centre for Configurational Studies, Open University.

Brown, S. (1998) What's the problem girls? CCTV and the gendering of public safety, in Norris, C., Morgan, J. and Armstrong, G. (eds) *Surveillance, Closed Circuit Television and Social Control*, Aldershot: Ashgate.

Brunn, S.D. (1987) A world of peace and military landscapes, *Journal of Geography*, **86**, 253–62.

Brunn, S.D. (1991) Peacekeeping missions and landscapes, in Rumley, D. and Minghi, J. (eds) *The Geography of Border Landscapes*, London: Routledge.

Burke, R.H. (1998) Begging, vagrancy and disorder, in Burke, R.H. (ed.) *Zero Tolerance Policing*, Leicester: Perpetuity Press.

Buttel, F., Hawkins, A. and Power, A. (1990) From limits to growth to global change: constraints and contradictions in the evolution of science and ideology, *Global Environmental Change*, **1**, 57–66.

Callan, H. (1970) *Ethology and Society: Towards an Anthropological View*, Oxford: Clarendon Press.

Calthorpe, P. (1993) *The Next American Metropolis: Ecology, Community and the American Dream*, New York: Princeton Architectural Press.

Campbell, R., Coleman, S. and Torkington, P. (1996) *Street Prostitution in Inner City Liverpool*, Final report of the Abercromby Prostitution Project, Applied Research Centre, Liverpool Hope University College.

Canniggia, G. (1979) *Composizione Architectettonica e Tiplologia Edilizia: 1. Lettura dell'Edilizia di Base*, Venice: Marsilio Editori.

Canniggia, G. (1984) *Composizione Architectettonica e Tiplologia Edilizia: 2. Il Progetto dell'Edilizia di Base*, Venice: Marsilio Editori.

Canter, D. (1987) Implications for 'new generation' prisons of existing psychological research into prison design and use, in Bottoms, A.E. and Light, R. (eds) *Problems of Long-term Imprisonment*, Aldershot: Gower.

Canter, D., Ambrose, I., Brown, J., Comber, M. and Hirsch, A. (1980) *Prison Design and Use Study*, Final Report, Department of Psychology, University of Surrey (mimeograph).

Carr, S., Francis, M., Rivlin, L.G. and Stone, A.M. (1992) *Public Space*, Cambridge: Cambridge University Press.

Carriage Park (n.d. a) *The Age of Discovery Isn't Over Yet*, Hendersonville, NC: Carriage Park.

Carriage Park (n.d. b) *Carriage Park: Don't Buy until You Get the Facts*, Hendersonville, NC: Carriage Park.

Casey, C. (1995) Lights on red, *Police Review*, **21**, 20–1.

Cauthen, M. (1997) The myth of divine election and Afrikaner ethnogenesis, in Hoskins, G. and Schopflin, G. (eds) *Myths and Nationhood and Ultra Nationalists*, London: Bloomfield.

Chaliand, G. and Rageau, J.P. (1985) *Strategic Atlas: A Comparative Geopolitics of the World's Powers*, New York: Harper & Row.

Champion Hills (n.d.) *Champion Hills Lifestyle*. Web page: http://www.championhills.com/ community.html

Chesshyre, R. (1997) Enough is enough, *Daily Telegraph Magazine*, 1 March.

Chestnut Hills (n.d.) *Chestnut Hill of Highlands: A Residential Retirement Community*. Web page: http://www.dnet.net/~chestnut/service.html

Christian, L. (1983) *Policing by Coercion*, London: GLC Police Committee Support Unit.

Christie, N. (1982) *The Limits to Pain*, Oxford: Robertson.

Christopherson, S. (1994) The fortress city: privatized spaces, consumer citizenship, in Amin, A. (ed.) *Post-Fordism: A Reader*, Oxford: Blackwell.

Churchill, W. (1994) *Indians are Us?: Culture and Genocide in Native North America*, Monroe, Minn.: Common Courage Press.

Clarke, R.V.G. (ed.) (1992) *Situational Crime Prevention: Successful Case Studies*, New York: Harrow and Heston.

Clarke, R.V.G. and Mayhew, P. (eds) (1980) *Designing Out Crime*, Home Office Research and Planning Unit, London: HMSO.

Cloke, P., Phillips, M. and Thrift, N. (1995) The new middle classes and the social con-structs of rural living, in Butler, T. and Savage, M. (eds) *Social Change and the Middle Classes*, London: UCL Press.

Cloke, P., Phillips, M. and Thrift, N. (1998) Class, colonisation and lifestyle strategies in Gower, in Boyle, M. and Halfacree, K. (eds) *Migration to Rural Areas*, London: Wiley.

Cohen, F. (1998) *Handbook of Federal Indian Law*, Buffalo, NY: Hein.

Coleman, A. (1985) *Utopia on Trial*, London: Hilary Shipman.

Coleman, R. and Sim, J. (1998) From the dockyards to the Disney store: surveillance, risk and security in Liverpool city centre, *International Review of Law, Computers and Technology*, **12**, 27–45.

COMEDIA, in association with the Calouste Gulbenkian Foundation (1991) *Out of Hours: A Study of Economic, Social and Cultural Life in Twelve Town Centres in the UK*, London: COMEDIA.

Conzen, M.P. (1990) Town-plan analysis in an American setting, cadastral processes in Boston and Omaha, 1630–1930, in Slater, T.R. (ed.) *The Built Form of Western Cities: Essays for M.R.G. Conzen on the Occasion of his Eightieth Birthday*, Leicester: Leicester University Press.

Conzen, M.R.G. (1958) The growth and character of Whitby, in Daysh, G.H.J. (ed.) *A Survey of Whitby and the Surrounding Area*, Windsor: Shakespeare Head Press.

Conzen, M.R.G. (1960) *Alnwick, Northumberland: A Study in Town-Plan Analysis*, Institute of British Geographers Publication No. 27, London: George Philip.

Conzen, M.R.G. (1962) The plan analysis of an English city centre, in Norborg, K. (ed.) *Proceedings of the IGU Symposium in Urban Geography, Lund 1960*, Lund: Gleerup.

Conzen, M.R.G. (1966) Historical townscapes in Britain: a problem in applied geography, in House, J.W. (ed.) *Northern Geographical Essays in Honour of G.H.J. Daysh*, Newcastle-upon-Tyne: Newcastle University.

Conzen, M.R.G. (1968) The use of town plans in the study of urban history, in Dyos, H.J. (ed.) *The Study of Urban History*, London: Edward Arnold.

Conzen, M.R.G. (1975) Geography and townscape conservation, in Uhlig, G. and Lienau, C. (eds) *Anglo-German Symposium in Applied Geography*, Giessen–Würzburg–München: Len.

Conzen, M.R.G. (1988) Morphogenesis, morphological regions and secular human agency in the historic townscape, as exemplified by Ludlow, in Denecke, D. and Shaw, G. (eds) *Urban Historical Geography: Recent Progress in Britain and Germany*, Cambridge: Cambridge University Press.

Cookson, H., Dowdeswell, P. and Murphy, M. (1994) *Control in Category C Prisons*, unpublished report to Prison Service Agency Directorate of Custody (Custody Group).

CORE (n.d.) *Thermal Oxide Reprocessing Plant: An In-depth Investigation*, Barrow-in-Furness: CORE.

Corbin, R. (1999) http://www.co.clark.nv.us/metro/CRIMPREV/96-2-2html

Cornerstone Properties (n.d.) *Grantham Place*. Web page: http://www.greensboro.com/repages/cornerstone/grantham.html

Cosgrove, D. (1978) Place, landscape and the dialectics of cultural geography, *Canadian Geographer*, **22**, 66–72.

Cosgrove, D. (1984) *Social Formation and the Symbolic Landscape*, London: Croom Helm.

Cosgrove, D. (1985) Prospect, perspective and the evolution of the landscape idea, *Transactions of the Institute of British Geographers*, NS **10**, 45–62.

Cosgrove, D. and Petts, G. (eds) (1990) *Water, Engineering and Landscape: Water Control and Landscape Transformation in the Modern Period*, London: Belhaven.

Cradon, L. (1997) Leadership is key factor in zero tolerance, *Police Review*, 20 June.

Crawford, A. (1995) Appeals to community and crime prevention, *Crime Law and Social Change*, **22**, 97.

Crawford, A. (1997) *The Local Governance of Crime: Appeals to Community and Partnership*, Oxford: Clarendon Press.

Crenson, M. (1971) *The Un-Politics of Pollution: A Study of Non-decisionmaking in the Cities*, Baltimore: Johns Hopkins University Press.

Cresswell, T. (1996) *In Place/Out of Place: Geography, Ideology, and Transgression*, Minneapolis: University of Minnesota Press.

Cresswell, T. (1997) Weeds, plagues and bodily secretions: a geographic interpretation of metaphors of displacement, *Annals of the Association of American Geographers*, **87**, 330–45.

Critchley, T.A. (1973) The idea of policing in Britain: success or failure? in Alderson, J.C. and Stead, P.J. (eds) *The Police We Deserve*, London: Wolfe Publishing.

Crowther, C. (1997) *Policing the Underclass: A Critical Consideration of the Current Agenda*, Studies in Crime, Order and Policing, Occasional Paper 12, Scarman Centre for the Study of Public Order, University of Leicester.

Currie, E. (1988) Two visions of community crime prevention, in Hope, T. and Shaw, M. (eds) *Communities and Crime Reduction*, London: HMSO.

Daniels, S.J. (1987) Marxism, culture and the duplicity of landscape, in Peet, R. and Thrift, N. (eds) *New Models in Geography*, Volume II, London: Unwin Hyman.

Daniels, S.J. (1993) Re-visioning Britain: mapping and landscape painting, 1750–1820, in Baetjer, K. and Rosenthal, M. (eds) *Glorious Nature: British Landscape Painting 1750–1850*, London: Zwemmer.

Davey, J. (1995) *The New Social Contract: America's Journey from Welfare State to Police State*, Westport, Conn.: Praeger.

Davies, S.G. (1996) *Big Brother: Britain's Web of Surveillance and the New Technological Order*, London: Pan Books.

Davis, M. (1990) *City of Quartz: Excavating the Future in Los Angeles*, London: Verso.

Davis, M. (1992) Fortress Los Angeles: the militarization of urban space, in Sorkin, M. (ed.) *Variations on a Theme Park: The New American City and the End of Public Space*, New York: Hill and Wang.

Davis, M. (1998) *Ecology of Fear: Los Angeles and the Imagination of Disaster*, London: Picador.

Davis, S.G. (1997) Space jam: family values in the entertainment city, paper presented at the American Studies Annual Meeting.

Dawson, T. (1994) Framing the villains, *New Statesman*, 28 January.

Dean, J. (1997) Can zero tolerance and problem oriented policing be part of the same philosophy? *Police Journal*, October, 5–7.

Dear, M. and Flusty, S. (1998) Postmodern urbanism, *Annals of the Association of American Geographers*, **88**, 50–72.

Debbage, K. and Rees, J. (1991) Company perceptions of comparative advantage by region, *Regional Studies*, **25**, 199–206.

de Certeau, M. (1984) *The Practice of Everyday Life*, Berkeley: University of California Press.

Deleuze, G. and Guattari, F. (1987) *A Thousand Plateaus*, Minneapolis: University of Minnesota Press.

Dennis, N. (ed.) (1998) *Zero-Tolerance: Policing a Free Society*, second edition, London: Institute of Economic Affairs.

Dennis, N. and Mallon, R. (1998a) Confident policing in Hartlepool, in Dennis, N. (ed.) *Zero-Tolerance: Policing a Free Society*, second edition, London: Institute of Economic Affairs.

Dennis, N. and Mallon, R. (1998b), Crime and culture in Hartlepool, in Dennis, N. (ed.) *Zero-Tolerance: Policing a Free Society*, second edition, London: Institute of Economic Affairs.

Di Genera, G. (ed.) (1975) *Prison Architecture, An International Survey of Representative Closed Institutions and Analysis of Current Trends in Prison Design*, United Nations Social and Defence Research Council, London: Architectural Press.

Dillon, M. (1996) *25 Years of Terror: The IRA's War against the British*, London: Bantam.

Dillon, M. (1998) *God and the Gun*, London: Orion.

Ditchfield, J. (1990) *Control in Prisons: A Review of the Literature*, Home Office Research Study 118, London: HMSO.

Ditchfield, J. (1995) *Assault in Category C Prisons*, unpublished report, Home Office Research and Planning Unit.

Ditton, J., Short, E., Phillips, S., Norris, C. and Armstrong, G. (1999) *The Effect of Closed Circuit Television Cameras on Recorded Crime Rates and Public Concern about Crime in Glasgow*, Edinburgh: Scottish Office Central Research Unit.

Dodgshon, R.A. (1987) *The European Past: Social Evolution and Spatial Order*, Basingstoke: Macmillan.

DoE (Department of the Environment) (1994) *Planning Out Crime*, London: HMSO.

DoE (Department of the Environment) (1997) *PPG1: General Policy and Principles*, London: HMSO.

Doel, M. and Clark, D.B. (1997) Transpolitical urbanism: suburban anomaly and ambient fear, *Space and Culture*, **2**, 13–37.

Domosh, M. (1992) Corporate cultures and the modern landscape of New York City, in Anderson, K. and Gale, F. (eds) *Inventing Places: Studies in Cultural Geography*, Melbourne: Longman.

Domosh, M. (1996) *Invented Cities: The Creation of Landscape in Nineteenth-century New York & Boston*, New Haven, Conn.: Yale University Press.

Donnelly, P.G. and Kimble, C.E. (1997) Community organizing, environmental change, and neighborhood crime, *Crime and Delinquency*, **43**(4), 493–511.

Donovan, B. (1999) Medicine man program: too little, too late? *Arizona Republic*, 15 March, B1–2.

Douglas, M. (1966) *Purity and Danger*, Harmondsworth: Penguin.

Douglas, N. and Shirlow, P. (1998) People in conflict in place: the case of Northern Ireland, *Political Geography*, **17**, 125–8.

Dowling, R. (1998) Neotraditionalism in the suburban landscape: cultural geographies of exclusion in Vancouver, Canada, *Urban Geography*, **24**, 105–22.

Duncan, N. (1996) Renegotiating gender and sexuality in public and private spaces, in Duncan, N. (ed.) *Bodyspace: Destabilising Geographies of Gender and Sexuality*, London: Routledge.

Duncan, S. and Goodwin, M. (1987) *The Local State and Uneven Development*, Cambridge: Polity Press.

Dunlap, R., Rosa, E., Baxter, R. and Mitchell, R. (1993) Local attitudes towards siting a high-level nuclear waste repository at Hanford, Washington, in Dunlap, R., Kraft, M. and Rosa, E. (eds) *Public Reactions to Nuclear Waste*, Durham, NC: Duke University Press.

Eder, K. (1993) *The New Politics of Class*, London: Sage.

Edwards, S.M. (1997) The legal regulation of prostitution: a human rights issue, in Scambler, G. and Scambler, S. (eds) *Rethinking Prostitution: Purchasing Sex in the 1990s*, London: Routledge.

Eisenberg, L. (1972) The *human* nature of human nature, *Science*, **176**, 123–8.

Elk River Development Corporation (n.d. a) *Elk River*, Elk River, NC: Elk River Development Corporation.

Elk River Development Corporation (n.d. b) *Elk River Airport*, Elk River, NC: Elk River Development Corporation.

Elk River Development Corporation (n.d. c) *Golf at Elk River*, Elk River, NC: Elk River Development Corporation.

Ellen, R. and Fukui, K. (1996) *Redefining Nature: Ecology, Culture and Domestication*, Oxford: Berg.

Ellin, N. (ed.) (1997) *Architecture of Fear*, Princeton, NJ: Princeton Architectural Press.

Erickson, R.C. (1996) Controlling chronic misconduct in city spaces: of panhandlers, skid rows and public space zoning, *Yale Law Review*, **105**, 1165–249.

Eskelson, D. (1999) http://www.clearwaterlandscapes.com/privacy_statement.html

Evans, D.J. (1992) Left realism and the spatial study of crime, in Evans, D.J., Fyfe, N.R. and Herbert, D.T. (eds) *Crime, Policing and Place: Essays in Environmental Criminology*, London: Routledge.

Fairweather, L. (1975) The evolution of the prison, in Di Genera, G. (ed.) *Prison Architecture, An International Survey of Representative Closed Institutions and Analysis of Current Trends in Prison Design*, London: Architectural Press.

Fairweather, L. (1989) Prisons: a new generation, *Prison Service Journal*, **76**, 15–21.

Fairweather, L. (1995) Does good design help those inside? *Prison Service Journal*, **101**, 19–24.

FBI (US Department of Justice, Federal Bureau of Intelligence) (1994), *Crime in the United States 1993: Uniform Crime Reports*, Washington, DC: US Department of Justice.

FBI (US Department of Justice, Federal Bureau of Intelligence) (1998), *Crime in the United States 1997: Uniform Crime Reports*, Washington, DC: US Department of Justice.

Featherstone, M. and Hepworth, M. (1995) Images of positive ageing, in Featherstone, M. and Wernick, A. (eds) *Images of Ageing*, London: Routledge.

Feher-Elston, C. (1988) *Children of the Sacred Ground: America's Last Indian War*, Flagstaff, Ariz.: Northland Publishers.

Feldman, A. (1991) *Formations of Violence*, Chicago: University of Chicago Press.

Findlay, J.M. (1992) An elusive institution: the birth of Indian reservations in Gold Rush California, in Castille, G.P. and Bee, R.L. (eds) *Perspectives of Federal Indian Policy*, Tucson: University of Arizona Press.

Finlayson, A. (1997) Discourse and contemporary loyalist identity, in Shirlow, P. and McGovern, M. (eds) *Who are the People? Unionism, Protestantism and Loyalism in Northern Ireland*, London: Pluto Press.

Fischer, B. and Poland, B. (1998) Exclusion, risk and social control: reflections on community policing and public health, *Geoforum*, 29, 187–97.

Flusty, S. (1994) *Building Paranoia: The Proliferation of Interdictory Space and the Erosion of Spatial Justice*, Los Angeles: Forum for Architecture and Urban Design.

Foot, M.R.D. (1990) *Art and War*, London: Headline.

Foucault, M. (1973) *The Birth of the Clinic*, London: Tavistock.

Foucault, M. (1977) *Discipline and Punish: The Birth of a Prison*, London: Allen Lane.

Franco, M. (1985) Killing priests, nuns, women and children, in Blonsky, M. (ed.) *Signs*, Baltimore: Johns Hopkins University Press.

Friendship Park. (n.d.) *A Senior Community: Friendship Park*, Jefferson, NC: Friendship Park.

Frey, W. (1995) Immigration and internal migration 'flight' from US metropolitan areas: toward a new demographic balkanization, *Urban Studies*, 32, 733–57.

Frey, W. (1996) Immigration, domestic migration, and demographic balkanization in America: new evidence for the 1990s, *Population and Development Review*, 22, 741–63.

Frey, W. and Johnson, K. (1998) Concentrated immigration, restructuring and the 'selective' deconcentration of the United States population, in Boyle, M. and Halfacree, K. (eds) *Migration to Rural Areas*, London: Wiley.

Funtowicz, S. and Ravetz, J. (1990) Post-normal science: a new science for new times, *Scientific European*, October, 20–2.

Furseth, O. (1996) The North Carolina business landscape: the view through foreign eyes, in Bennett, G.C. (ed.) *Snapshots of the Carolinas: Landscapes and Cultures*, Charlotte, NC: Association of American Geographers.

Fyfe, N.R. (1995) Policing the city, *Urban Studies*, 32, 759–78.

Fyfe, N.R. (ed.) (1998) *Images of the Street*, London: Routledge.

Fyfe, N.R. and Bannister, J. (1996) City watching: closed circuit television surveillance in public spaces, *Area*, 28, 37–46.

Fyfe, N.R. and Bannister, J. (1998) The 'eyes upon the street': closed circuit television surveillance and the city, in Fyfe, N.R. (ed.) *Images of the Street: Planning, Identity and Control in Public Space*, London: Routledge.

Gagen, E. (2000) 'An example to us all': children's bodies and identity construction in early twentieth century playgrounds, *Environment and Planning A*, in press.

Gallaher, C. (1997) Identity politics and the religious Right: hiding hate in the landscape, *Antipode*, 29, 256–77.

Gamble, A. (1994) *The Free Economy and the Strong State*, second edition, London: Macmillan.

Gardiner, S. (1995) Criminal Justice Act 1991: management of the underclass and the potentiality of community, in Noaks, L., Levi, M. and Maguire, M. (eds) *Contemporary Issues in Criminology*, Cardiff: University of Wales Press.

Garland, D. (1990) *Punishment and Modern Society*, Oxford: Clarendon Press.

Garland, D. (1996) The limits of the sovereign state: strategies of crime control in contemporary society, *British Journal of Criminology*, 36, 445–71.

Garland, D. (1997) 'Governmentality' and the problem of crime: Foucault, criminology and sociology, *Theoretical Criminology*, **1**, 173.

Garland, D. (1998) A review of 'The Local Governance of Crime' by Adam Crawford, *British Journal of Criminology*, **38**, 516–19.

Garland, D. (1999) 'Governmentality' and the problem of crime, in Smandych, R. (ed.) *Governable Places: Readings on Governmentality and Crime Control*, Aldershot: Ashgate Dartmouth.

Gellner, E. (1964) *Thought and Change*, London: Weidenfeld & Nicholson.

Gellner, E. (1983) *Nations and Nationalism*, Oxford: Basil Blackwell.

Gerber, M. (1992) *On the Home Front: The Cold War Legacy of the Hanford Nuclear Site*, Lincoln: University of Nebraska Press.

Gibbons, S. (1995) Trading places, *Police Review*, **21**, 22–3.

Gibbons, S. (1996) Reclaiming the streets, *Police Review*, **22**, 13.

Gibbons, S. (1997) Zero intolerance, *International Police Review*, May/June.

Giddens, A. (1988) *The Consequences of Modernity*, Cambridge: Polity Press.

Giddens, A. (1991) *Modernity and Self-Identity: Self and Society in the Late Modern Age*, Cambridge: Polity Press.

Giddens, A. (1992) *The Transformation of Intimacy: Sexuality, Love and Eroticism in Modern Societies*, Cambridge: Polity Press.

Gill, H.B. (1972) Correctional philosophy and architecture, in Carter, R.M. (ed.) *Correctional Institutions*, New York: Lippincott.

Gilling, D. (1996) Policing, crime prevention and partnerships, in Leishman, F., Loveday, B. and Savage, S. (eds) *Core Issues in Policing*, London: Longman.

Gilling, D. (1997) *Crime Prevention: Theory, Policy and Politics*, London: UCL Press.

Giuliani, R.W. and Bratton, W.J. (1994) *Police Strategy No.5: Reclaiming the Public Spaces of New York*, New York: New York Police Department.

Glasgow, N. and Reeder, R.J. (1990) Economic and fiscal implications of non-metro retirement migration, *Journal of Applied Gerontology*, **9**, 433–51.

Godlewska, A. and Smith, N. (1994) *Geography and Empire*, Oxford: Blackwell.

Gold, J.R. (1982) Territoriality and human spatial behaviour, *Progress in Human Geography*, **6**, 44–67.

Gold, J.R. and Gold, M.M. (1995) *Imagining Scotland: Tradition, Representation and Promotion in Scottish Tourism since 1750*, Aldershot: Scolar Press/Ashgate Press.

Gold, J.R. and Gold, M.M. (2000) Representing Culloden: social memory, heritage and landscapes of regret, paper presented to symposium on 'Memory/Identity/Landscape', Annual Meeting, Association of American Geographers, Pittsburgh, Penn.

Gold, J.R. and Revill, G.E. (1999) Landscapes of defence, *Landscape Research*, **29**(3), 229–39.

Gold, J.R. and Ward, S.V. (eds) (1994) *Place Promotion: The Use of Publicity and Marketing to Sell Towns and Regions*, Chichester: Wiley.

Golomstock, I. (1990) *Totalitarian Art in the Soviet Union, the Third Reich, Fascist Italy and the People's Republic of China*, London: Collins Harvill.

Goodwin, M. (1998) The governance of rural areas: some emerging research issues and agendas, *Journal of Rural Studies*, **14**, 5–12.

Goodwin, M., Johnstone, C. and Williams, K. (1998a) Redefining the spaces of deterrence: CCTV and the policing of public space, paper presented to RGS/IBG conference, Guildford, 5–8 January 1998.

Goodwin, M., Johnstone, C. and Williams, K. (1998b) CCTV, public behaviour, and the policing of public space, paper presented to the Association of American Geographers Annual Meeting, Boston, 25–29 March 1998.

Goss, J. (1993) The magic of the mall, _Annals of the Association of American Geographers_, **83**, 18–47.

Graham, B. (1997) Ulster: a representation of place yet to be imagined, in Shirlow, P. and McGovern, M. (eds) _Who are the People? Unionism, Protestantism and Loyalism in Northern Ireland_, London: Pluto Press.

Graham, S., Brookes, J. and Heery, D. (1996) _Towns on the Television: Closed circuit Television Surveillance in British Towns and Cities_, Newcastle, working paper, Department of Town and Country Planning, University of Newcastle-upon-Tyne.

Graham, S. and Marvin, S. (1996) _Telecommunications and the City: Electronic Spaces, Urban Places_, London: Routledge.

Graves, R.M. (1949) _Experiment in Anarchy_, London: Victor Gollancz.

Green, M.K. (1995) Cultural identities: challenges for the twenty-first century, in Green, M.K. (ed.) _Issues in Native American Cultural Identity_, Portland, Ore.: Sussex Academic Press.

Greenbie, B.B. (1992) The landscape of social symbols, in Nasar, J.L. (ed.) _Environmental Aesthetics: Theory, Research and Applications_, Cambridge: Cambridge University Press.

Gregory, D. (1994) _Geographical Imaginations_, Oxford: Blackwell.

Gresty, B. (1996) The New York Police Department: managing the crime rates, _Focus_, **7**, March.

Gribb, W.J. (1992) Taos Pueblo's struggle for Blue Lake, in Janelle, D. (ed.) _Geographical Snapshots of North America: Commemorating the 27th Congress of the International Geographical Union and Assembly_, New York: Guilford Press.

Grogger, J. and Weatherford, M.S. (1995) Crime, policing and the perception of neighbourhood safety, _Political Geography_, **14**, 521–41.

Guessoum-Benderbouz, Y. (1997) Planning for women's safety in city centres, unpublished PhD thesis, University of Nottingham.

Haas, W.H. and Serow, W.J. (1990) The influence of retirement immigration on local economic development, Asheville, NC: North Carolina Center for Creative Retirement.

Habermas, J. (1991) _The Structural Transformation of the Public Sphere_, Cambridge, MA: Harvard University Press.

Hagan, W.T. (1993) _American Indians_, Chicago: University of Chicago Press.

Halfacree, K. (1993) Locality and social representation: space, discourse and alternative definitions of the rural, _Journal of Rural Studies_, **9**, 1–15.

Hall, S., Critcher, C., Jefferson, T., Clarke, J. and Roberts, B. (1978) _Policing the Crisis_, Basingstoke: Macmillan.

Hall, S. (1980) _Drifting into a Law and Order Society_, London: Cobden Trust.

Hall, S. and Jacques, M. (1983) _The Politics of Thatcherism_, London: Lawrence & Wishart.

Hamlin, D. (1997) _Commentaries_, Hendersonville, NC: Carriage Park.

Ham-Rowbottom, K.A., Gifford, R. and Shaw, K.T. (1999) Defensible space theory and the police: assessing the vulnerability of residences to burglary, _Journal of Environmental Psychology_, **19**(2), 117–29.

Hamshaw, K. (1998) Living with the 24 hour economy, paper presented at the English Historic Towns Forum Living in the City Conference, Brighton and Hove.

Harper, S. (1997) Contesting later life, in Cloke, P. and Little, J. (eds) _Contested Countryside Cultures: Otherness, Marginalisation and Rurality_, London: Routledge.

Harries, M. (1983) _The War Artists: British Official War Art of the Twentieth Century_, London: Joseph in association with the Imperial War Museum and the Tate Gallery.

Hart, J.F. (1995) Reading the landscape, in Thompson, G.F. (ed.) _Landscape in America_, Austin: University of Texas Press.

Harvey, D. (1985) *Consciousness and the Urban Experience: Studies in the History and Theory of Capitalist Urbanisation*, Baltimore: Johns Hopkins University Press.

Harvey, D. (1989) *The Condition of Postmodernity*, Oxford: Basil Blackwell.

Hay, C. (1996) *Restating Social and Political Change*, Milton Keynes: Open University Press.

Hayes, B. (1998) Applying Bratton to Britain: the need for sensible compromise, in Weatheritt, M. (ed.) *Zero Tolerance: What Does it Mean and Is it Right for Policing in Britain?* London: Police Foundation.

Heath, T. (1997) The twenty-four hour city concept: a review of initiatives in British cities, *Journal of Urban Design*, **2**, 193–204.

Heffernan, M. (1995) For ever England: the Western Front and the politics of remembrance in Britain, *Ecumene*, **2**, 293–323.

Helgerson, R. (1986) *The Land Speaks: Cartography, Chorography and Subversion in Renaissance England*, Berkeley: University of California Press.

Henderson, M. (1992) American Indian reservations: controlling separate space, creating separate environments, in Dilsaver, L. and Colten, C. (eds) *The American Environment: Interpretations of Past Geographies*, Savage, Md.: Rowman & Littlefield.

Herbert, D.T. (1993) Neighbourhood incivilities and the study of crime in place, *Area*, **25**, 45–54.

Hetherington, K. (1997) *The Badlands of Modernity: Heterotopia and Social Ordering*, London: Routledge.

Hewlett, S.A. and West, C. (1998) *The War Against Parents: What We Can Do for America's Beleaguered Moms and Dads*, Boston and New York: Houghton Mifflin.

Hill, R. (1996) Mission possible: a new role for the local state, *Renewal*, **4**(2), May, 21.

Hillier, T. (1994) web site: www.emergency.com/carbomb.html

Hinchcliffe, S. (1997) Locating risk: energy use, the ideal home and the non-ideal world, *Transactions of the Institute of British Geographers*, **22**, 197–209.

Hirsch, E. and O'Hanlon, M. (1995) *The Anthropology of Landscape: Perspectives on Place and Space*, Oxford: Clarendon Press.

HMCIP (1993) *Report of an Inquiry by Her Majesty's Chief Inspector of Prisons for England and Wales into the Disturbance at HM Prison Wymott on 6 September 1993*, London: HMSO.

HM Prison Service (1990) *Report on the Seminar for Category C establishments 24–26 January 1990*, Regime Management Report No. 2, May (unpublished).

Home Office (1983) *Home Office Circular*, 114/1983, London: Home Office.

Home Office (1984a) *Home Office Circular*, 8/1984, London: Home Office.

Home Office (1984b) *Managing the Long-term Prison System, the Report of the Control Review Committee* (The CRC Report), London: HMSO.

Home Office (1985) *New Directions in Prison Design: Report of a Home Office Working Party on American New Generation Prisons* (The Platt Report), London: HMSO.

Home Office (1989) *Home Office Standing Conference Report on the Fear of Crime*, London: Home Office.

Home Office (1994a) *Closed Circuit Television: Looking Out for You*, London: Home Office.

Home Office (1994b) *Operation of Certain Police Powers Under PACE*, Statistical Bulletin, 15/94, 24 June.

Home Office (1996) *Press Release*, 007/96, London: Home Office.

Home Office (1997) *Tackling Youth Crime*, London: Home Office.

Home Office (1998) *Crime and Disorder Bill 1998: Guidance on Statutory Crime and Disorder Partnerships*, London: Home Office.

Home Office (1999a) Crime reduction programme: CCTV initiative, available from Home Office at www.homeoffice.gov.uk, Crime Reduction Unit home page.

Home Office (1999b) *Operation of Certain Police Powers Under PACE*, Statistical Bulletin, 2/99, 22 January.

Home Office (1999c) Press release 168/99 at http:/www.nds.coi.gov.uk/coi/coipress

Homel, R., Hauritz, M., McLiwain, G., Wortley, R. and Carvolth, R. (1997) Preventing drunkenness and violence around night-clubs in a tourist resort, in Clarke, R.V.G. (ed.) *Situational Crime Prevention: Successful Case Studies*, second edition, New York: Harrow and Heston.

Homer-Dixon, T.F. (1999) *Environmental, Scarcity, and Violence*, Princeton, NJ: Princeton University Press.

Horowitz, C. (1995) The end of crime as we know it, *New York*, 14 August.

Horvath, R.J. (1974) Machine space, *Geographical Review*, **64**, 167–88.

Hough, J.M., Clarke, R.V.G. and Mayhew, P. (1980) Introduction, in Clarke, R.V.G. and Mayhew, P. (eds) *Designing Out Crime*, Home Office Research and Planning Unit, London: HMSO, pp. 1–17.

Hound Ears Club (n.d.) *Hound Ears Club*. Web page: http://www.houndears.com/houndears2.html

House of Lords (1999) *Management of Nuclear Waste*, third report of the Select Committee on Science and Technology, Session 1998–99, London: HMSO.

Howard, H.E. (1920) *Territory in Bird Life*, London: John Murray.

Howard, M. (1996) A letter from Michael Howard, *New Scientist*, **149**, 13 January, 47.

Hubbard, P. (1997) Red-light districts and toleration zones: changing geographies of female street prostitution in England and Wales, *Area*, **29**, 129–40.

Hubbard, P. (1999) *Sex and the City: Geographies of Prostitution in the Urban West*, London: Ashgate Press.

Hughes, G. (1997) Policing late modernity: changing strategies of crime management in contemporary Britain, in Jewson, N. and MacGregor, S. (eds) *Transforming Cities: Contested Governance and New Spatial Divisions*, London: Routledge.

Hughes, R. (1987) *The Fatal Shore: A History of the Transportation of Convicts to Australia, 1787–1868*, London: Folio Society.

Hunter, J. (1999) Native American perceptions of higher education employment experiences: a case study of Northern Arizona University, unpublished Masters thesis, Department of Geography and Public Planning, Northern Arizona University.

Husain, S. (1988) *Neighbourhood Watch in England and Wales: A Longitudinal Analysis*, Crime Prevention Unit Paper 12, London: HMSO.

Immigration and Naturalization Service (1998) *Fact Sheet: Operation Gatekeeper: New Resources, Enhanced Results*, http://www.ins.usdoj.gov/graphics/publicaffairs/factsheets/OpGateFS.htm

Immigration and Naturalization Service (1999a) *Fact Sheet: Border Management*, http://www.ins.usdoj.gov/graphics/publicaffairs/factsheets/borderfs.htm

Immigration and Naturalization Service (1999b) INSPASS, http://www.ins.usdoj.gov/graphics/lawenfor/bmgt/inspect/inspass.htm

Ingold, T. (1996) Hunting and gathering as ways of perceiving the environment, in Ellen, R. and Fukui, K. (eds) *Redefining Nature: Ecology, Culture and Domestication*, Oxford: Berg.

Jackson, P. (1998) Domesticating the street: the contested spaces of the high street and the mall, in Fyfe, N. (ed.) *Images of the Street*, London: Routledge.

Jacobs, J. (1961) *The Death and Life of Great American Cities: The Failure of Town Planning*, London: Peregrine Books in association with Jonathan Cape.

Jarman, N. (1998) *Material Conflicts*, Oxford: Berg.

Jessop, B., Bonnet, K., Bromley, S. and Ling, T. (1988) *Thatcherism*, Cambridge: Polity Press.

Johnson, N. (1973) *The Human Cage: A Brief History of Prison Architecture*, New York: Walker.

Johnston, L. (1996) Policing diversity: the impact of the public–private complex on policing, in Leishman, F., Loveday, B. and Savage, S. (eds) *Core Issues in Policing*, London: Longman.

Johnstone, C., Goodwin, M. and Williams, K. (1999) An evaluation of the introduction of CCTV to Aberystwyth and Cardigan town centres, unpublished report prepared for Ceredigion Regional Council.

Jones, P., Hillier, D. and Turner, D. (1999) Towards the 24 hour city, *Town and Country Planning*, **68**(5), 164–5.

Joss, S. (1998) Danish consensus conferences as a model of participatory technology assessment: an impact study of consensus conferences on Danish parliament and Danish public debate, *Science and Public Policy*, **25**, 2–22.

Judd, D. (1995) The rise of the new walled cities, in Liggett, H. and Perry, D. (eds) *Spatial Practices*, Thousand Oaks, Calif.: Sage.

Judge, J. (1972) Israel: the seventh day, *National Geographic*, **142**, 6.

Kaplan, S. (1992) Perception and landscape: conceptions and misconceptions, in Nasar, J.L. (ed.) *Environmental Aesthetics: Theory, Research and Applications*, Cambridge: Cambridge University Press.

Kaplan, S. (1992) Perception and landscape: conceptions and misconceptions, in Nasar, J.L. (ed.) *Environmental Aesthetics: Theory, Research and Applications*, Cambridge: Cambridge University Press.

Katz, C. (1993) Growing girls/closing circles, in Katz, C. and Monk, J. (eds) *Full Circles: Geographies of Women over the Lifecourse*, London: Routledge.

Keith, M. and Pile, S. (eds) (1993) *Place and the Politics of Identity*, London: Routledge.

Kellerman, A. (1996) Settlement myth and settlement activity: interrelationships in the Zionist land of Israel, *Transactions of the Institute of British Geography*, **21**, 363–78.

Kelling, G.L. (1998) The evolution of broken windows, in Weatheritt, M. (ed.) *Zero Tolerance: What Does it Mean and Is It Right for Policing in Britain?* London: Police Foundation.

Kelly, O. (1994) By all means necessary, *Police Review*, **20**, 15 April, 14–16.

Kelsey, J. (1993) *Rolling Back the State: Privatisation of Power in Aotearoa/New Zealand*, Wellington: Bridget Williams Books.

Kenmure Enterprises Inc. (n.d. a) *Kenmure*. Web page: http://www.kenmure.com/community.htm (as accessed December 1997).

Kenmure Enterprises Inc. (n.d. b) *Kenmure: A Mountain Legacy*, Flat Rock, NC: Kenmure.

Kidron, M.H. and Smith, D. (1991) *The New State of War and Peace*, New York: Simon & Schuster.

Kimmerer, R. (1998) Intellectual diversity: bringing the native perspective into natural resources education, *Winds of Change*, Summer.

King, A.D. (1976) *Colonial Urban Development: Culture, Social Power and Environment*, London: Routledge & Kegan Paul.

King, R.D. (1972) *An Analysis of Prison Regimes*, unpublished report to the Home Office.

King, R.D. (1985) Control in prisons, in Maguire, M., Vagg, J. and Morgan, R. (eds) *Accountability and Prisons: Opening up a Closed World*, London: Tavistock.

Klein, B.S. (1998) Politics by design: remapping security landscapes, *European Journal of International Relations*, **4**, 327–46.

Kliot, N. (1994) *Water Resources and Conflict in the Middle East*, London: Routledge.

Koch, B. (1998) *The Politics of Crime Prevention*, Aldershot: Ashgate Press.

Kowinski, W.S. (1982) *The Malling of America*, New York: William Morrow.

KPM (Klynveld Peat Marwick) (1991) *Counting Out Crime: The Nottingham Crime Audit*, Nottingham: Nottingham City Council.

Krannich, R. (1985) Rapid growth and fear of crime: a four community comparison, *Rural Sociology*, **49**, 193–209.

Kristeva, J. (1982) *Powers of Horror*, New York: Columbia University Press.

Kristeva, J. (1991) *Strangers to Ourselves*, New York: Columbia University Press.

Kropf, K. (1993) An inquiry into the definition of built form in urban morphology, unpublished PhD thesis, School of Geography, University of Birmingham.

Kropf, K. (1994) Typological zoning and the maintenance of local and historical character, paper presented to the Urban Morphology Research Group, School of Geography, University of Birmingham, October.

Kropf, K. (1995) Interpretations of 'type' in urban morphology and architectural typology, paper presented to the Urban Morphology Research Group, School of Geography, University of Birmingham, 30 October.

LaGrange, R.L., Ferraro, K.F. and Supanic, M. (1992) Perceived risk and fear of crime: role of social and physical incivilities, *Journal of Research in Crime and Delinquency*, **29**, 31–4.

Lake Toxaway Company (n.d.) *Lake Toxaway*. Web page: http://www.laketoxaway.com

Larsen, E.N. (1992) The politics of prostitution control: interest group politics in four Canadian cities, *International Journal of Urban and Regional Research*, **16**, 169–89.

Lash, S. and Urry, J. (1994) *Economies of Signs and Spaces*, London: Sage.

Lavender, D. (1980) *The Southwest*, New York: Harper & Row.

Law, J. (1994) *Organising Modernity*, Oxford: Blackwell.

Law, L. (1997) Dancing on the bar, in Pile, S. and Keith, M. (eds) *Geographies of Resistance*, London: Routledge.

Laws, G. (1994) Oppression, knowledge and the built environment, *Political Geography*, **13**, 7–32.

Laycock, G. and Tilley, N. (1995) *Policing and Neighbourhood Watch: Strategic Issues*, Crime Detection & Prevention Series Paper 60, Police Research Group, London: Home Office.

Leeds City Council (1993) *Unitary Development Plan*, revised draft, Leeds: Leeds City Council.

Leeds City Council (1998) *Leeds City Centre Briefings, City Centre Management*, Leeds: Leeds City Council.

Leeds City Council (1999) *City Centre First Annual Report*, Leeds: Leeds City Council.

Leuthold, S. (1995) Native American responses to the western, *American Indian Culture and Research Journal*, **19**, 153–89.

Lew, A.A. (1998) American Indians in state tourism promotional literature, in Lew, A.A. and Van Otten, G. (eds) *Tourism and Gaming on American Indian Lands*, New York: Cognizant Communication Corporation.

Lew, A.A. (1999) Managing tourist space in Pueblo villages of the American southwest, in Singh, T.V. (ed.) *Tourism Development in Critical Environments*, New York: Cognizant Communications Corporation.

Lew, A.A. and Van Otten, G. (eds) (1998) *Tourism and Gaming on American Indian Lands*, New York: Cognizant Communication Corporation.

Lewis, D.R. (1995) Native Americans and the environment: a survey of twentieth-century issues, *American Indian Quarterly*, **19**, 423–50.

Lilly, J.R., Cullen, F.T. and Ball, R.A. (1995) *Criminological Theory: Context and Consequences*, second edition, Thousand Oaks, Calif.: Sage.

Livingstone, D.N. (1992) *The Geographical Tradition: Episodes in the History of a Contested Enterprise*, Oxford: Blackwell.

Loeb, P. (1986) *Nuclear Culture: Living and Working in the World's Largest Atomic Complex*, Philadelphia: New Society Publishers.

London Chamber of Commerce and Industry (1996) *Invest in London: An International City*, London: EMP Publications.

Lovatt, A. (1994) *More Hours in the Day*, Manchester: Manchester Institute of Popular Culture.

Lowman, J. (1992) Street prostitution control: some Canadian reflections on the Finsbury Park experience, *British Journal of Criminology*, **32**, 1–16.

Lukes, S. (1974) *Power: A Radical View*, London: Macmillan.

Lyon, D. (1994) *The Electronic Eye: The Rise of Surveillance Society*, London: Polity Press.

MacCannell, D. (1994) Tradition's next step, in Norris, S. (ed.) *Discovered Country: Tourism and Survival in the American West*, Albuquerque: University of New Mexico Press.

MacCormac, R. (1983) Urban reform: MacCormac's manifesto, *Architects' Journal*, 15 June, 59–72.

Macpherson of Cluny, Sir W. (1999) *The Stephen Lawrence Inquiry: Report of an Inquiry by Sir William Macpherson of Cluny*, Cm 4262-I, London: HMSO.

Maffesoli, M. (1996) *The Time of the Tribes*, London: Sage.

Maguire, M., Morgan, R. and Reiner, R. (1997) *The Oxford Handbook of Criminology*, second edition, Oxford: Clarendon Press.

Mallory, J. and McNeill, T. (1991) *The Archaeology of Ulster*, Belfast: Institute of Irish Studies.

Malmberg, T. (1980) *Human Territoriality: A Survey of the Behavioural Territories of Man with Preliminary Analysis and Discussion of Meaning*, Chicago: Aldine.

Mansfield, M. (1993) *Presumed Guilty*, London: Heinemann.

March, L. (1972) Elementary models of built forms, in Martin, L. and March, L. (eds) *Urban Space and Structures*, Cambridge: Cambridge University Press.

Marshall, S.W. (1994) An undesirable institution? the changing location of prisons in England and Wales, *Conference Proceedings of the First Societies in Transition Conference*, Vol. 2, Herriot-Watt University, Edinburgh, June.

Marshall, S.W. (1995) *Control in Category C Prisons: Built-Environment and Regime Study*, final report for the Home Office Research and Planning Unit on behalf of HM Prison Service Custody Group, December 1995 (unpublished).

Marshall, S.W. (1997a) Always greener on the other side of the fence? order and control in medium security prisons in England and Wales, unpublished doctoral thesis, University of Birmingham.

Marshall, S.W. (1997b) *Control in Category C Prisons*, Home Office Research and Statistics Directorate Research Findings No. 54, London: Home Office.

Martin, L. and March, L. (eds) (1972) *Urban Space and Structures*, Cambridge: Cambridge University Press.

Marx, L. (1967) *The Machine in the Garden: Technology and the Pastoral Ideal in America*, London: Oxford University Press.

Massey, D. (1995) Thinking radical democracy, *Environment and Planning D: Society and Space*, **13**, 283–8.

Matley, R. (1999) 24 hour city: the Leeds experience, paper given at the 24 Hour City Economy Conference, University of North London, 31 March.

Matless, D. (1998) *Landscape and Englishness*, London: Reaktion Books.

Matthews, R. (1986) *Policing Prostitution: A Multi-agency Approach*, Centre for Criminology, Middlesex Polytechnic.

Matthews, R. (1993) *Kerb-Crawling, Prostitution and Multi-agency Policing*, Series Paper No. 43, London: Police Research Group Crime Prevention Unit.

Mayr, E. (1935) Bernard Altum and the territorial theory, *Proceedings of the Linnaean Society of New York*, **45–6**, 24–38.

McCann, E. (1995) Neo-traditional developments: the anatomy of a new urban form, *Urban Geography*, **16**, 210–33.

McConville, M., Sanders, A. and Leng, R. (1991) *The Case for The Prosecution*, London: Routledge.

McCormick, K. and Hamilton, D.P. (1991) *Images of War: The Artist's Vision of World War II*, London: Cassell.

McDowell, L. (1997) *Capital Culture: Gender at Work in the City*, Oxford: Blackwell.

McGarry, J. and O'Leary, B. (1996) *Explaining Northern Ireland*, Oxford: Blackwell.

McGuire, T. (1994) Federal Indian policy: a framework for evaluation, in Wells, R. (ed.) *Native American Resurgence and Renewal: A Reader and Bibliography*, London: Scarecrow Press.

McKay, G. (1996) *Senseless Acts of Beauty*, London: Verso.

McKenzie, E. (1994) *Privatopia: Homeowners Associations and the Rise of Residential Private Government*, New Haven, Conn.: Yale University Press.

McLaughlin, E. and Muncie, J. (eds) (1996) *Controlling Crime*, London: Sage/Open University.

McRobbie, A. (1994) Folk devils fight back, *New Left Review*, **203**, January–February, 107–16.

McSorley, J. (1990) *Living in the Shadow*, London: Pan Books.

Medvedev, Z. (1979) *Nuclear Disaster in the Urals*, London: Angus & Robertson.

Medvedev, Z. (1990) *The Legacy of Chernobyl*, Oxford: Blackwell.

Menzie, K. (1998) Home, safe home: security is a high priority, *Baltimore Sun*, 15 March, 9J.

Mihesuah, D. (1996) *American Indians: Stereotypes and Realities*, Canada: Clarity International.

Mitchell, D. (1993) State intervention in landscape production: the wheatland riot and the California Commission of Immigration and Housing, *Antipode*, **25**(2), 91–113.

Mitchell, D. (1994) Landscape and surplus value: the making of the ordinary in Brentwood, CA, *Environment and Planning D: Society and Space*, **12**, 7–30.

Mitchell, D. (1995) The end of public space? People's park, definitions of the public and democracy, *Annals of the Association of American Geographers*, **85**, 108–33.

Mitchell, D. (1996) Introduction: public space and the city, *Urban Geography*, **17**, 127–31.

Mitchell, D. (1997) The annihilation of space by law: the roots and implications of anti-homeless laws in the United States, *Antipode*, **29**, 303–35.

Mitchell, W.J.T. (1994) Imperial landscape, in Mitchell, W.J.T. (ed.) *Landscape and Power*, Chicago: University of Chicago Press.

Moffat, C.B. (1903) The spring rivalry of birds, *Irish Naturalist*, **12**, 156–66.

Montgomery, J. (1994) The evening economy of cities, *Town and Country Planning*, **63**, 302–7.

Montgomery, J. (1995) Urban vitality and the culture of cities, *Planning Practice and Research*, **10**, 101–9.

Montgomery, R.H. (1974) A measurement of inmate satisfaction/dissatisfaction in selected South Carolina institutions, unpublished doctoral thesis, University of South Carolina.

Moos, R.H. (1968) Assessment of the social climates of correctional institutions, *Journal of Crime and Delinquency*, **5**, 174–88.

Morgan, R. (1994) Thoughts about control in prisons, *Prison Service Journal*, **93**, 57–60.

Morgan Committee (1991) *Safer Communities: The Local Delivery of Crime Prevention through the Partnership Approach*, London: HMSO.

Morris, A.E.J. (1994) *History of Urban Form: Before the Industrial Revolution*, Harlow, Essex: Longman.

Morris, N. and Rothman, D.J. (eds) (1996) *The Oxford History of the Prison: The Practice of Punishment in Western Society*, Oxford: Oxford University Press.

Morrow, D. (1997) Suffering for righteousness' sake? Fundamentalist Protestantism and Ulster politics, in Shirlow, P. and McGovern, M. (eds) *Who are the People? Unionism, Protestantism and Loyalism in Northern Ireland*, London: Pluto Press.

Mott, J. (1985) *Adult Prisons and Prisoners in England and Wales 1970–1982*, Home Office Research Study 84, London: HMSO.

Mountbatten, L. (1966) *Report of the Inquiry into Prison Escapes and Security by Admiral of the Fleet, the Earl Mountbatten of Burma*, Cmnd 3175, London: HMSO.

Muir, R. (1999) *Approaches to Landscape*, Basingstoke, Hampshire: Macmillan.

Muncie, J. (1996) The construction and deconstruction of crime, in Muncie, J. and McLaughlin, E. (eds) *The Problem of Crime*, London: Sage/Open University.

Muratori, S. (1959) *Studi per una Operante Storia Urbana di Venezia*, Palladio: Rivista di Storia dell' Architettura, Nuova Serie IX.

Muratori, S. (1963) *Studi per una Operante Storia Urbana di Roma*, Rome: Instituto Poligrafico dello Stato.

Nagel, W.G. (1973) *The New Red Barn: A Critical Look at the Modern American Prison*, New York: Walker.

Nash, R. (1967) *Wilderness and the American Mind*, New Haven, Conn.: Yale University Press.

Nelson, A.L. (1997) Public perceptions of the electronic eye, *Town and Country Planning*, **66**, July/August.

Nelson, A.L. (1998) Public perceptions of the effectiveness of closed circuit television, unpublished discussion paper presented at Worcester College of Higher Education.

Newman, O. (1972) *Defensible Space*, New York: Macmillan.

Newman, O. (1996) *Creating Defensible Space*, Washington, DC: Department of Housing and Urban Development.

Newman, O. and Franck, K.A. (1982) The effects of building size on personal crime and fear of crime, *Population and Environment*, **5**, 203–20.

Neyroud, P. (1998) Intelligent zero policing, *Police Research and Management*, Spring.

Niedersachsen (1993) *International Hearing on Final Disposal of Nuclear Waste*, Niedersachsisches Umweltministerium (English version).

Nirex (1997) *Nirex Approach to Publication and Peer Review*, Harwell, UK: Nirex Ltd, January.

Norris, C. and Armstrong, G. (1997) *The Unforgiving Eye: CCTV Surveillance in Public Spaces*, Centre for Criminology, University of Hull.

Norris, C. and Armstrong, G. (1998) 'Introduction: power and vision', in Norris, C., Morgan, J. and Armstrong, G. (eds) *Surveillance, Closed Circuit Television and Social Control*, Aldershot: Ashgate Press.

Norris, C. and Armstrong, G. (1999) *The Maximum Surveillance Society*, London: Berg.

NYPD (New York Police Department) (1995) *Managing for Results: Building a Police Organization that Can Dramatically Reduce, Crime, Disorder and Fear*, New York: NYPD.

Nyström, L. (ed.) (1999) *City and Culture: Cultural Processes and Urban Sustainability*, Karlskrone: Swedish Urban Environmental Council, National Board of Housing Building and Planning.

Oc, T. and Trench, S. (1993) Planning and shopper security, in Bromley, R.D.F. and Thomas, C.J. (eds) *Retail Change*, London: UCL Press.

Oc, T. and Tiesdell, S. (eds) (1997) *Safer City Centres: Reviving the Public Realm*, London: Paul Chapman.

Oc, T. and Tiesdell, S. (1998) City centre management and safer city centres: approaches in Coventry and Nottingham, *Cities*, **15**, 85–103.

O'Connor, J. (1993) Towards the twenty-four hour city, in *Transcript of the First National Conference on the Evening Economy*, Manchester: Manchester Institute of Popular Culture.

Old North State Club (n.d.) *Old North State Club at Uwharrie Point*, New London, NC: Old North State Club.

O'Malley, P. (1987) Marxist theory and Marxist criminology, *Crime and Social Justice*, **29**, 70–87.

Olwig, K.R. (1996) Recovering the substantive nature of landscape, *Annals of the Association of American Geographers*, **86**, 630–53.

Orians, G.H. (1986) An ecological and evolutionary aproach to landscape aesthetics, in Penning-Rowsell, E.C. and Lowenthal, D. (eds) *Landscape Meaning and Values*, London: Allen & Unwin.

Osborne, D. and Gaebler, T. (1992) *Reinventing Government: How the Entrepreneurial Spirit is Transforming the Public Sector*, Reading, Mass.: Addison-Wesley.

Ó Tuathail, G. (1986) The languages and nature of 'new' geopolitics, *Political Geography*, **11**, 190–204.

Page, V. (1994) Reservation development in the United States: peripherality in the core, in Wells, R. (ed.) *Native American Resurgence and Renewal: A Reader and Bibliography*, London: Scarecrow Press.

Painter, K. (1996) It's safety first and last, *Landscape Design*, March, 52.

Pavlich, G. (1999) Preventing crime: 'social' versus 'community' governance in Aotearoa/New Zealand, in Smandych, R. (ed.) *Governable Places: Readings on Governmentality and Crime Control*, Aldershot: Ashgate Press.

Pawson, E. (1992) Two New Zealands: Maori and European, in Anderson, K. and Gale, F. (eds) *Inventing Places: Studies in Cultural Geography*, Melbourne: Longman.

Pease, K. (1992) Preventing burglary on a British public housing estate, in Clarke, R. (ed.) *Situational Crime Prevention: Successful Case Studies*, New York: Harrow and Heston.

Perkins, D.D., Brown, B.B. and Taylor, R.B. (1996) The ecology of empowerment: predicting participation in community organizations, *Journal of Social Issues*, **52**(1), 85–110.

Perkins, D.D. and Taylor, R.B. (1996) Ecological assessments of community disorder: their relationship to fear of crime and theoretical implications, *American Journal of Community Psychology*, **24**(1), 63–107.

Phillips, M. (2000) Exclusivity and anxiety in the promotion of private ruralities: a study of private communities in the North Carolina countryside, *Rural Sociology*, (forthcoming).

Pile, S. (1997) Opposition, political identities and spaces of resistance, in Pile, S. and Keith, M. (eds) *Geographies of Resistance*, London: Routledge.

Pollard, C. (1998) Zero tolerance: short term fix, long term liability, in Dennis, N. (ed.) *Zero-tolerance: Policing a Free Society*, second edition, London: Institute of Economic Affairs.

Porter, B. (1987) *Origins of the Vigilant State*, London: Weidenfeld & Nicolson.

Pred, A. (1990) *Making Histories and Constructing Human Geographies*, Boulder: Westview Press.

Pugh, S. (1988) *Garden – Nature – Language*, Manchester: Manchester University Press.

Pugh, S. (ed.) (1990) *Reading Landscape: Country – city – capital*, Manchester: Manchester University Press.

Punter, J. (1990) The privatisation of the public realm, *Planning Practice and Research*, **5**, 17–21.

Ramsey, M. (1990) *Lager land lost? an experiment in keeping drinkers off the streets in central Coventry and elsewhere*, Crime Prevention Unit Paper 22, London: Home Office.

RCCP (Royal Commission on Criminal Procedure) (1981) Cmnd 8092, London: HMSO.

RCPPP (Royal Commission on Police Powers and Procedure) (1929), Cmnd 3297, London: HMSO.

Rees, A.L. and Borzello, F. (1986) *The New Art History*, London: Camden.

Rees, J. and Debbage, K. (1996) Economic change in metropolitan Carolina, in Bennett, G.C. (ed.) *Snapshots of the Carolinas: Landscapes and Cultures*, Charlotte, NC: Association of American Geographers.

Reeve, A. (1996) The private realm of the managed town centre, *Urban Design International*, **1**, 61–80.

Rhodes, R.A.W. (1996) *Understanding Governance: Policy Networks, Governance, Reflexivity and Accountability*, Buckingham: Open University Press.

Richardson, D. (1938) *About Policemen*, London: Ginn.

Riger, S., Gordon, M.T. and Le Bailly, R.K. (1982) Coping with crime: women's use of precautionary behaviours, *American Journal of Community Psychology*, **10**, 369–86.

Roberts, B.K. (1987) *Rural Settlement*, London: Macmillan.

Robertson, R. (1992) Globality and modernity, *Theory, Culture and Society*, **9**, 153–61.

Romeanes, T. (1998) A question of confidence: zero tolerance and problem-oriented policing, in Burke, R.H. (ed.) *Zero Tolerance Policing*, Leicester: Perpetuity Press.

Rose, D. (1996) *In The Name of the Law: The Collapse of Criminal Justice*, London: Jonathan Cape.

Rose, G. (1992) Geography as a science of observation: the landscape, the gaze and masculinity, in Driver, F. and Rose, G. (eds) *Nature and Science: Essays in the History of Geographical Knowledge*, Institute of British Geographers Research Series 28, 8–18.

Rose, G. (1993) *Feminism and Geography: The Limits of Geographical Knowledge*, Cambridge: Polity Press.

Roseberry, W. (1989) *Anthropologies and Histories*, New Brunswick, NJ: Rutgers University Press.

Rossi, A. (1970) *L'Analisi Urbana e la Progettazione Architettonica*, Milan: Cooperativa Libraria Universaria Politeccnico.

Rossi, A. (1982) *The Architecture of the City*, Cambridge, Mass.: MIT Press.

Rossi, G. (1999) Tribes deal themselves in: Indian gaming money and political influence. Intellectual Capital.com 2(34): http://www.intellectualcapital.com/issues/97/0821/icpolicy1.html, accessed 1 October.

Routledge, P. (1997a) A Spatiality of Resistance, in Pile, S. and Keith, M. (eds) *Geographies of Resistance*, London: Routledge.

Routledge, P. (1997b) Pollock Free State and the practice of postmodern politics, *Transactions of the Institute of British Geographers*, **22**, 359–77.

Rudzitis, G. (1993) Nonmetropolitan geography: migration, sense of place, and the American west, *Urban Geography*, **14**, 574–85.

Russell, G. (1993) *The American Indian Digest*, Phoenix, Ariz: Thunderbird Enterprises.

RWMAC (Radioactive Waste Management Advisory Committee) (1997) *Response on: Nirex's Proposals for Publication and Peer Review*, January.

RWMAC (Radioactive Waste Management Advisory Committee) (1998) *Initial Recommendations on the Long Term Management of Intermediate Level Radioactive Waste following Rejection of the UK Nirex Rock Characterisation Facility Planning Application: 'Rethinking Disposal'*, January.

RWMAC (Radioactive Waste Management Advisory Committee) (1999) *The Establishment of Scientific Consensus on the Interpretation and Significance of the Results of Science Programmes into Radioactive Waste Disposal*, advice to Ministers, April.

RWMAC/ACSNI (Radioactive Waste Management Advisory Committee/Advisory Committee on the Safety of Nuclear Installations) (1995) *Site Selection for Radioactive Waste Disposal Facilities and the Protection of Human Health*, report of a study group drawn from members of the Radioactive Waste Management Advisory Committee and the Advisory Committee on the Safety of Nuclear Installations, March.

Sack, R.D. (1980) *Conception of Space in Social Thought: A Geographic Perspective*, Baltimore: Johns Hopkins University Press.

Sack, R.D. (1986) *Human Territoriality*, Cambridge: Cambridge University Press.

Sack, R.D. (1997) *Homo Geographicus*, Baltimore: Johns Hopkins University Press.

Sadnicki, M., Barker, F. and MacKerron, G. (1999) *THORP: The Case for Contract Renegotiation*, London: Friends of the Earth.

Saleh, M.A.E. (1999) Reviving traditional design in modern Saudi Arabia for social cohesion and crime prevention purposes, *Landscape and Urban Planning*, **44**, 43–62.

Samuels, I. (1985) *Theories des Mutations urbaines en Pays developpes*, Joint Centre for Urban Design, Oxford Brookes University, unpublished monograph.

Scarman, Lord (1981) *The Brixton Disorders 10–12 April 1981: Report of an Inquiry by the Rt Honourable the Lord Scarman OBE*, Cmnd 8427, London: HMSO.

Schama, S. (1995) *Landscape and Memory*, London: HarperCollins.

Scott, J. (1992) Domination, acting and fantasy, in Martin, A. (ed.) *The Paths to Domination, Resistance and Terror*, Berkeley: University of California Press.

Scruton, R. (1982) *A Dictionary of Political Thought*, London: Pan.

Sennett, R. (1990) *The Conscience of the Eye: The Design and Social Life of Cities*, London: Faber.

Serow, W.J. and Haas, W.H. (1992) Measuring the economic impact of retirement migration: the case of western North Carolina, *Journal of Applied Gerontology*, **11**, 200–15.

Sharpe, K. (1998) *Red Light, Blue Light: Prostitutes, Punters and Police*, London: Ashgate.

Shields, R. (1991) *Places on the Margins: Alternative Geographies of Modernity*, London: Routledge.

Shirlow, P. and McGovern, M. (1998) Language, discourse and dialogue: Sinn Fein and the Irish peace process, *Political Geography*, **17**, 171–86.

Shoard, M. (1987) *This Land is Our Land*, London: Paladin.

Short, E. and Ditton, J. (1996) *Does Closed Circuit Television Prevent Crime? An Evaluation of the Use of CCTV Surveillance Cameras in Airdrie Town Centre*, Glasgow: Central Research Unit, Scottish Office.

Short, E. and Ditton, J. (1998) Seen and now heard: talking to the targets of open street CCTV, *British Journal of Criminology*, **38**, 404–28.

Shotter, J. (1993) *Cultural Politics of Everyday Life*, Milton Keynes: Open University Press.

Sibley, D. (1995) *Geographies of Exclusion: Society and Difference in the West*, London: Routledge.

Siting Task Force (1995a) *Deep River Community Agreement-in-Principle*, Siting Task Force and the Corporation of the Town of Deep River, July.

Siting Task Force (1995b) *A Community Volunteers*, Siting Task Force, Final Report, November.

Skogan, W.G. (1990) *Disorder and Decline: Crime and the Spiral of Decay in American Neighborhoods*, New York: Free Press.

Slater, T.R. (1981) The analysis of burgage patterns in medieval towns, *Area*, **13**, 211–16.

Slater, T.R. (ed.) (1990) *The Built Form of Western Cities: Essays for M.R.G. Conzen on the Occasion of his Eightieth Birthday*, Leicester: Leicester University Press.

Slovic, P., Fischoff, B. and Lichtenstein, S. (1980) Facts and fears: understanding perceived risk, in Schwing, R. and Albers, W. (eds) *Societal Risk Assessment: How Safe Is Safe Enough?* New York: Plenum Press.

Smandych, R. (ed.) (1999) *Governable Places: Readings on Governmentality and Crime Control*, Aldershot: Ashgate Press.

Smith, A.D. (1991) *National Identity*, London: Penguin.

Smith, A.D. (1997) The 'Golden Age' and national renewal, in Hosking, G. and Schöpflin, G. (eds) *Myths & Nationhood*, London: Hurst.

Smith, B. (1984) *European Vision and the South Pacific*, New Haven, Conn.: Yale University Press.

Smith, D.H. (1994) The issue of compatibility between cultural integrity and economic development among Native American tribes, *American Indian Culture and Research Journal*, **18**, 177–205.

Smith, N. (1996) *The New Urban Frontier: Gentrification and the Revanchist City*, New York: Routledge.

Snipp, M. (1994) The changing political and economic status of the American Indians: from captive nations to internal colonies, in Wells, R. (ed.) *Native American Resurgence and Renewal: A Reader and Bibliography*, London: Scarecrow Press.

Soffer, A. and Minghi, J.V. (1986) Israel's security landscapes: the impact of military considerations on land uses, *Political Geographer*, **38**, 28–41.

Sorkin, M. (ed.) (1995) *Variations on a Theme Park: The New American City and the End of Public Space*, New York: Hill and Wang.

Southwood, M. and Ben-Joseph, E. (1997) *Streets and the Shaping of Towns and Cities*, New York: McGraw-Hill.

Sparks, W., Bottoms, A.E. and Light, R. (1996) *Prisons and the Problem of Order*, Oxford: Clarendon Press.

Spradley, P. and McCurdy, D. (1987) *Conformity and Conflict*, Boston: Little, Brown and Co.

Staeheli, L. and Thompson, A. (1997) Citizenship, community and struggles for public space, *Professional Geographer*, **49**, 28–38.

Steadman, J.P. (1994) Built forms and building types: some speculations, *Environment and Planning B: Planning and Design*, **21**, 7–30.

Sternberg, E. (1997) A case of iconographic competition: the building industry and the postmodern landscape, *Journal of Urban Design*, **1**, 145–64.

Sternleib, G. (1990) Things aren't what they used to be, *Journal of the American Planning Association*, **56**, 494.

Stevens, P. and Yach, D.M. (1995), *Community Policing in Action: A Practitioner's Guide*, London: Kenwyn, Juta & Co.

Stoker, G. (1996) *Governance as Theory: Five Propositions*, mimeographed report, Department of Government, University of Strathclyde.

Stoker, G. (1997) *Public–Private Partnerships and Urban Governance*, mimeographed report, available from author at Department of Government, University of Strathclyde.

Strang, V. (1997) *Uncommon Ground: Cultural Landscapes and Environmental Values*, Oxford: Berg.

Sullivan, G.A. (1998) *The Drama of Landscape: Land, Property and Social Relations on the Early Modern Stage*, Stanford, Calif.: Stanford University Press.

Sullivan, R.R. (1997) Bill Bratton's zero tolerance, *City Security*, **13**, 26–9.

Straw, J. (1998) Introduction, in *Crime and Disorder Bill 1998: Guidance on Statutory Crime and Disorder Partnerships*, London: Home Office.

Sutton, M. (1996) *Implementing Crime Prevention Schemes in a Multi-Agency Setting: Aspects of Process in the Safer Cities Programme*, Home Office Research Study 160, London: HMSO.

Swanson, G. (1995) Drunk with the glitter: consuming spaces and sexual geographies, in Watson, S. and Gibson, K. (eds) *Postmodern Cities and Spaces*, Oxford: Blackwell.

Swentzell, R. (1990) The Bureau of Indian Affairs landscape within Santa Clara pueblo, in Groth, P. (ed.) *Vision, Culture, and Landscape: Working Papers from the Berkeley Symposium on Cultural Landscape Interpretation*, University of California at Berkeley.

Swyngedouw, E. (1997) Neither global nor local: 'glocalisation' and the politics of scale, in Cox, K. (ed.) *Spaces of Globalization*, New York: Guilford Press, 137–66.

Symanski, R. (1981) *The Immoral Landscape: Female Prostitution in Western Societies*, London: Butterworth.

Takahashi, L.M. and Dear, M.J. (1997) The changing dynamics of community opposition to human service facilities, *Journal of the American Planning Association*, **63**, 79–93.

Taos Pueblo (n.d.) *Guide for Visitors*, Taos Pueblo, Ariz.: Taos Pueblo Community.

Taylor, I. (1996) Fear of crime, urban fortunes and suburban social movements: some reflections from Manchester, *Sociology*, **30**, 317–37.

Theroux, P. (1988) *Riding the Iron Rooster: By Rail through China*, New York: Ivy Books.

Theroux, P. (1995) *The Pillars of Hercules: A Grand Tour of the Mediterranean*, Harmondsworth: Penguin Books.

Thornton, D. (1985) Intra-regime variation in inmate perception of prison staff, *British Journal of Criminology*, **24**, 24–7.

Thrift, N. and Forbes, D. (1983) A landscape with figures: political geography within human conflict, *Political Geography*, **2**, 247–63.

Tiesdell, S. and Oc, T. (1998) Beyond 'fortress' and 'panoptic' cities: towards a safer public realm, *Environment and Planning B: Planning and Design*, **25**, 639–55.

Tijerino, R. (1998) Civil spaces: a critical perspective on defensible space, *Journal of Architectural and Planning Research*, **15**, 321–37.

Tilbury, N. (1992) *Lonely Planet: Israel, a Travel Survival Kit*, Hawthorn, Australia: Lonely Planet Publications.

Till, K. (1993) Neo-traditional towns and urban villages: the cultural production of a geography of 'otherness', *Environment and Planning D: Society and Space*, **11**, 709–32.

Tilley, N. (1992) *Safer City and Community Safety Strategies*, Home Office Police Research Group, Crime Prevention Series Paper 38, London: Home Office.

Tilley, N. (1993) *Understanding Car Parks, Crime and CCTV: Evaluation Lessons from Safer Cities*, Police Research Group, Crime Prevention Series Paper 42, London: Home Office.

Tonkin, E. (1992) *Narrating Our Past: Social Construction of Oral History*, Cambridge: Cambridge University Press.

Tosh, H. (1977) *Living in Prison: The Ecology of Survival*, New York: Free Press.

Towle, L. (1995) Addressing for success, *North Carolina*, June, 19–20, 24–5.

Trench, S. (1996) Safer transport and parking, in Oc, T. and Tiesdell, S. (eds) *Safer City Centres: Reviving the Public Realm*, London: Paul Chapman.

Trench, S., Oc, T. and Tiesdell, S. (1992) Safer cities for women – perceived risks and planning measures, *Town Planning Review*, **63**, 279–96.

Trosper, R.L. (1995) Traditional American Indian economic policy, *American Indian Culture and Research Journal*, **19**, 65–95.

Trosper, R.L. (1998) Hunters-gatherers in late modernity: is survival becoming easier? Paper presented at the Seventh Common Property Conference, International Association for the Study of Common Property, 10–14 June, Vancouver.

Tuan, Y.-F. (1974) *Topophilia: A Study of Environmental Perception, Attitudes and Values*, Englewood Cliffs: NJ: Prentice-Hall.

Tuan, Y.-F. (1979) *Landscapes of Fear*, Oxford: Blackwell; New York: Pantheon Books.

Tyler, S.L. (1973) *A History of Indian Policy*, Washington, DC: Bureau of Indian Affairs, US Department of the Interior.

UIPA (United Indian Planners Association) (1977) *National Indian Planning Assessment*, Washington, DC: US Government Printing Office for the US Department of Housing and Urban Development and the Economic Development Administration.

UKCEED (1999) *Radioactive Waste Management*, UK National Consensus Conference, Citizens' Panel Report, 21–4 May.

Urban Task Force (1999) *Towards an Urban Renaissance*, London: HMSO.

USGAO (United States General Accounting Office) (1987) More federal efforts needed to improve Indian's standard of living through business development, in *Report of the Comptroller* (15 February), CED-78-50, Washington, DC: General Accounting Office.

Valentine, G. (1993) Heterosexing space: lesbian perceptions and experiences of everyday spaces, *Environment and Planning D: Society and Space*, **9**, 395–413.

Valentine, G. (1997) 'Oh yes I can,' 'Oh no you can't': children and parents' understanding of kids' competence to negotiate public space safety, *Antipode*, **29**, 65–89.

Van Otten, G.A. (1985) A geographer's perception of land use planning in Arizona's Native American reservations, *Papers and Proceedings of the Applied Geography Conferences*, **8**, 307–13.

Varenne, H. (1993) The question of European nationalism, in Wilson, T. and Smith, M. (eds) *Cultural Change and the New Europe*, Boulder, Colo: Westview Press.

Virilio, P. (1986) *Speed and Politics: An Essay on Dromology*, trans. M. Polizzotti, New York: Columbia University Press/Semiotext(e).

Visser, J. (1997) Zoning street prostitution, unpublished paper presented to 'Sexuality Across Cultures' conference, Amsterdam.

Walker, C., Starmer, K., Schiff, D. and Nobles, R. (1994) *Evaluation of Urban Crime Funding in West Yorkshire*, final report to the West Yorkshire Police Authority, Centre for Criminal Justice Studies, University of Leeds.

Walkowitz, J. (1992) *The City of Dreadful Delight: Narratives of Sexual Danger in Victorian England*, London: Virago.

Walmsley, D.J. and Lewis, G.J. (1993) *People and Environment: Behavioural Approaches in Human Geography*, Harlow: Longman.

Ward, H., Day, S., Mezzone, J., Dunlop, L., Donegan, C., Farrar, S., Whitaker, L., Harris, J. and Miller, D. (1993) Prostitution and risk of HIV: female prostitutes in London, *British Medical Journal*, **307**, 356–8.

Warnke, M. (1994) *Political Landscape: The Art History of Nature*, London: Reaktion Books.

Watts, M. (1997) Black gold, white heat, in Pile, S. and Keith, M. (eds) *Geographies of Resistance*, London: Routledge.

Wells, R. (1994) Transforming Native American education: the long road from acculturation to cultural self-determination, in Wells, R. (ed.) *Native American Resurgence and Renewal: A Reader and Bibliography*, London: Scarecrow Press.

Wekerle, G.R. and Whitzman, C. (1995) *Safe Cities: Guidelines for Planning, Design and Management*, New York: Van Nostrand Reinhold.

West, M. and Reid, B. (1999) Ageing on the Navajo and Hopi reservations . . . old and alone: rural Native American elders struggle in land without young, *Arizona Republic*, 5 September, A1 and A16.

Whitaker, B. (1964) *The Police*, Harmondsworth: Penguin Books.

Whitehand, J.W.R. (1981) Background to the urban morphogenetic tradition, in Whitehand, J.W.R. (ed.) *The Urban Landscape: Historical Development and Management*, London: Academic Press.

Whitehand, J.W.R. (1987a) Urban Morphology, in M. Pacione (ed.) *Historical Geography: Progress and Prospect*, London: Croom Helm.

Whitehand, J.W.R. (1987b) M.R.G. Conzen and the intellectual parentage of urban morphology, *Planning History Bulletin*, **9**(2), 35–41.

Whitehand, J.W.R. (1987c) *The Changing Face of Cities: A Study of Development Cycles and Urban Form*, Institute of British Geographers Special Publication 21, Oxford: Blackwell.

Whitehand, J.W.R. and Larkham, P.J. (eds) (1992) *Urban Landscapes: International Perspectives*, London: Routledge.

Whitehouse, P. (1997) *Report by the Chief Constable: Positive Street Policing*, Hove: Sussex Police Authority.

Whitt, L.A. (1995) Cultural imperialism and the marketing of Native America, *American Indian Culture and Research Journal*, **19**, 1–31.

Wilkinson, C. (1992) *The Eagle Bird: Mapping a New West*, New York: Pantheon.

Williams, K.S. (1997) *Textbook on Criminology*, third edition, London: Blackstone.

Williams, R. (1973) *The Country and the City*, London: Chatto & Windus.

Williamson, T. (1998) *Polite Landscapes: Gardens & Society in Eighteenth-Century England*, Stroud: Alan Sutton.

Wilson, A. (1992) *The Culture of Nature: North American Landscape from Disney to the Exxon Valdez*, Cambridge, Mass.: Blackwell.

Wilson, A. and Charlton, K. (1997) *Making Partnerships Work*, York: Joseph Rowntree Foundation.

Wilson, J.Q. and Kelling, G.L. (1982) Broken windows, *The Atlantic Monthly*, March, 29–38.

Wilson, R. (1996) Indian gaming in California. Public Law Research Institute, U.C. Hastings College of Law Website: http://www.uchastings.edu/plri/spr96tex/indgam.html, accessed 1 October 1999.

Wilton, R. (1998) The constitution of difference: space and psyche in landscapes of exclusion, *Geoforum*, **29**, 173–85.

Winters, H.A. (1998) *Battling the elements: weather and terrain in the conduct of war*, Baltimore: Johns Hopkins University Press.

WISE (1994) *Wastes from Reprocessing Foreign Spent Fuel at La Hague*, World Information Service on Energy, Paris, January.

WISE (1999) *Plutonium Investigation: Russia, Nuclear Superpower's Inheritance*, World Information Service on Energy, Paris, March/April.

Wolfensberger, W. (1970) The principle of normalisation and its implications for psychiatric services, *American Journal of Psychiatry*, **127**, 291–7.

Woolf, Lord (1991) *Prison Disturbances April 1990: Report of An Inquiry by the Rt Hon Lord Justice Woolf (Parts I and II) and His Honour Judge Stephen Tumin (Part II), presented to Parliament by the Secretary of State for the Home Department by Command of Her Majesty February 1991*, Cmnd 1456, London: HMSO.

Worpole, K. and Greenhalgh, L. (1996) *The Freedom of the City*, London: Demos.

Wright, M. (1995) Victims, medication and criminal justice, *Criminal Law Review*, **187**, 187–99.

Wright, P. (1985) *On Living in an Old Country: The National Past in Contemporary Britain*, London: Verso.

Wynne, B. (1992) Misunderstood misunderstandings: social identities and public uptake of science, *Public Understanding of Science*, **1**, 19–46.

Wynne, B., Waterton, C. and Grove-White, R. (1993) *Public Perceptions and the Nuclear Industry in West Cumbria*, Centre for the Study of Environmental Change, Lancaster University.

Yaar, E. (1998) Between Masada and Goa: personal, social and national attitudes of Israeli youth, unpublished paper, conference co-sponsored by the Rabin Centre and the Israeli Institute for Social Research, Tel Aviv University.

Young, E. (1992) Hunter-gatherer concepts of land and its ownership in remote Australia and North America, in Anderson, K. and Gale, F. (eds) *Inventing Places: Studies in Cultural Geography*, Melbourne: Longman.

Young, I.M. (1990) *Justice and the Politics of Difference*, Princeton, NJ: Princeton University Press.

Zonabend, F. (1993) *The Nuclear Peninsula*, Cambridge: Cambridge University Press.

Zukin, S. (1991) *Landscapes of Power*, Berkeley: University of California Press.

Zukin, S. (1995) *The Culture of Cities*, Oxford: Blackwell.

FILM

Lucas, P. (1979) *Images of Indians*: Part 5 'The movie reel Indians'. KCTS-9, Seattle, Washington, in cooperation with the United Indians of All Tribes Foundation.

INDEX